GUIDELINES FOR SAFE AND RELIABLE INSTRUMENTED PROTECTIVE SYSTEMS

This book is one in a series of process safety guideline and concept books published by the Center for Chemical Process Safety (CCPS). Please go to www.wiley.com/go/ccps to see the full list of titles.

GUIDELINES FOR SAFE AND RELIABLE INSTRUMENTED PROTECTIVE SYSTEMS

Center for Chemical Process Safety

CCPS
CENTER FOR CHEMICAL PROCESS SAFETY
An AIChE Industry Technology Alliance

WILEY-INTERSCIENCE
A JOHN WILEY & SONS, INC., PUBLICATION

A Joint Publication of the Center for Chemical Process Safety of the American Institute of Chemical Engineers and John Wiley & Sons, Inc.

Published by John Wiley & Sons, Inc., Hoboken, New Jersey.
Published simultaneously in Canada.

Library of Congress Cataloging-in-Publication Data is available.

ISBN 978-0-471-97940-1

Printed in the United States of America.

10 9 8 7 6 5 4 3 2 1

ACKNOWLEDGEMENTS

The American Institute of Chemical Engineers (AIChE) wishes to thank the Center for Chemical Process Safety (CCPS) and those involved in its operation, including its many sponsors whose funding made this project possible and the members of the Technical Steering Committee who conceived of and supported this Guidelines project. The members of the Guidelines for Safe and Reliable Instrumented Protective Systems Subcommittee deserve special recognition for their dedication and technical contributions leading to the creation of this useful addition to the CCPS process safety Guidelines series.

The members of the Subcommittee were:

Dave A. Deibert, Chair	Air Products and Chemicals, Inc.
Arthur J. Schwartz	Bayer (retired)
A. Doug Cates	Celanese
Bob Roubion	Degussa
Gregory Schultz	The Dow Chemical Company
Helmut Bezecny	The Dow Chemical Company
Richard R. Dunn	Dupont
Jan Windhorst	NOVA Chemicals, Inc.
Dallas L. Green	Rohm and Haas Company
Randy Freeman	Safety & Property Protection Consulting

Adrian L. Sepeda was the CCPS staff liaison and was responsible for overall project administration. Adrian also wrote the book preface.

The task of preparing the text from Subcommittee input was entrusted to Dr. Angela Summers, President, SIS-TECH Solutions LP, and principal book author. She and selected members of her capable staff organized and drafted the concepts and emphasis areas requested. Dr. Michela Gentile and Susan Wiley assisted with book drafting and review. Laurie Mayes-Fisher was responsible for typesetting and graphic design.

A special thanks and appreciation to Ken Bond, Shell (retired), Bernard Michaux, Total (retired), and Vic Maggioli, IEC 61511 Chairman, for input and council during the drafting of the book.

CCPS also gratefully acknowledges and thanks those who peer reviewed the draft and offered meaningful comments and suggestions. These peer reviewers were:

Hal Thomas	Air Products and Chemicals, Inc.
Robert Zittleman	Degussa
Vern Darling	The Dow Chemical Company
Richard Stougie	The Dow Chemical Company
Wayne Chastain	Eastman Chemicals
Robert Blanco	Fibertel
William Olsen	Merck & Company , Inc.
Tony Thompson	Monsanto
Brian Smith	Nova Chemicals
Lisa Morrison	PPG Industries, Inc.
Art Dowell	Rohm & Haas Company
John Alderman	RRS Engineering
Bryan Zachary	SIS-TECH Solutions, LP
Kevin Klein	Solutia

Their insights, comments, and suggestions helped ensure a balanced perspective for the Guideline.

Lastly, we wish to express our special appreciation for the guidance, counsel, and contributions of Dr. Arthur Schwartz, committee member, who passed away before the book was published. Art was a scholar and a gentleman.

PREFACE

The American Institute of Chemical Engineers (AIChE) has a more than 50 year history of involvement with process safety as it relates to chemical processing facilities. Through its strong ties with process designers, builders, operators, safety professionals and academia, AIChE has enhanced communication and fostered improvement in the already high safety standards of the industry. AIChE publications and symposia have become a recognized valuable information resource for the engineering profession on the causes of accidents and means of prevention.

The Center for Chemical Process Safety (CCPS), an Industry Technology Alliance of AIChE, was established in 1985 to develop and disseminate technical information for use in the prevention of major chemical accidents. CCPS is supported by a diverse group of industrial sponsors in the chemical industry and related industries who provide the necessary funding and professional guidance for its projects. The CCPS Technical Steering Committee and the technical subcommittees oversee individual projects selected by CCPS. Professional representatives of the sponsoring companies staff the subcommittees, with a member of the CCPS staff coordinating subcommittee activities.

Since its founding, CCPS has published many volumes in its "Guidelines" series and in its smaller "Concept" series texts. These CCPS books address not only scientific techniques, practices and issues faced by engineers in plant design, operations and maintenance, they also cover the broader subject of chemical process safety management. Successful process safety programs and management systems are the products of committed and active participation of personnel at all levels who apply a systematic approach to process safety as an integral part of operations management.

This Guideline explains the decision-making processes for the management of instrumented protective systems (IPS) throughout a project's life cycle. It uses IEC and ISA standards as a basis for the work processes used to achieve safe and reliable process operation. It establishes a framework for a protective management system that can be used to design and manage those specific instrumented systems. By walking the reader through a project's life cycle, engineering, maintenance, and operations disciplines can easily focus on their responsibilities and duties. Using this approach, the book is useful as a primer, guidelines reference and resource manual. Examples are used to provide "real-world" experience applications. This book is a companion publication to the earlier published CCPS book, "Guidelines for Safe Automation of Chemical Processes."

CCPS hopes that the guidance and examples provided herein will aid in promoting safer and more reliable IPSs.

TABLE OF CONTENTS

LIST OF FIGURES

LIST OF TABLES

1

INTRODUCTION

Instrumented Protective Systems (IPS) implement protective functions that detect abnormal or unacceptable operating conditions and take action on the process to achieve or maintain a safe state. IPSs are used to reduce the process risk associated with health and safety effects, environmental impacts, loss of property and business interruption costs.

Safe operation cannot be achieved in isolation. The risk reduction strategy must also consider the owner/operator's business needs. Personnel are expected to operate process units to achieve target production rates, product quality, and cost performance. Balancing safety and production goals can be challenging when the IPS design and management does not adequately address the operational needs. The following can add significantly to this challenge:

- High initiating cause frequency results in frequent loss of control and process shutdown,
- High frequency of spurious IPS operation leads to:
    - Lack of trust in the IPS (leading cause of improper bypassing)
    - Frequent process equipment shutdown with subsequent process unit impact
    - Frequent process unit start-up which may have significant inherent risk
- High frequency of IPS equipment failure results in high operating and maintenance costs,
- Ignoring functionality requirements leads to an IPS design which does not adequately support the various process operating modes and potentially causes excessive IPS equipment bypassing, alarms, and shutdowns, and
- Ignoring maintainability requirements leads to inadequate maintenance resources and facilities and potentially failure of the mechanical integrity program.

It is well understood that plant productivity and operability improves when quality control processes are applied to process equipment operation. Given the potential problems associated with IPS implementation, it simply makes sense to apply the same quality control processes across the IPS lifecycle.

Quality control processes rely on the use of appropriate metrics to verify compliance with the work process expectations. For IPS design and management, these metrics are associated with core attributes that are considered essential for an instrumented safeguard to be classified as an IPS. Seven core attributes should be achieved by the IPS design and supported by appropriate management practices:

1. Independence,
2. Functionality,
3. Integrity,
4. Reliability,
5. Auditability,
6. Access security, and
7. Management of change.

These core attributes are periodically assessed to determine the degree to which they are being maintained and improved. Quality control processes, such as verification, assessment, auditing, and validation, are necessary to ensure the required attributes are achieved throughout the IPS life. The level of rigor employed in the quality control limits the performance which can be reasonably achieved by the IPS.

IPS implementation and continuous improvement involve the effort of many stakeholders, e.g., management, process safety, process, instrumentation and electrical, operations, maintenance, and manufacturers. Projects are often iterative processes requiring careful consideration of each discipline's needs and the core attributes.

This guidelines book intends to:

- Clarify the essential role of the various personnel responsible for IPSs,
- Establish a protective management system framework for IPS design and management,
- Provide the work processes to be followed for IPS development from risk assessment through its implementation and transfer to operations,
- Discuss essential on-going, day-to-day activities necessary to maintain the core attributes, and
- Challenge owner/operators to continuously evaluate opportunities for improvement.

1.1 PURPOSE

The process industry has made great strides toward improving process unit performance and safe operation. It has made and continues to make significant investment to address process risk using a variety of approaches aimed at identifying and controlling risk. These approaches often must fit within a regulatory framework, which relies on the use of recognized and generally accepted good engineering practices to define the minimum requirements.

Many governments (e.g., the United States of America, the European Union, the United Kingdom, Germany, The Netherlands, Korea, Taiwan, and Brazil) have regulations concerning the prevention of releases of hazardous chemicals that pose serious injury or life threatening consequences. Although each government uses unique terminology to describe such events, the concept of process safety management is well known throughout the world. It is widely supported even by governments that do not have specific regulations mandating its implementation. Most require, at a minimum, that an owner/operator demonstrate compliance with the good engineering practices applicable to the manufacturing process and its associated hazards.

The application of control and shutdown equipment to manage hazardous events was first discussed in *Guidelines for Safe Automation of Chemical Processes* (CCPS/AIChE 1993, referred to as *Safe Automation*). In particular, *Safe Automation* provided information for the design and implementation of the Basic Process Control System (BPCS) and the SIS. It established for the process industry many of the fundamental concepts used today, such as independent protection layer (IPL), safety integrity level (SIL), separation and diversity of the BPCS and SIS, access security, and fault tolerance.

Safe Automation was later referenced by the Instrumentation, Systems and Automation (ISA) society standard, ANSI/ISA 84.01-1996, *Application of Safety Instrumented Systems (SIS) for the Process Industry*. This standard provided good engineering practices for the SIS lifecycle, starting with the design phase and continuing through decommissioning.

The globalization of the process industry resulted in demand for international practices. Numerous good engineering practices, previously considered national or regional, are being modified, updated, harmonized, and issued as international practices. One such standard is IEC 61511, *Functional Safety: Safety Instrumented Systems for the Process Industry Sector*, which expanded the requirements of ANSI/ISA 84.01-1996.

IEC 61511 is the first sector standard issued using the lifecycle framework established by IEC 61508, *Functional Safety of Electrical/ Electronic/Programmable Electronic Safety Related Systems* and covers the complete SIS lifecycle for the process sector. It was developed and is maintained by the International Electrotechnical Commission (IEC) with volunteer support from organizations worldwide, including ISA and CCPS/AIChE.

IEC 61511 was accepted in 2004 by the European Committee for Electrotechnical Standardization (CENELEC) as EN IEC 61511 and the American National Standards Institute (ANSI) as ANSI/ISA 84.00.01-2004 Parts 1-3. In 2005, ISA published, *Guidelines on the Implementation of ANSI/ISA 84.00.01-2004*, to provide guidance to owners/operators concerning the application of the SIS standard to new and existing equipment. To recognize the contribution of both ISA and IEC to the documentation of good engineering practices for SIS, this book refers to the standard as ISA 84.01/IEC 61511.

ISA 84.01/IEC 61511 uses the SIL concept to benchmark the integrity of the instrumentation and controls used to achieve the required performance from the SIS. The required SIL is defined during a risk assessment process, which examines the process risk and identifies IPLs. ISA 84.01/IEC 61511 requires that the SIL be quantitatively verified using estimates of the random hardware failure rate of the SIS components in the intended operating environment.

Since ISA 84.01/IEC 61511 is an instrumentation and controls standard, it places a great deal of emphasis on the functionality and integrity of the hardware. The assignment and verification of SIL establishes a robust relationship between hardware design and risk reduction. It also provides justification for separation, fault tolerance, and proof test intervals. However, the SIS's capability to achieve or maintain a safe state is dependent on more than the sum of its hardware components.

Integrity and functionality are essential performance attributes, but excess attention on these can result in a loss of focus on other core attributes. While weak links in the hardware design may be identified during a numerical analysis of the SIS equipment, the ability of the installed SIS to achieve the SIL is generally limited by human performance against practices and procedures. Independence, reliability, auditability, access security, and management of change must receive as much, if not more, attention to detail.

The core attributes support the SIS throughout its life by ensuring appropriate focus on minimizing the potential impact of human error on the SIS performance. The absence of a rigorous management system can lead to discrepancies between the desired functionality and integrity and what is achievable in actual operation.

As process units become increasingly automated, integrated and complex, the deliberate and intentional act of implementing IPLs becomes more important. SISs are only one IPL of many that can be used to achieve and maintain safe operation. Other IPLs, such as relief devices and protective alarms, may be identified and should be managed appropriately. The management system ensures that protective equipment are designed, inspected, maintained, tested, and operated in a safe manner. Many incidents in the process industry have been caused by poor management systems that allowed systematic errors to erode safe operation to the point of catastrophic release.

This book uses the seven core attributes, namely, independence, functionality, integrity, reliability, auditability, access security, and management of change, to define the required performance for the human and equipment systems necessary for safe operation. Following in the footsteps of *Safe Automation*, this book is intended for use by people who are familiar with the manufacture and use of chemicals. It expands the work processes to cover the major activities executed by the various disciplines supporting the SIS lifecycle.

With such an encompassing scope, no single book can possibly cover all of the detailed tasks required for safe and reliable operation. Instead, this book concentrates on the overall work processes, task intent, input information, considerations, and output deliverables. When necessary, the book provides references to other technical publications for greater detail and guidance on specific topics.

1.2 TARGET AUDIENCE

A performance-based management system relies on metrics to support prudent business decisions. Performance-based systems only work in a safety culture nurtured and directed by top management. Sustainable performance requires long-term vision, consistent focus and attention, and financial commitment from senior management. Consensus and participation of personnel are necessary to support the operational and strategic objectives, as well as foster a safe working environment.

The target users of this book are the various disciplines responsible for safe and reliable operation in the process industry. At any given facility, these disciplines may be represented by individuals, departments, or organizations. At some facilities, one person may be responsible for the activities listed for multiple disciplines. The site management system should specify the individuals, departments, or organizations responsible for work activities.

Table 1.1 provides the essential knowledge to be gained by reading this book for seven disciplines. Chapters 3 through 7 include a target audience section that identifies the essential tasks to be discussed for each discipline:

- **Management** includes personnel responsible for establishing policies related to safe and reliable operation and for oversight of the management system. Includes corporate and site organizations,

- **Process safety** includes personnel responsible for process safety management. Includes environmental, health, and process safety management organizations,

- **Process** includes personnel responsible for the process design and operation. Includes research and development, process, and process control,

- **Instrumentation and Electrical (I&E)** includes personnel responsible for instrumentation and control design and implementation. Includes I&E, process control and reliability,

- **Operations** include personnel responsible for the operation of the process. Includes process operations and operations management,

- **Maintenance** includes any personnel responsible for inspecting, testing, and maintaining IPS equipment. This may include personnel from maintenance, process control, I&E, and reliability (equipment), and

- **Manufacturers** include any entity that develops, markets, and sells a product for IPS use.

Target Audience	Will Gain Essential Knowledge On
Everyone	Role and responsibility Risk criteria and affect on IPS requirements Core attributes of IPLs and IPSs Effect of IPS classification on design and management
Management	Management system and its fundamental features Activities, training, tasks, and systems required to support IPSs Communication of risk criteria and expectations
Process Safety	Activities, training, tasks, and systems required to support IPSs Risk criteria and affect on hazard and risk analysis and IPL requirements
Process	Protective requirements specification How functionality, operability, maintainability, and reliability affect design and operating basis Content of I&E requirements specification
Instrumentation and Electrical	Content of process requirements specification I&E requirements specification User approval of equipment How equipment selection, subsystem architecture, diagnostic capability, and proof test interval affect the integrity and reliability
Operations	Administrative procedures--access security management of change, bypass management, and event reporting Operating procedures--hazardous event description, failure response, compensating measures, when to execute a safe shutdown, and what to do when a shutdown fails
Maintenance	Administrative procedures--access security, management of change, bypass, configuration management, and failure reporting Maintenance procedures--hazardous event description, failure response, allowable repair time, inspection, preventive maintenance, and proof tests
Manufacturers	How functionality, operability, maintainability, and reliability affect safe operation

Table 1.1. Target Audience and Essential Knowledge.

1.3 BOOK ROAD MAP

The book is organized using a project lifecycle with six major phases:

- Planning,
- Risk Assessment,
- Design,
- Engineering, Installation, Commissioning and Validation,
- Operational and Mechanical Integrity, and
- Continuous Improvement.

These phases are shown in Table 1.2, which also provides a road map for the book listed by the target audience (see Section 1.2). An “X” is an indication that the chapter contains material that is important to that resource, e.g., a discussion of fundamental principles or specific task responsibility.

The protective management system discussed in Chapter 2 reduces the systematic errors through quality management processes and good engineering practices. Chapter 3 provides an overview of the risk assessment phase, which uses a variety of hazard and risk analysis techniques to identify and classify IPSs.

Chapters 4 and 5 address work processes for IPS design and implementation. Chapter 4 discusses the development of the design basis, which must achieve the intent of the risk assessment and the core attributes defined for each IPS. Chapter 5 covers the engineering, installation, commissioning and validation phase, where the design basis is turned into an installed and operational IPS.

Long-term operational and mechanical integrity is supported by an operating basis, discussed in Chapter 6, which addresses IPS operating and maintenance procedures and personnel training. Chapter 6 also discusses the importance of bypass management, compensating measures, periodic proof testing, and configuration management in achieving the core attributes. Finally, in Chapter 7, long-term performance is monitored and options for improvement are periodically considered.

Audience	Chapter 2 Planning	Chapter 3 Risk Assessment	Chapter 4 Design	Chapter 5 Engineering, Installation, Commissioning and Validation	Chapter 6 Operational and Mechanical Integrity	Chapter 7 Continuous Improvement
Management	X					X
Process Safety	X	X				X
Process	X	X	X	X		X
Instrumentation and Electrical	X	X	X	X	X	X
Operations	X	X		X	X	X
Maintenance	X			X	X	X
Manufacturers			X	X	X	X

Table 1.2. Road Map by Target Audience.

1.4 MANAGEMENT COMMITMENT

Management must make it a priority to develop a protective management system that ensures safe operation of their facilities. Management must support and approve the documentation of policies, practices, and procedures, which provide the work processes and metrics essential to effective risk management. Global competition also demands that the end result of these work processes yield reliable and cost effective IPS operation. Work processes should address significant classes of business risk, that is, safety, environmental, business interruption, and asset, to obtain the greatest value.

Many different management systems are used in the process industry to achieve safety and business risk goals. Consistent performance is directly related to the relevance of the prescribed practices to actual work tasks and the rigor employed to achieve quality workmanship.

In general, the management system should:

- Establish risk criteria and risk management philosophy,
- Verify work activities and associated documentation are in alignment with this philosophy,
- Establish processes to identify and classify IPS,
- Identify personnel or departments who are responsible for IPS work activities and specialists who support the management system elements,
- Verify competence of those assigned responsibility for the IPSs,
- Establish a process to evaluate whether existing IPSs meet applicable good engineering practices, and
- Verify that the behavior and actions of its personnel are consistent with a culture that encourages continuous improvement in IPSs.

Management must be fully committed and engaged in the development and implementation of the management system. Management responsibility includes establishing safe and reliable operation as a priority and providing the resources, tools and training required to get the job done. Successful execution requires decision criteria be clearly stated and consistently followed. For efficient implementation, these criteria should be embedded into each project and operational phase so that safety and reliability issues are considered a normal part of doing business. The best results are obtained when personnel who are responsible for safe and reliable operation:

- Understand their individual responsibility and authority within the management system,
- Understand the owner/operator risk criteria and how it is applied in the design and management of IPSs,
- Are expected to consistently apply internal practices related to safe and reliable operation,
- Feel support to do what is right (even if it requires changing the way things are done now),
- Have the competency and necessary resources to accomplish their assigned responsibilities, and
- Understand the boundaries of acceptable design, operation and mechanical integrity.

2

PLANNING

In the book, *Lessons from Disaster: How organizations have no memory and accidents recur* (Kletz 1993), Trevor Kletz states that "listing...human error as the cause of an accident is about as helpful as listing gravity as the cause of a fall. It may be true, but it does not lead to constructive action." When a bridge collapses, the incident investigation report does not say that the incident was the result of the force of gravity. It is understood that gravity is a fundamental property considered in the bridge's design and construction. The incident report will refer to improper steel specification, inadequate support structure, etc. Gravity is not listed as the cause, because it is obvious that given the right conditions all things succumb to gravity.

Similarly, it should be recognized that given the right conditions all things succumb to human error, such as:

- Actuator is undersized such that the valve does not close as required,
- Hazard is not recognized, so it is not managed,
- Hazard is identified but deemed non-credible,
- Operator opens valves in wrong sequence,
- Transmitter is left in the "test" mode following repair,
- Incorrect kinetic data results in poor selection of set point for the IPS,
- Configuration error results in the closure of the wrong valve,
- Management makes decision to bypass IPS to allow continued production, and
- Inadequate staffing leads to poor maintenance.

The process industry has adopted automation to improve product quality and production rates, to reduce the potential for operator error, and to decrease resource requirements. Process automation includes many different instrumented systems, such as process controls, alarms, instrumented protective systems (IPS), and emergency shutdown systems. These systems range from simple hardwired systems to complex programmable electronic systems.

While the latest in instrumentation and controls often requires less support once implemented, the increased complexity may require a significantly greater level of attention during assessment, design, inspection, testing, and maintenance. Unforeseen failures can occur during long-term operation, rendering the system incapable of executing its specified action. This is especially problematic in programmable electronic systems (PESs) due to the large number of potential failure combinations between the various software and hardware elements that comprise the system. PESs also tend to require more frequent upgrades, with the potential for new failures resulting from conflicts with interconnected new and old equipment.

Good design leads to safe operation by anticipating potential errors and failures and implementing means to detect their presence, so that they can be corrected. Good design ensures that the core attributes of independence, functionality, integrity, auditability, reliability, access security, and management of change are addressed consistent with owner/operator requirements. Poorly implemented design can lead to significant hazardous events, involving impact to people, the environment, and plant assets.

Much has been done to encourage, guide, and regulate safety through identification of hazardous events and management of significant risk. However, catastrophic incidents continue to occur around the world. Some of these incidents are discussed in this guidelines book as examples of how both simple and complex events lead to catastrophic consequences. There are common threads woven throughout these incidents that point to cultural issues, where decisions were made that were contrary to what was appropriate for safety.

Developing a safety culture requires that values and policies are converted to practices and behaviors. The policy of "Safe operation is our mission" must be backed with the authority, resources, and tools necessary to achieve the policy. Everyone from the front-line operator to the board of directors must understand and believe that safe operation is a core operating principle. Periodic audits of actual practices and behaviors should be used to verify that internal practices and procedures are being followed. It should also be clearly communicated that safe operation is rewarded and that unsafe operation is not tolerated.

IPS design is a team effort requiring knowledge and skills from a variety of disciplines. Some of these skills may be rarely used, but when required, high levels of expertise must be available. Massive restructuring and mergers have become commonplace within the process industry. Corporate technical departments that once served as clearinghouses for new technologies and new ideas no longer exist. Responsibility has shifted in many companies from specialized corporate personnel to site personnel, who often find it difficult to maintain the required expertise. In some facilities, there is an increasing reliance on manufacturers and other third party contractors to provide the missing expertise. At the plant site, numerous departments and many different individuals may be responsible for various, yet limited, aspects of safe operation.

Without strong oversight, it is easy for simple slips of reason, errors, loss of knowledge, loss of skill, or bad behavior to become the predominant cause of process safety incidents. A safety and reliability driven culture can only exist when the owner/operator documents the information, practices and procedures necessary to achieve the attributes important to safe and reliable operation. The goal is to maintain the IPS equipment in an "as good as new" condition throughout the process equipment life. To accomplish this, random failures and systematic errors that cause IPS equipment to fail must be identified and corrected, as early as possible.

The protective management system recommended in this chapter provides a lifecycle process that encourages the early identification and correction of failures and errors. It provides a framework for the implementation of quality control processes for the development, implementation, and management of a risk reduction strategy that reduces the risk below the owner/operator risk criteria. The protective management system is intended to ensure:

- Protective functions and required risk reduction are identified for each mode of process operation,
- IPSs are designed, installed, commissioned, and validated to meet the design basis,
- IPSs are inspected, tested, maintained, and operated to ensure the equipment operates in an "as good as new" condition,
- The process is adequately protected during periods where the process is being operated with a known IPS failure,
- Changes to the IPSs are evaluated through a management of change process,
- Access to the IPS is controlled administratively and physically, and
- IPS equipment failures and the occurrence of process demands are tracked and periodically assessed to ensure prompt response and resolution of any identified inadequacy.

The management system described in this chapter is applicable to all classes of IPSs, including those preventing safety, environmental, asset losses, or business interruption risks. Further, it stresses the importance of process reliability, operability, functionality, and maintainability to safe operation. A generalized lifecycle is provided, where internal practices are established to support performance requirements. This lifecycle serves as the outline for the remainder of this book.

2.1 PROTECTIVE MANAGEMENT SYSTEM LIFECYCLE

The protective management system supports IPSs throughout their lifecycle. It starts with the identification of IPS requirements in the risk assessment phase and ends when the IPS is decommissioned as the result of continuous improvement activities. When the lifecycle is integrated with a project management process, it provides an overall work process that can be applied to any IPS. Figure 2.1 provides an example of such a work process, covering seven major phases:

- Planning—establish a management system for how IPSs will be identified, designed, inspected, maintained and operated to meet the core attributes,
- Risk assessment—perform hazard and risk analysis to identify and classify IPSs and to define the required risk reduction,
- Design--develop a design basis to achieve the core attributes,
- Engineering—execute detailed design and software development to achieve design basis,
- Installation, commissioning, and validation—implement the IPS following the design basis and detailed design documentation,
- Operational and mechanical integrity—define what is required of operation and maintenance personnel to maintain the core attributes for the IPSs, and
- Continuous Improvement--ensure the IPS performs as intended by assessing the operation, inspection, testing, and maintenance practices for the installed equipment.

The management system can be used to address any risk, whether safety, environmental, asset or business related. Many governments require the use of good engineering practices in the design and management of the instrumented systems that maintain safe operation in the process industry. Local regulations, applicable codes, or insurance practices may require the use of specific good engineering practices. Consequently, the management system should incorporate relevant good engineering practices to provide a comprehensive program.

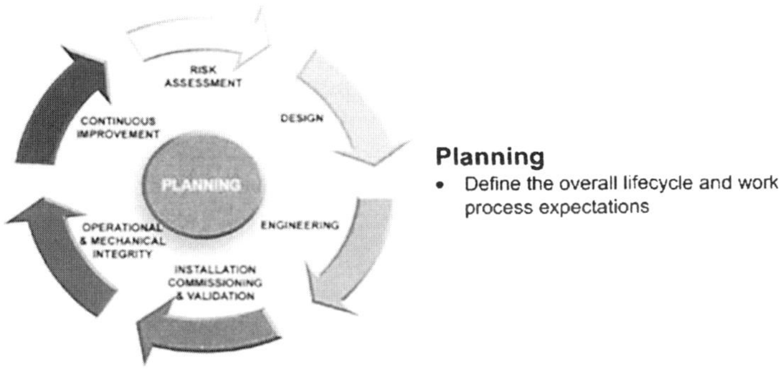

Figure 2.1. Planning Phase.

The protective management system will be different for each owner/operator, and possibly for different plant locations. This occurs for many reasons including local government requirements, safety culture, risk management philosophy, process age, resource availability, and personnel experience. Resource requirements vary among owner/operators due to:

- Internal corporate structure and culture,
- Intended management system rigor,
- Administrative control philosophy,
- Process complexity and maturity,
- Control system complexity,
- Protection layer complexity, and
- Existing resource skills and experience.

Some owner/operators establish "classes" of IPSs, providing different management systems for safety, environmental, asset and business risks. The management system establishes priority for the implementation of and adherence to the core attributes for each class. The level of rigor employed in the management system is typically based on the event consequence severity and the risk reduction expected from the IPS. The main difference between IPS and non-IPS is the level of rigor required to support the core attributes, especially management of change, access security, and auditability.

Knowledge-based management systems (see Figure 2.2) rely on the communication of the owner/operator risk management philosophy to its personnel through broad objectives. It provides maximum flexibility for managing the process risk because each solution can be custom tailored as desired by the individuals involved with the process.

Knowledge-based systems are used throughout the process industry. This system relies on knowledgeable, skilled, and experienced personnel to yield the required human, equipment, and process performance. Knowledge-based systems rely heavily on personnel skills and experience to correctly interpret management goals and implement IPSs. The performance achieved by a knowledge-based system can be highly affected by personnel and organizational changes.

The greatest caveat of a knowledge-based system is the uncertainty of consistent quality workmanship. Without strong oversight, each person's unique perspective on what is the "right" thing to do becomes a variable in process performance. The culmination of numerous independent decisions can erode performance over time. With this management system, verification, assessment and audit activities are very important to ensure personnel behavior matches owner/operator expectations.

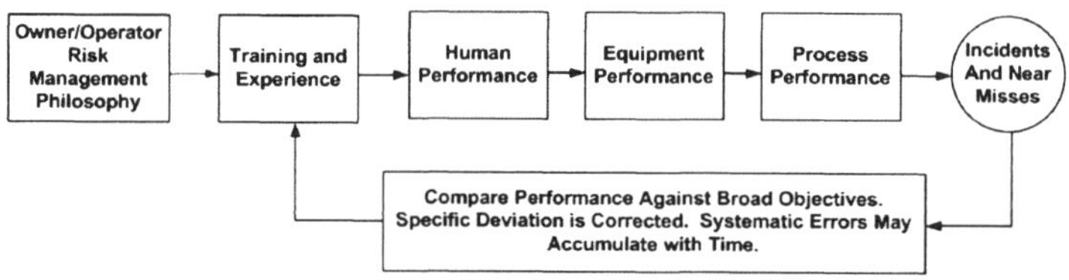

Figure 2.2. System Relying on Personnel Training and Experience. (adapted From Reason 1997)

The management system illustrated in Figure 2.3 communicates the owner/operator's risk management philosophy using polices, procedures, and practices. These prescriptive practices cover the human, equipment, and process lifecycles. Personnel knowledge and experience are not as important, because work activities and tasks are of sufficient detail that random variation in personnel knowledge does not significantly influence the work outcome.

This management system is not flexible in managing the process risk, because it dictates the solution. Maintaining the procedures requires discipline. Incremental technology changes can render existing procedures irrelevant. The natural tendency is to allow the behaviors to evolve organically over time, as a mixture of procedures that are followed to varying degrees and on-the-job training that provides the "here's how we actually do it" instructions.

While prescriptive systems require substantial effort to develop and maintain, these systems achieve very consistent performance when personnel are expected to comply with the practices and procedures. Periodic evaluation of practices and procedures is necessary to ensure that they remain up-to-date and relevant to current work expectations and requirements. Work practices and procedures should also be periodically benchmarked against industry published good engineering practices to ensure that internal practices keep pace with industry guidance and requirements.

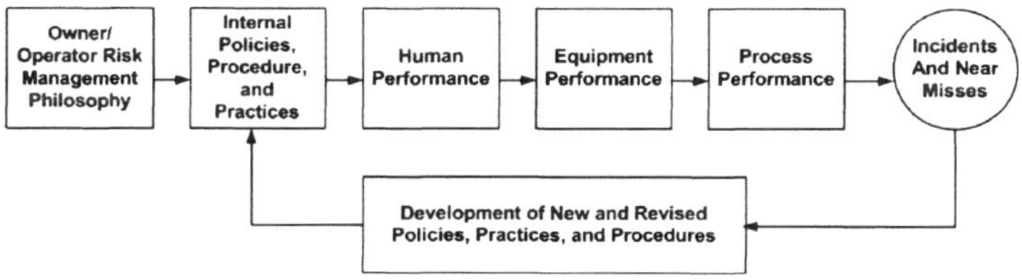

Figure 2.3. System Relying on Procedures and Practices. (adapted from Reason 1997)

For the management systems represented in Figures 2.2 and 2.3, unacceptable deviations are identified through incidents and near misses. Many accidents are complex and often involve multiple errors, lapses, and failures. Very rarely, does a single deviation cause an incident; it is usually the accumulation of deviations over time. Only a small percentage of unacceptable deviations are identified through incidents and near misses.

The effectiveness of incident and near miss investigations on correcting behavior and ensuring reliable performance is limited if the root cause analysis does not seek to identify the systemic problems. It is normal human behavior for personnel to believe that, due to the complex nature of the event, the incident or near miss is unique. The natural tendency is to focus on personnel errors or equipment malfunction, because these errors or failures are inherently linked to the chain of events leading to the incident. However, effective feedback requires examination of the management system and a determination of how the deviations went unchecked.

Incidents and near misses are lagging indicators of safe operation, because they do not allow the early identification and correction of the individual deviations. In general, by the time a near miss or incident manifests itself, the systemic problems that lead to the event may be (and often are) pervasive throughout the human, equipment, and process systems.

The management system recommended by this guidelines book establishes a set of attributes which are considered fundamental characteristics of a quality IPS management system. The management system (see Figure 2.4) uses a combination of internal practices and personnel knowledge, skills, and experience to achieve and maintain the core attributes throughout the lifecycle.

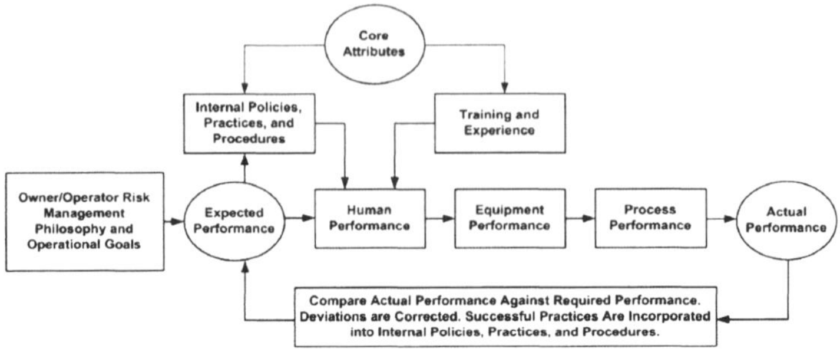

Figure 2.4. System Relying on Establishing and Monitoring Core Attributes. (adapted from Reason 1997)

The level of rigor used to control and monitor the core attributes affects the actual performance which can be achieved from personnel, the equipment, and the process. Actual IPS performance is monitored through various leading indicators, such as detected faults, dangerous and safe failures, process demands, and personnel conformance to management system requirements. These leading indicators allow an owner/operator to determine more quickly when the operational and mechanical integrity of equipment is insufficient to ensure consistent safe and reliable operation.

Risk management is an area where many technical personnel have limited training and skills, yet personnel are often expected to make risk based decisions with little guidance. Personnel responsible for process safety must understand the hazards associated with the process and the consequences of abnormal operation. Without this knowledge, personnel cannot be expected to make risk management decisions.

The attribute priority and management rigor typically reflect an owner/operator's risk management philosophy and operational goals. These attributes should be an inherent part of internal policies, practices, and procedures and conformance to these attributes should be monitored. The rigor required to maintain the attributes is influenced by the site culture. Lessons learned in applying the attributes should be incorporated into the work processes and requirements so that risk is continuously reduced.

Rather than waiting for an incident to identify a systemic problem, the recommended management system examines the reports and records created during day-to-day activities affecting safe and reliable operation. For example, how many pieces of IPS equipment are behind schedule for proof testing? Why are they behind? Is it a resource limitation or priority issue?

Unlike the incident-driven management systems represented by Figures 2.2 and 2.3, the management system proposed by Figure 2.4 acknowledges that sustainable performance is obtained when:

- Important work processes are documented,
- Personnel with the appropriate skill and training are assigned responsibility,
- Personnel are trained on the work processes,
- Compliance to approved practices is expected and audited, and
- Continuous improvement is considered important to business viability.

Sustainable performance results in decreased business losses, less rework, and higher plant performance. The protective management system is intended to:

- Outline the general work processes and deliverables required to properly manage risk,
- Identify key resource needs and gain management commitment to fulfill these needs,
- Outline the essential criteria for the various decision-making processes that occur throughout the life of a process unit,
- Provide a clear definition of risk criteria in terms of safety, environmental, and economic protection,
- Define an engineering approach to prevent process incidents, especially those that involve the catastrophic release of hazardous chemicals or energy, and
- Incorporate process and equipment reliability goals in terms of on-line time, spurious trip rate and on-line repairs.

2.2 WHY IT MAKES GOOD BUSINESS SENSE

A protective management system makes good business sense because it is the most efficient way to achieve consistent, predictable results from process operation, while providing:

- More consistent application of the core attributes to each protection layer,
- More opportunity to achieve business objectives such as improved process availability, fewer incidents, and increased market share,
- Greater chance to motivate, educate, and retain personnel,
- Greater likelihood to be viewed as a good neighbor, a good employer, a reliable supplier, and a strong competitor, and
- Better opportunity to gain a competitive advantage and lead the industry.

An effective management system takes a systematic approach to safe and reliable operation. The objective of the management system is to define the overall lifecycle and work process expectations. Many owner/operators are already executing many of the activities recommended by these guidelines, but in a somewhat fragmented manner. The task is to tie the existing policies, procedures, and practices into a comprehensive program, streamlining work execution and reducing implementation cost.

With a systematic approach, the process design, IPS design, operation and maintenance procedures, training programs, change management, and continuous improvement activities are considered appropriately during each lifecycle stage. When the management system incorporates quality control checkpoints, verifications, and validations, there is greater assurance that the process design and construction achieves risk management goals and regulatory compliance. A management system produces positive effects on the process operation and offers significant benefits.

Achieving safe and reliable operation requires resources to create a sustainable management system, to audit performance, and to initiate improvement, when necessary. It requires a substantial commitment from the top to provide the resources necessary for the effort and commitment from the bottom to make it work. However, many owner/operators have recognized significant benefits from implementing a management system, including:

- Reduced manufacturing costs,
- Increased reliability,
- Increased production, product yield and product quality,
- Reduced incident rate and severity,
- Reduced liability,
- Improved operability, functionality and maintainability,
- Increased efficiency in assessing changes,
- Improved design and management processes that lower project costs, and
- More effective regulatory compliance.

2.3 DOCUMENTATION

It is beyond the scope of this guidelines book to discuss everything needed to achieve safe and reliable operation for every process throughout the significantly diverse process industry. For IPSs, information is collected and documentation is generated throughout the lifecycle. Information concerning the hazardous events, the process technology and the process equipment is required at a minimum.

Hazardous events are identified by examining information concerning the process chemicals and their hazards:

- Physical property data
- Chemical data sheets, such as material safety datasheets,
- Chemical and thermal stability, ignitability, reactivity, corrosivity, and toxicity data,
- Permissible exposure limits, and
- Affects of inadvertent mixing of different materials.

The process is operated to produce a product at a target production rate and product quality. The process technology may be described by:

- Process description,
- Process chemistry
- Process flow diagrams,
- Operating modes (e.g., start-up, normal, reduced rates, maintenance, and shutdown),
- Maximum intended inventory,
- Hazard and risk analysis, and
- Safe operating limits.

The process technology is implemented and its hazards are controlled using equipment that is designed and constructed in accordance with specified good engineering practices. The equipment may be described by the following:

- Piping and Instrumentation Diagrams,
- Material and energy balances,
- Materials of construction,
- Electrical area classification,
- IPS design basis,
- Relief system design basis,
- Ventilation system design basis,
- Process equipment specification sheets, and
- Protective equipment specification sheets.

Operator procedures are also developed to define how the process is safely operated under normal, abnormal, and emergency conditions. The procedures should define when it is acceptable to operate the process with a known IPS failure and when and how bypasses can be used for start-up and/or equipment maintenance. Operators should be trained on the procedures and tested on their comprehension.

To ensure mechanical integrity of the IPS, its equipment must be inspected, tested, and maintained according to written procedures. These procedures should define how to safe conduct maintenance, including the use of any bypasses. Trips and equipment failure should be investigated and findings reported. Maintenance records, as well trip and equipment failure reports, should be maintained for a sufficient period to allow equipment performance to be tracked and trended.

Near misses and incidents should be investigated and reported to others within the owner/operator organization. This alerts other locations to potential problems that may need correction.

Access to IPS equipment should be controlled using safe working practices that restrict access to only trained and authorized personnel changes to the hardware, software, or procedures associated with IPS should be controlled under a management of change (MOC) process. Approved MOC records should define the documentation or procedures requiring update related to the change. Approved substitution for hardware and software should be provided on an Approved Equipment List.

An audit process should be used to demonstrate that personnel understand the procedures and are following the intent of the procedures related to safe and reliable operation. Gaps should be addressed through audit reports that provide findings, recommendations, and an action plan.

Each lifecycle phase is listed in Table 2.1 with its objective, input information and output documentation. While not specifically listed, work processes are also necessary to support the day-to-day activities of project engineering and on-going plant operation and maintenance. Detailed work processes and activities are provided in each subject-related chapter. The resources necessary to develop and maintain the required information and documentation is not trivial. Responsibility should be formally assigned to individual(s) or department(s).

Objective	Input	Output
	Planning	
1. Define the overall lifecycle and work process expectations	1. Government requirements 2. Risk criteria 3. Project management process 4. Site organization 5. Good engineering practices	1. Overall lifecycle and plan 2. Applicability 3. Responsibilities for major disciplines 4. Requirements for competency, verification, independent review, functional assessment, management of change (MOC), and auditing
	Risk Assessment	
1. Identify hazardous events 2. Develop risk reduction strategy	1. Government requirements 2. Risk criteria 3. Process safety information 4. Personnel who have knowledge of the process operation and its hazards 5. Hazard and risk analysis procedures	1. Hazardous events, describing initiating causes, consequence severity, and frequency 2. Independent protection layers including protective functions and risk reduction requirements 3. Action plans regarding residual risk
	Design	
1. Develop IPS design basis that implements the risk reduction strategy	1. Process safety information 2. Internal technical practices 3. Hazardous events, describing initiating causes, consequence severity, and frequency 4. Independent protection layers including allocated protective functions and risk reduction requirements 5. Operability, functionality, maintainability, and reliability requirements 6. Cost benefit criteria 7. Approved Equipment List	1. IPF Design Basis 2. Verification that IPF meets risk reduction
	Engineering, Installation, Commissioning and Validation	
1. Execute detailed design and application programming 2. Install, commission, and validate IPS 3. Specify operations and maintenance requirements	1. Process safety information 2. Hazardous events, describing initiating causes, consequence severity, and frequency 3. IPL Analysis report, including allocated protective functions and risk reduction requirements 4. Operability, functionality, maintainability, and reliability requirements 5. Approved equipment list 6. Internal technical practices	1. Detailed engineering specification 2. Installation, commissioning, and validation plan 3. Validation documents (software and hardware) 4. Pre-startup Safety Review 5. Operating and maintenance constraints 6. Complete, as built engineering and process safety information

Operation and Maintenance		
1. Operations and Maintenance Planning 2. Develop Operating Basis including operating, inspection, maintenance, and MOC procedures	1. Design Basis 2. Detailed specification 3. Bypass management process 4. Training requirements 5. Access permit requirements 6. Configuration management process 7. MOC 8. Failure tracking process 9. Trip tracking process 10. Audit process	1. IPS equipment list shall identify equipment by a unique tag number, classification, and risk reduction and schedule inspection, preventive maintenance, and proof tests 2. Operating procedures, including hazardous event description, failure response, compensating measures, when to execute a safe shutdown, what to do when a shutdown fails, access security, and MOC 3. Maintenance procedures, including hazardous event description, failure response, bypass, access security, and MOC 4. Fault alarm, protective alarm, and trip records 5. Bypass records 6. Inspection, maintenance and proof test records 7. MOC records 8. IPS equipment failure reports 9. Trip reports 10. Audit reports
Continuous Improvement		
1. Monitor performance and close gap between expected and actual performance 2. Make investments to reduce risk further when practical	1. As built documentation 2. Training records 3. Authorized bypass records 4. MOC records 5. Proof test records with failure investigation reports 6. Incident and near miss investigation reports 7. Trip reports 8. Verification, assessment, and audit reports	1. Recommended changes to equipment, practices, or training 2. Incorporation of lessons learned as appropriate into personnel training, equipment selection, installation practices, and mechanical integrity practices 3. Improved process safety and reliability performance 4. Fewer incidents and near misses 5. Regulatory and good engineering practice compliance 6. Better owner/operator reputation and shareholder value 7. Better personnel participation

Table 2.1. Objectives, Inputs and Outputs by Lifecycle Phase.

2.4 GOOD ENGINEERING PRACTICES

Technology is evolving at a rapid pace, bringing tremendous productivity, quality, and cost benefits. For instrumented systems, technology improvements have brought more capability and flexibility, but these improvements have also added complexity to the system design due to new, sometimes unpredictable, failures. Interoperability, compatibility, increased upgrades, and obsolescence are major issues for project execution, because the lifespan of some advanced technologies has decreased to less than 10 years.

When a new technology is deployed, owner/operators should incorporate lessons learned during its operation into design and management practices. In this way, the same mistakes can be avoided when equipment is used for other services. A culture of continuous improvement includes the documentation of successful practices and, when necessary, the explicit exclusion of bad practices or inappropriate technologies. With rapid technology change, it is sometimes difficult to gain this experience fast enough to realize the full benefit of the technology without operational impact or problems. With complex technology, the number of potential problems is great and learning each lesson, one at a time, is resource intensive.

Active participation in industrial trade organizations is important than ever, serving as a forum for open, technical exchange. Organizations, such as the American Society of Mechanical Engineers (ASME), American Petroleum Institute (API), American National Standards Institute (ANSI), National Fire Protection Association (NFPA), Instrumentation, Systems, and Automation Society (ISA), International Electrotechnical Commission (IEC), Institute of Electrical and Electronics Engineering (IEEE), International Organization for Standardization (ISO), International Society of Pharmaceutical Engineering (ISPE), and American Institute of Chemical Engineers (AIChE), publish good engineering practices covering a wide variety of subjects and applications related to safe and reliable operation.

Good engineering practice is a continuously moving target, as old practices are modified or replaced to meet the needs of today's process operation and technology. Applicable good engineering practices should be reviewed to gain an understanding of the consensus approaches and prescriptive methods for achieving safe operation. Few governments mandate specific industrial codes, standards, or practices, instead most require that the owner/operator:

- Review current good engineering practices to determine intent and applicability, and
- Incorporate applicable techniques and approaches into internal work practices.

Finally, relevant good engineering practices should be identified, evaluated, and incorporated into internal work practices. The good engineering practices used during the process design become part of its design basis. If the design deviates from, or conflicts with, referenced good engineering practices, the design basis should provide the rationale behind the deviation.

2.4.1 Conformance

The owner/operator is responsible for determining the good engineering practices that are applicable to the process and its equipment. The latest revision of a practice should always be considered when designing a new facility, expanding an existing facility, or upgrading existing equipment or controls.

Each owner/operator should have a process for evaluating how changes in good engineering practices affect the operation of, or requirements for, IPSs throughout the process life. Management should establish policies and provide necessary resources to support this evaluation and to ensure that internal practices keep pace.

Upgrading existing facilities to meet the requirements of newer good engineering practices may be accelerated when systems are found to no longer fulfill the requirements for safe operation. This determination may be made in any number of ways, such as:

- Hazard and risk analysis findings,
- Test and inspection reports,
- Management of change activity,
- Mechanical integrity program findings,
- Near miss or incident investigation reports,
- Internal audit reports,
- Insurance audits,
- Inspection agency audits, and
- Operation and maintenance reports.

2.4.2 Application-Specific Practices

Many industry organizations, such as NFPA, API, and ASME, publish good engineering practices that cover specific process equipment applications. These application-specific practices tend to focus on process equipment that is widely used within the process industry. In these cases, there are stakeholders, e.g., owner/operators, manufacturers, insurers, and government agencies, interested in undertaking the effort of developing a consensus practice.

Application practices tend to be prescriptive in the design requirements, but are often generic and limited to very specific hazards that are common to the majority of applications. When these practices are applied to a particular piece of equipment, consideration should be given to the hazards associated with the actual process design, which may be more extensive than what is covered by the application practice. In general, hazard and risk analysis should be applied to the process equipment to ensure that hazards are appropriately identified and managed.

2.4.3 SIS Good Engineering Practices

The first international standard for safety-related systems was published by the International Electrotechnical Commission located in Geneva, Switzerland. This standard, IEC 61508, *Functional Safety of Electrical/Electronic/Programmable Electronic Safety Related Systems*, is a seven part document and establishes the lifecycle and requirements for safety-related systems. Sector-specific standards, such as IEC 61511 for the process sector, follow the fundamental structure and requirements of IEC 61508, but reflect its intent in sector-specific terminology.

IEC 61508 also provides specific criteria for the evaluation of hardware and software comprising a manufactured product intended for use in a safety-related system. Such an evaluation provides evidence that equipment can be considered for user approval in safety applications. Equipment compliance to IEC 61508 requirements is only one form of evidence for user approval. Actual field experience should be gained with a manufactured product before use in a safety application. IEC 61508 covers many industrial sectors in addition to the process industry, such as transportation, manufacturing, and medical. Equipment under review may not be appropriate for the operating environment in all sectors or in all process applications within a sector.

IEC 61511, *Functional Safety: Safety Instrumented Systems for the Process Sector*, was released internationally as the first sector-specific standard developed using the IEC 61508 framework. The US ISA SP84 committee adopted IEC 61511 as the replacement for its existing standard ANSI/ISA 84.01-1996. IEC 61511 was accepted in 2004 by the European Committee for Electrotechnical Standardization (CENELEC) as EN IEC 61511 and the American National Standards Institute (ANSI) as ANSI/ISA 84.00.01-2004 Parts 1-3 (IEC 61511 Mod.) For the purposes of this book, IEC 61511 and ANSI/ISA 84.00.01-2004 (IEC 61511 Mod.) will be jointly referred to as ISA 84.01/IEC 61511.

In 2005, the Occupational Safety and Health Administration (OSHA) acknowledged the US version as representing recognized and generally accepted good engineering practice for safety instrumented system (OSHA 2005). This letter was similar to one issued in 2000 related to ANSI/ISA 84.01-1996 (OSHA 2000). Other governments have acknowledged the international standard or language-translated versions as good engineering practice.

The only significant difference between ANSI/ISA 84.00.01-2004 and IEC 61511 is a "grandfather clause" included in the US standard for existing safety instrumented systems (SIS) designed and constructed in accordance with other codes, standards, or practices prior to the issuance of and ANSI acceptance of the new standard. This grandfather clause provides minimum criteria to determine the adequacy of the existing SIS. This issue is discussed further in Chapter 2.6.5 and in ANSI/ISA TR84.00.04 Clause 3.0 and Annex A.

2.4.4 ISA 84.01/IEC 61511 Overview

ISA 84.01/IEC 61511 requires a management system for identified SIS. This management system defines how an owner/operator intends to assess, design, engineer, verify, install, commission, validate, operate, maintain, and continuously improve their SIS. The essential roles of the various personnel assigned responsibility for the SIS should be defined and procedures developed, as necessary, to support the consistent execution of their responsibilities.

ISA 84.01/IEC 61511 uses an order of magnitude metric, the safety integrity level (SIL), to establish the necessary performance. A hazard and risk analysis is used to identify the required safety functions and risk reduction for specified hazardous events. Safety functions allocated to the SIS are safety instrumented functions; the allocated risk reduction is related to the SIL. The design and operating basis is developed to ensure that the SIS meets the required SIL. Field data is collected through operational and mechanical integrity program activities to assess actual SIS performance. When the required performance is not met, action must be taken to close the gap.

ISA 84.01/IEC 61511 is applicable to many different technologies, e.g., electrical, electronic, and programmable electronic systems. ISA 84.01/IEC 61511 does not address product development, such as the use or modification of a manufactured product in a manner not intended by the original manufacturer. Manufacturer claims related to product performance should be made in accordance with IEC 61508.

ISA 84.01/IEC 61511 also does not apply to programming in full variability language. This language is generally used in advanced process control applications, rather than safety instrumented systems. Programming in full variability language should be executed according to IEC 61508 Part 4 to ensure adequate use of techniques and measures which reduce systematic error. This is also true for embedded software development, which should follow IEC 61508 Part 4 requirements. Finally, for SIL 4, which is the highest safety integrity level, ISA 84.01/IEC 61511 refers the owner/operator to IEC 61508 for equipment selection, for the development of the application program, and for determination of the requirements for fault tolerance against dangerous failure.

The lifecycle and its requirements can be presented in many different ways. Figure 2.5 is an illustration of the ISA 84.01/IEC 61511 lifecycle process. Due to its broad scope, the standard has many general requirements, which are organized in 19 clauses. The ten steps of the lifecycle are summarized as follows:

1. Development of a management system (Clauses 5 through 7)

The intent of the ISA 84.01/IEC 61511 management system is the identification and correction of systematic errors that potentially impact safe operation. Effective safety planning is accomplished using a management system that includes procedures or practices to ensure that:

- Safety functions and required risk reduction are identified for each mode of process operation,
- Identified SISs are designed, installed, commissioned, and validated to meet the safety requirements specification,
- Identified SISs are inspected, tested, maintained, and operated to maintain equipment in the "as good as new" condition,
- The process is adequately protected during periods where the process is being operated with a known SIS fault,
- Changes to the SIS are evaluated through a change management process, and
- Dangerous SIS equipment failures and the occurrence of process demands are tracked and periodically assessed to ensure prompt response and resolution of any identified SIS inadequacy.

The first step in minimizing systematic error is the management of human errors. Management must identify essential skills, experience, and knowledge necessary to execute the activities associated with each lifecycle phase. Then, personnel are assigned responsibility for the lifecycle activities based on these requirements.

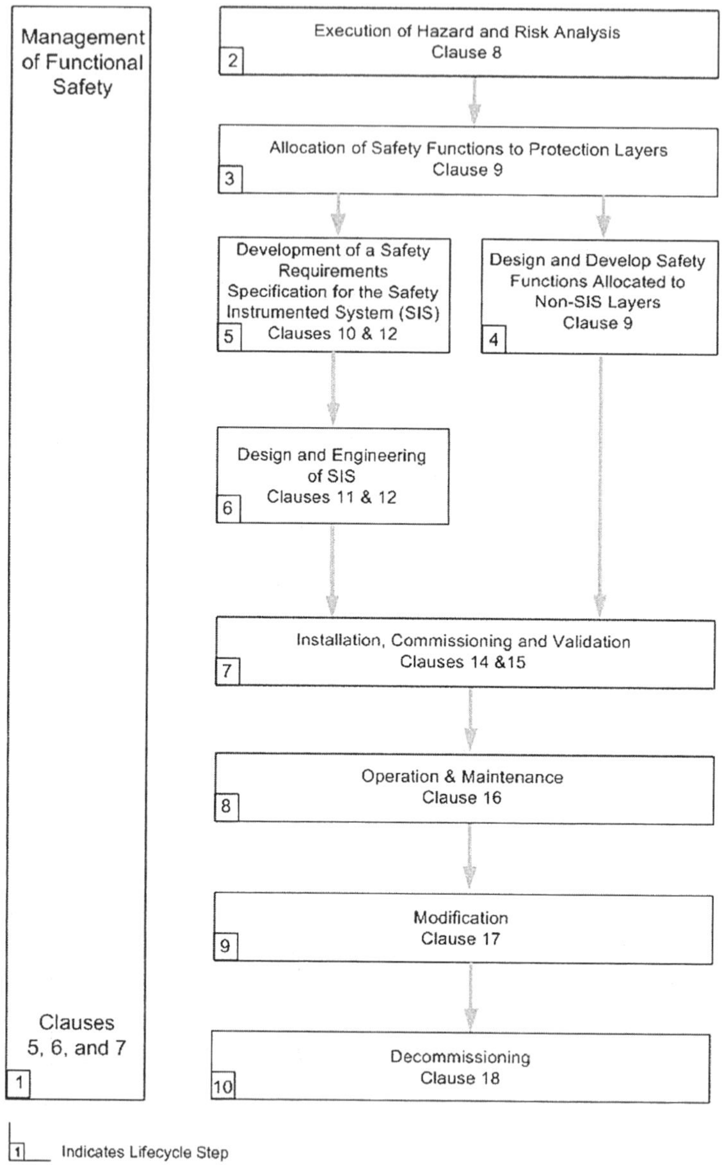

Figure 2.5. ISA 84.01/IEC 61511 Lifecycle.

The second step is the establishment of review processes throughout the SIS lifecycle to identify and correct systematic errors. A multi-layered approach is necessary that requires:

- Verification of individual tasks for consistency with purpose,

- Functional safety assessment at major lifecycle stages to ascertain whether the planned SIS can meet or is meeting expected performance, and
- Audits to determine consistency of the actual performance against expectations.

2. Execution of hazard and risk analysis (Clause 8)

A hazard and risk analysis team identifies potential hazardous events using an accepted hazard and risk analysis methodology. The initiating causes leading to each hazardous event are identified and their frequency of occurrence is estimated. The dangerous failure rate (or hazard rate) of a basic process control system that is not designed and managed according to ISA 84.01/IEC 61511 may not be assumed to be lower than 10-5/hr. The consequence severity of the hazardous event is determined based on site specific issues. The initiating cause frequency and the consequence severity define the process risk for the identified hazardous event.

The team compares the process risk to the risk criteria; if the process risk is greater than the risk criteria deems suitable, safety functions are identified that reduce the process risk. Inherently safer design should be considered, where possible, to eliminate or minimize process risk. Each safety function is allocated risk reduction as necessary to achieve the risk criteria.

3. Allocation of safety functions to protection layers (Clause 9):

Once the safety functions are identified, each is allocated to a protection layer that is capable of executing the required function and achieving the required risk reduction. The risk reduction allocated to a basic process control system that does not conform to ISA 84.01/IEC 61511 must be less than 10. The safety functions allocated to the safety instrumented system (SIS) are safety instrumented functions (SIF). The risk reduction is the safety integrity level (SIL) for each SIF.

4. Design and develop safety functions allocated to non-SIS layers (Clause 8 and 9):

IEC 61511/ISA 84.01 concerns SIS lifecycle management. However, all identified safety functions must achieve their allocated risk reduction. Consequently, each safety function should have a design and operating basis defining its requirements. The potential for common cause, common mode, and dependent failures between protection layers and the BPCS should be assessed. These failures should be sufficiently low compared to the overall risk reduction requirements.

5. Development of a safety requirements specification for the SIS (Clauses 10 and 12).

The safety requirements specification defines how the SIFs are designed and implemented in an SIS to meet the SIL for each process operating mode, such as:

- Requirements for the detection of and response to potential hazardous events,
- Requirements for fault detection, such as diagnostics and proof testing,
- Requirements for fault tolerance against dangerous failures,
- Provisions for safe bypass for maintenance and testing including maximum length of time that the SIS can be in bypass prior to management of change action,
- Provisions for safe operation when continuing process equipment operation with a detected SIS fault,
- Provision for safe shutdown if SIS fails to take action when required, and
- Requirements for start-up and shutdown.

Quantitative SIL verification is important to ensure that the design can achieve the required risk reduction. Reliability Block Diagrams, Markov models and Fault Tree Analysis are acceptable techniques for SIL verification. ISA TR84.00.02 provides an overview of SIL verification.

6. Design and engineering of the SIS (Clauses 11 and 12):

These clauses provide various requirements for the detailed design and engineering package. Clause 11 specifically addresses hardware requirements from equipment selection through engineering. These requirements are intended to address random hardware failures that can cause the SIS to fail to operate as required. Of special significance are the requirements for determining whether equipment is appropriate for use in an SIS and for determining the requirements for fault tolerance against dangerous failure. Clause 12 addresses the application program and provides the lifecycle requirements directed at minimizing the systematic errors that can cause software to perform incorrectly. It also emphasizes the importance of software specification, verification, and validation.

7. Installation, commission, and validation (Clause 14 and 15):

These clauses shift focus from design and engineering to the field installation and validation activities. Clause 14 covers the installation and commissioning of the SIS, while Clause 15 addresses validation. Validation involves proof testing the SIS to demonstrate that the installed SIF operates as required for the various process operating modes; that test, bypass, and manual shutdown facilities operate correctly; and that any fault alarms are generated as required.

8. Operation and maintenance (Clause 16):

This clause addresses the long-term performance of the SIS. It outlines the activities required during operation, including operator and maintenance response to detected faults and the execution of periodic proof testing. The clause requires the development of procedures to support these activities and personnel training. Requirements for auditing of operation and maintenance personnel performance against the procedures and SIS performance against expectations are also identified.

9. Modification (Clauses 5 and 17):

This clause defines the basic issues to be addressed when determining the impact of proposed changes to the process or its equipment on the expected SIS performance. These changes can be as obvious as substituting a feed chemical with different physical properties and hazards into the process or as subtle as changing how equipment is configured to fail on detection of equipment failure. This latter issue requires a special management of change process addressing equipment configuration. The standard refers to this as configuration management (Clause 5).

Changes should be evaluated and approved prior to implementation by assessing impact of the change on the SIS functional and integrity requirements. Follow-up activities are required to ensure that:

- Changes were successfully completed,
- Change did not impact any other functions,

- Documentation is updated, and
- Personnel are trained on the change.

Change control should also be supported through administrative controls (e.g., signed and dated approvals required), access security provisions, and auditing.

10. Decommissioning (Clause 18):

This clause requires an analysis be performed to determine the effect of SIS decommissioning on safe operation. Typically, decommissioning is done in conjunction with the installation of an upgraded SIS. Part or all of an existing SIS may be decommissioned when upgrade is required.

2.5 KEY MANAGEMENT SYSTEM ELEMENTS

Safe and reliable operation cannot be achieved without an effective management system. Lifecycle cost is lowest when trained resources are provided with work processes that address significant consequence events, including safety, environmental, asset protection, and business interruption risks. Process-and site-specific criteria should also be considered, such as operability, functionality, maintainability, and reliability requirements.

Quality design is achieved in an environment that fosters a safety and reliability-focused workforce. Quality systems address existing equipment through user approval practices and configuration management procedures. All successful quality programs have one thing in common: a documented system with defined metrics for monitoring, measuring, and improving performance. Personnel activities and tasks should be monitored using clear and unambiguous metrics, e.g., is equipment being fully proof tested as required by the operating basis?

The risk reduction achievable by any IPS is limited by the administrative processes used to determine the resource requirements, to assign work to personnel, and to review work against expectations. Human error must be reduced to the point that it has a negligible impact on IPS performance. This can only be achieved by applying a management system in a culture where safe and reliable operation is expected. Deviations from approved practices and procedures should require justification and independent approval.

2.5.1 Staffing

Management should identify the resources required to fulfill the lifecycle requirements. Competent personnel should be assigned to the various tasks necessary to design, inspect, maintain, test, and operate the IPSs. Operating plans should include provision for additional staffing to support commissioning, validation, and start-up activities and to cover any periods of process operation with detected IPS faults.

Consideration should be given to the independence of personnel assigned responsibility for the various IPSs to reduce potential common cause, e.g., assigning different departments responsibility for the BPCS and SIS. Also, the independence of personnel conducting verification activities should be assessed to ensure objective review.

2.5.2 Responsibility and Competence

Management should establish internal practices and procedures, outlining the work processes used to design and manage IPS. The intent is to minimize the potential for human error that causes IPS failure. The level of detail employed when writing practices and procedures is related to the allowable variability in the final output. Risk assessment, design, engineering, installation, and commissioning are often covered by general practices, allowing the physical systems to be tailored to meet specific process risk and business needs. The operation and maintenance activities are typically covered by prescriptive procedures, identifying tasks that should be executed as directed.

Competence is closely related to the individual's understanding of their role in safe operation. Competency is affected by the availability of information and procedures, as well as data integrity, consistency and quality. Many activities associated with the protective management system are task oriented. Focusing on day-to-day task execution can lead to a narrow understanding of facts and procedures. When compliance expectations are not well understood, it is easy for actual practices to deviate significantly from documented practices.

Without documented practices and procedures, the purpose and specifics of essential activities often become known to only a few individuals within an organization. When personnel changes occur, internal practices provide a safety net, ensuring work tasks are consistently executed and information is not lost. Internal practices are the most effective means of communicating minimum requirements, including required work activities and the expected outcome of those activities.

Personnel assigned responsibility for a work process activity should understand how to execute assigned tasks according to the owner/operator's practices and procedures. Personnel should be trained on the overall management system and the specific work processes related to their job function. Whether the "person" part of personnel is an engineering contractor, a manufacturer, construction company representative, consultant, contractor, or direct-hire employee, personnel should have the ability to use their knowledge, understanding, practical and logical skills to execute their work effectively to the standards required for employment.

Competence can be gained through general education, work process training, skills-based training, and hands-on experience. As procedures are developed to support the various lifecycle activities, key skills and knowledge required for work execution should be identified and documented. Government agencies, insurers, or the owner/operator may encourage or even require that personnel who perform certain tasks have specific types of training or credentials. For example, many owner/operators require that operators have a two-year process technology degree.

When project assignments are made, the required skills and knowledge should be compared to the individual's capabilities. Lack of necessary skills or knowledge should be addressed through training, increased supervision and review, mentoring programs, and/or project oversight. Competent trainers and mentors should be identified and formally assigned to monitor project and plant personnel activities, as necessary. Specialized training should be considered to focus on job responsibilities at the discipline and department level.

In training personnel, it is important to recognize that there is generally a lag between a statement of corporate intention and a change in an individual's belief system. In a pressure situation, it is the individual's belief system that is most likely to determine the individual's actions, not simple knowledge. Sustainable performance is achieved when the management system is taken seriously and treated as a license to operate.

2.5.3 Independent Review

Throughout the IPS life, verifications, functional assessments, management of change reviews, and audits are performed to objectively determine the appropriateness of its design and management. Depending on the task under review, objectivity can be attained by peer, supervisory, external department, or external site review. Independent reviews are performed by qualified personnel who are not directly associated with the work activity being verified, assessed, or audited.

Quite naturally, once the reviewer becomes involved in the work process, absolute "independence" is lost. Further, in many organizations, the independent reviewer will have mentoring and training responsibility, so independence can not be judged absolutely.

In a more general sense, consider independence as a mental attitude reflecting a freedom from bias that allows the reviewer to objectively look at the information and make a judgment as to its acceptability against owner/operator standards. The reviewer(s) should have sufficient knowledge, skills, and experience to understand the basis for work execution and to determine whether the final result meets expectations.

2.5.3.1 Verification

Verifications generally involve review of available information and documents by an alternate member of the project team or department. Verifications evaluate the consistency of the input (scope) and output (deliverables) documents. For example, project documents are reviewed to determine whether the scope for the specific work process step was met, e.g., does the engineering design package meet the established design basis.

Verification is the act of determining and documenting whether the outcome of work processes, activities or tasks conform to specified requirements and traceable input information. Verification activities are quality control checks performed to reduce the potential for human error to cause IPS failure. These activities include a wide variety of activities, such as reviewing, inspecting, checking, and testing.

Verification can be conducted by peers, supervisors, or managers, as deemed appropriate for the work process activity and the IPS classification. Anyone responsible for a verification activity should have sufficient skills, knowledge and experience to determine whether the outcome of the task is consistent with expectations.

The review rigor may be varied by the IPS classification, design complexity, and experience with the equipment being implemented. Complex or highly sequenced systems need more detailed review than those that follow standardized or prescriptive designs. When unfamiliar hardware and software is implemented, additional reviews may be needed to ensure the equipment is designed and managed appropriately to meet perhaps similarly unfamiliar requirements.

2.5.3.2 Functional Assessment

Functional assessments are quality control measures used to ensure the design, maintenance, inspection, testing, and operating strategy is consistent with practices and expectations. Functional assessments are recommended at five stages of the IPS lifecycle, as shown in Figure 2.6:

- Design,
- Engineering,
- Installation, Commissioning, and Validation,
- Operational and Mechanical Integrity, and
- Continuous Improvement.

Significant changes between the design and installation phases are less likely on short duration projects. When a project has a significant time gap between design and implementation, numerous changes may be identified. Some may be triggered by the identification of better practices. Or perhaps, there are changes in the approved equipment list.

As time passes, many changes may occur, which affect the IPS design and operating basis. These changes may be as fundamental as owner/operator expectations or the process requirements. They may also be as detailed as the operating environment specification, the available technology or the personnel assigned responsibilitv.

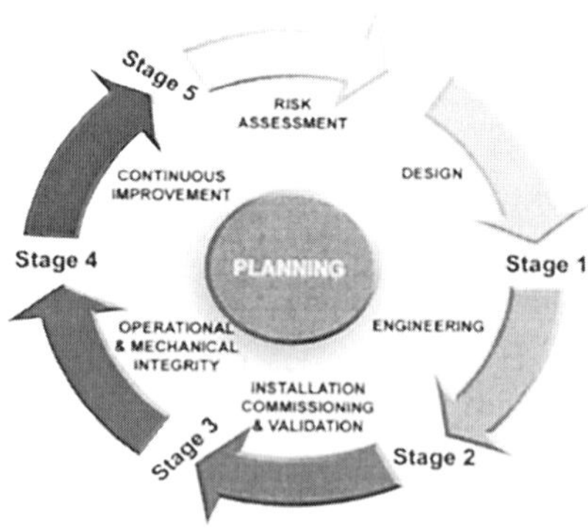

Figure 2.6. Lifecycle Illustrating Functional Assessment Stages.

Functional assessments are executed to ensure the cumulative outcome of the various changes still meets the original design intent and is compliant with owner/operator practices and expectations. These assessments ensure:

- Equipment is installed and demonstrated to operate according to design basis,
- Adequate safety, operating, maintenance, and emergency procedures are in place,
- Appropriate hazard and risk analysis or management of change reviews have been conducted and their recommendations addressed, and
- Training of relevant personnel is complete.

The stage 1 and 2 assessments are typically included as part of normal project execution. In many project management processes, stage 1 occurs at the end of front-end loading, while stage 2 occurs toward the completion of detailed engineering. Since the assessments may result in recommendations for change, they should be conducted early enough to allow resolution without significant disruption to the project schedule or budget.

The stage 3 functional assessment is executed for any new or modified IPS after installation, commissioning, and validation is completed. This assessment is equivalent to the pre-startup safety review required by many governments, among them Germany, the United Kingdom, and the United States. The pre-startup safety review is discussed further in Chapter 5.12.5 and in the CCPS/AICHE book, *Guidelines for Performing Effective Pre-Startup Safety Reviews*.

The stage 4 assessment evaluates how existing IPS equipment, procedures, and personnel are performing relative to expectations. IPS performance is monitored through fault alarm, protective alarm, trip, bypass, inspection, maintenance, and proof test records. Stage 4 seeks to identify and close gaps between the expected and actual performance as indicated by these records. Operational and mechanical integrity is achieved when the process operates in a safe and reliable manner and IPS equipment is maintained in the "as good as new" condition. The Stage 4 assessment often identifies opportunities for improvement of existing systems.

Continuous improvement seeks to improve equipment, procedures, and personnel performance. The need to change the process or the IPS equipment often results in action items to be tracked to completion. Periodic hazard and risk analyses identify new hazards and new opportunities for risk reduction. Lessons are learned at other facilities and shared throughout the organization.

A stage 5 assessment benchmarks existing IPS design and management practices to current owner/operator expectations and good engineering practices. It also determines whether failure and trip reports demonstrate that IPS equipment is suitable for continued use or should be replaced. Stage 5 should examine various reports and records, such as hazard and risk analysis and management of change action plans, audit reports, and near miss/incident investigation reports. An active continuous improvement program should result in efficient and timely implementation of action plans, improvement in adherence to practices and procedures, and a reduction of facility near misses and incidents.

Assessments may be conducted by one person or a team, depending on the amount of information and documentation being reviewed and the number of personnel to be consulted and interviewed. The assessment team should include personnel with knowledge and experience in the management system, as well as documentation and record expectations. When possible, the team should include personnel, who are sufficiently independent of the work being reviewed to ensure an objective assessment.

The rigor of the functional assessment depends on many factors, including the project size, IPS complexity, project duration, design standardization, and previous experience. Deficiencies discovered during the functional assessment should be should be tracked to completion using an appropriate management of change process.

2.5.3.3 Management of Change

Properly managing changes to process chemicals, design, technology, equipment, and facilities is essential to safe and reliable operation. Changes should be identified, reviewed, and approved through a management of change process prior to implementation. Changes include any modification to the design, equipment, procedures, diagnostics, testing, and installation, other than "replacement in kind." Typical drivers for changing the IPS are reduced maintenance cost, increased reliability, impending obsolescence, and compliance with applicable regulations or good engineering practices. The management of change process ensures that:

- Those assigned responsibility for safe operation agree that the change is appropriate,
- The change does not increase the risk reduction requirements,
- The change does not reduce the risk reduction capability of the IPS,
- Personnel only implement approved changes,
- Changes are recorded, and
- Changes are traceable to the design basis.

Changes in process documents, such as piping and instrumentation diagrams, operating procedures, mechanical integrity programs, electrical classifications, etc., should be noted, so that other related drawings and procedures can be updated. Copies of approved process changes should be kept in an accessible location. Design changes should be available to operating personnel, as well as to process hazards analysis (PHA) teams during any relevant hazards analysis.

Changes deemed “temporary” have caused a number of catastrophes over the years, and owner/operators need to establish ways to detect and track such temporary changes, as well as those that are intended to be permanent. It is important that a time limit for temporary changes be established and monitored since, without control, these changes tend to become permanent through inaction.

Temporary changes should be subject to management of change. Proper change management should ensure the equipment and operating procedures are returned to their original (or as intended) condition at the suspension of these temporary modifications. Thorough documentation and review of these changes is invaluable in assuring that the safety and health considerations are being incorporated into the operating procedures and the process.

Employers may wish to develop a form or clearance sheet to facilitate the processing of changes through the management of change procedures. A typical change form may include the following:

- Description and the purpose of the change,
- Technical basis for the change,
- Safety and health considerations,
- Requirements for updating the design and operating basis,
- Impact to electrical classification,
- Requirements for communication to affected personnel,
- Requirements for training,
- Requirements for pre-startup safety review,
- Allowable duration, if a temporary change, and
- Approval or authorization.

Management of change should also be practiced at the configuration level for IPSs. Configuration management is the process of identifying the components of an evolving (hardware and/or software) system for the purposes of controlling changes to those components and maintaining continuity and traceability throughout the lifecycle.

Configuration management typically becomes an owner/operator responsibility after the pre-start-up safety review has confirmed that the IPS equipment is operating per design intent. To support configuration management, the management system should include procedures to uniquely identify all constituent parts of the IPS, including hardware and software components. Configuration management is intended to ensure appropriate replacement in kind evaluation and to prevent unauthorized equipment from being placed in an IPS application.

Controlling configuration at this level requires not only a detailed procedure, but also the expertise to make the replacement in kind assessment. For example, when a replacement transmitter is purchased, the software and, perhaps even the hardware, may have been modified by the manufacturer. A knowledgeable person should review the transmitter changes to ensure that these changes do not affect the design and operating basis requirements. For example, does the new version have different requirements for specification, configuration, installation, inspection, operation, maintenance or proof test? This review includes assessment of the transmitter and interconnected equipment. Manufacturers can assist with this assessment by reporting how changes to the equipment affect the functionality, integrity, or reliability.

2.5.3.4 Auditing

Audits determine and document whether management system work processes and procedures are in place, up-to-date, and being followed. Audits are an important quality assurance review for the management system. Personnel should not audit themselves or their own department. Auditors should understand management system expectations related to input information, work processes, and output documentation.

The person (or team of people) performing the audit should have the experience necessary to evaluate the appropriateness of the design and management of the identified IPSs. The number of people required for the audit is dependent on the amount of equipment under review, the number of potential hazards involved, and the complexity of IPS design, operation, and maintenance. At least one person on the audit team should be independent of the facility being audited

An audit is an independent review of the available information and documentation, inspection of physical facilities, and interviews with plant personnel to:

- Determine the adequacy and effectiveness of management system,
- Ensure compliance with policy, practices, and procedures, and
- Recommend any necessary changes.

The audit interval should be defined by the management system. Those sites demonstrating poor performance related to the protective management system should be audited more frequently. The interval is typically affected by:

- Site safety culture,
- Required risk reduction,
- Expected frequency of IPS change,
- Number of changes made to the IPS, and
- Government or jurisdictional requirements.

2.6 SPECIAL TOPICS

2.6.1 Non-safety Risks

Many governments have regulations regarding the prevention of catastrophic incidents. Throughout the lifecycle from the earliest stages of process design to decommissioning, the potential for hazardous events causing catastrophic impact is analyzed and means to reduce this potential are implemented. In some applications and market sectors, the IPS used to prevent such events are subject to inspection by third parties.

Asset risk is different. While insurance requirements may play a role, there are no government mandates to minimize economic, asset or business interruption losses. These non-safety risks should be considered during the development of the risk reduction strategy. It makes good business sense to minimize incidents that move the process equipment away from normal and efficient operation.

Industrial boilers, furnaces, and rotating equipment are highly integrated with the process unit, providing essential utilities, such as heat, steam, and power. The risk assessment findings for this equipment often identify unacceptable process conditions that can result in extensive internal damage with significant potential downtime. While the internal damage can be very serious, the event may not result in immediate safety consequences. If the IPS were defined by the safety risk alone, significant economic risks would remain unchecked.

Any major incident, even one with no immediate safety impact, results in emergency response (exposing emergency personnel to hazards), in demolition and rebuild (with hazard to workers and surrounding process units), and a demand for committed resources, which would normally be supporting other activities, until restart of the process unit. Simple calculations can show that incidents cost money. Owner/operators who abuse their franchise to operate often suffer a significant financial impact.

The work processes outlined in this book are applicable to the analysis and control of most identified hazards. Non-safety risks can be managed through the same day-to-day work processes that are recommended for safety risks. The specific application to non-safety risk is dependent on many factors, such as insurance requirements, process equipment criticality, and loss potential. However, the cost associated with extending the protective management system to cover these risks is generally negligible, while the derived return is quite substantial.

A classification system can be used to delineate between safety, environmental, and asset protective systems. Many governments require the identification and management of the risk associated with catastrophic safety and environmental consequences. Delineation of those systems reducing catastrophic safety and environmental risk versus that addressing asset risk allows the owner/operator to establish different management systems for "safety" versus "non-safety" risk. Asset protective systems can be implemented based on prescriptive requirements, performance targets, or cost versus benefit analysis. Refer to Chapter 2.6.4 for more information on classification.

2.6.2 Lifecycle Costs

Lifecycle cost should be carefully considered. Manufacturers provide a wide range of choices for IPS design. These choices also come with a wide variety of capital and long-term operation and maintenance costs. The choice between low redundancy and high redundancy should be influenced not only by the required performance, but also by short- and long-term costs and resource requirements.

For example, it is possible to design a low capital cost, low redundancy IPS that achieves a high degree of risk reduction. From a capital budget perspective, this low redundancy system may appear to be the best choice. However, to achieve its high integrity, this system may require extensive diagnostics (with a high potential for fault alarms and spurious trips) and frequent proof testing. Resources must be identified to respond to the expected fault alarm load and to perform the required proof tests.

In contrast, using higher redundancy may result in higher capital cost, but lower overall operation and maintenance costs. A fault tolerant, high integrity system requires less frequent proof testing and provides more options for on-line diagnostics and repair. With redundancy, it may be possible to extend the proof test interval to the process unit turnaround interval, reducing the requirements for on-line tests.

2.6.3 Managing Organization Changes

Organizational shifts can affect safe operation due to changes in the roles and responsibilities of individual personnel, organizational reporting structure, and allocated resources. When departments and organizations are re-organized, right sized, or re-engineered, management system activities can get lost in the shuffle. These transitions must be reviewed to ensure that trained resources with the proper experience are identified and retained. The importance of managing organizational changes should not be underestimated, because these changes can lead to the following problems:

- Excessive work load leading to increased systematic errors,
- Loss of competency within the organization,
- Loss of communication structure to share lessons learned,
- Conflicting priorities where day-to-day issues overwhelm long-term risk reduction priorities, and
- Decreased morale with reduced focus on essential work activities.

Change is a normal part of business life in the process industry, so policies and procedures should be in place to deal explicitly with change. Organizational changes affecting personnel assigned responsibility for safe operation should be evaluated to ensure work activities continue as required. Management of change processes can be used to assess change impact and to develop an action plan that addresses any gaps and manages any transition.

2.6.4 IPS Classification

Proper equipment classification is important. Classification reduces systematic error through a clear and unambiguous distinction of what equipment is safety and/or protective and falls under the protective management system and what equipment is associated with control and does not.

All safety equipment should be designed, inspected, maintained, tested, and operated in accordance with applicable good engineering practices. Other protective equipment, such as those responsible for reducing business risk, including production loss, diminished product quality, and equipment damage, may be managed in a less rigorous manner if implemented sufficiently independent of the safety equipment.

Safeguards consist of a wide variety of measures deployed to protect the process equipment throughout its life. Some of these safeguards are defined as independent protection layers during the risk assessment phase. Instrumentation and controls which stop hazardous events posing safety risk are instrumented safety systems (ISS).

Within the process industry, ISS can be implemented as various types of systems. Appendix B of this book provides a discussion of various protection layers, including those relying on instrumentation and controls. Figure 2.7 illustrates various terms, which can be used to further classify ISS. Other terms are used within the process industry.

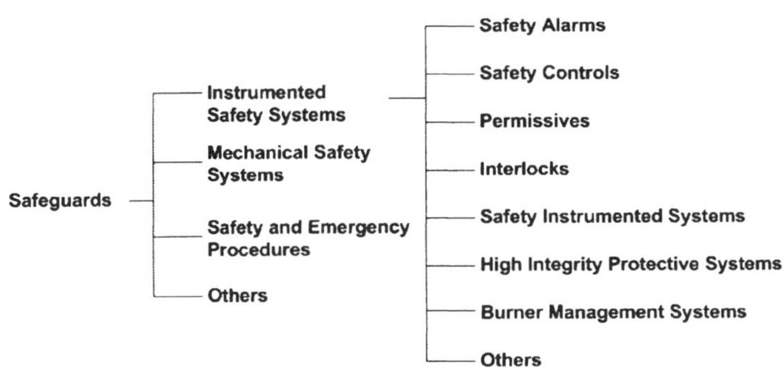

Figure 2.7. Example Instrumented Safety System Classification.

Classification has been accomplished using a wide variety of methods. The risk assessment phase, as presented in Chapter 3, results in the identification and classification of each ISS. Hazard analysis employs proven techniques for identify potential hazardous events with undesirable consequences. Risk analysis is used to make decisions on the risk reduction strategy. The classification system typically addresses undesirable consequences, which pose safety, environmental, asset protection, or business interruption risks.

During hazard and risk analysis, protective functions are allocated risk reduction as necessary to reduce the hazardous event risk in accordance with specified risk criteria. When the consequence involves significant harm to humans, such as serious injuries and/or fatalities or serious impact to public health, the protective function is a safety function. Likewise, when it involves significant damage to the environment, it is an environmental function, and so on. This yields a clear, rigorous, and consistent rationale for the IPS classification.

However, if the IPSs are not currently classified, it takes time to go through an existing facility and perform a hazard and risk analysis. Consequently, it may be necessary to use screening processes to define the equipment classification in an existing facility. Simple screening methods are commonly used to make an initial classification.

Screening examines existing functions to identify the hazardous event(s) being protected against. The team then determines the consequence and its severity for each identified event, e.g., what can happen and how widespread. If the event exceeds a consequence threshold, additional analysis is performed to determine the risk reduction requirements. Regardless of the technique used for classification, the rationale should demonstrate appropriate consideration for the potential risk and its allocation to the various IPLs.

2.6.5 Existing SIS

There is one significant difference between IEC 61511 and ISA 84.01-2004. The latter adds a single clause to Part 1, Clause 1y. The clause is known as the "grandfather clause" and states the following:

"For existing SIS designed and constructed in accordance with codes, standards, or practices prior to the issuance of this standard (e.g. ANSI/ISA 84.01-1996), the owner/operator shall determine that the equipment is designed, maintained, inspected, tested, and operating in a safe manner."

The grandfather clause provides essential criteria that must be considered when evaluating whether an SIS should be upgraded to the new requirements or can be left "as is." ISA TR84.00.04 is a technical report developed by the ISA SP84 committee to provide users with information and guidance related to the ISA 84.01-2004. According to ISA TR84.04, there are two essential steps to determine the applicability of the grandfather clause:

- Confirm that a hazard and risk analysis has been done to determine qualitatively or quantitatively the level of risk reduction needed for each SIF in the SIS, and
- Confirm that an assessment of the existing SIF has been performed to determine that it delivers the needed level of risk reduction.

It is further recommended in ISA TR84.04 that, if the above activities have not been done, they should be scheduled for review at the next appropriate opportunity. It is the responsibility of the owner/operator to:

- Determine that existing SISs meet the intent of the grandfather clause, and
- Document the operating, testing, inspection and maintenance conditions under which this will remain true.

"Grandfathering" does not apply to the management system or documentation requirements. Shortcomings in the process safety information and management system should be addressed prior to "grandfathering" the SIS.

The most important step in addressing the applicability of the "grandfather" status is the development of a methodology to identify existing SIS and to evaluate their appropriateness against current practices. An action plan should be established for evaluating existing SISs, considering, among other factors, the risk potential and anticipated gaps with the standard.

Each owner/operator should develop internal guidance for its facilities to follow with regard to existing SIS. Management of change activities and periodic process hazards analysis drive the evaluation of existing process risk and can challenge the appropriateness of a grandfather claim. Various study findings require prioritization and actions plans. Work processes and procedures developed for process safety management should be leveraged.

Good engineering practice dictates that any process posing a significant risk to personnel, the environment or the community should undergo a periodic evaluation of its risk reduction strategy. In the United States, this periodic review must occur at a minimum of every five years. Likewise, in the United Kingdom, "Top Tier" COMAH (Control of Major Accident Hazards) sites are required to review and revise their Safety Report every five years. Similar requirements exist throughout the European Union, as well as in many other countries.

During the hazard and risk analysis, the team should confirm that the demonstrated SIS performance meets the requirements and that the SIS design, maintenance, inspection, testing and operation are consistent with the required risk reduction. This judgment should also consider the frequency of process demands on the SIS and the history of incidents and near misses associated with the SIS. The maintenance, testing, and inspection records should also be reviewed to ensure that the frequency of these activities is sufficient and that all critical equipment is performing as intended.

The upgrade of existing facilities should be accelerated when existing equipment is found to no longer meet the required risk reduction. This determination may be made through a variety of activities, such as the hazard and risk analyses, inspection and proof tests, failure and trip investigations, audits, and gap analyses.

Checklists may be used to support or supplement the SIS evaluation. Checklists may be developed to address major technology, architecture, design, and testing practices. A significant deviation from accepted practice would indicate potential exclusion from the "grandfather" clause.

Once an owner/operator determines that the existing SIS does not meet the intent of the "grandfather" clause, there should be a defined process to address the identified deviations. When the risk is greater than desired, these gaps should be addressed with an action plan which often establishes priority based on the size of the gap and the nature of process risk (e.g., frequency and consequence) associated with the potential event. Many companies use a risk ranking matrix to rank process hazards analysis recommendations. Similar work processes can be used to prioritize action plans for closing the gaps.

3

RISK ASSESSMENT

As catastrophic incidents occurred in the process industry, communities demanded regulations to protect workers, the environment, and the community itself. Many governments responded by issuing regulations that require the identification and management of risk associated with hazardous events posing catastrophic consequences. While each government has its own definition, catastrophic consequences typically result in losses with major consequences and unacceptable lasting effects. Such consequences usually involve significant harm to humans, substantial damage to the environment, and/or loss of community trust with possible loss of franchise to operate.

Achieving compliance requires a risk assessment process, where a formal hazard and risk analysis process identifies potential hazardous events with undesirable consequences and makes decisions on the risk reduction strategy for the identified events. Risk management requires the systematic application of management policies, procedures and practices to the task of preventing these hazardous events.

Although governments generally require risk assessment for catastrophic incidents only, most owner/operators consider potential economic impacts as well. It makes good business sense to minimize events that move the process unit away from normal operation. The derived return associated with properly addressing business risk can be substantial. Consequently, the risk assessment discussed in this chapter addresses significant consequence events, which may cause significant harm to humans, damage to the environment, loss of property, and business interruption.

When designing process equipment, there is always the question of just how much to invest in resources, equipment, instrumentation and controls, time, etc. to reduce the risk further. Consider the following examples of risk decisions suggested in *Tools for Making Acute Risk Decisions with Chemical Process Safety Applications* (CCPS/AIChE 1995):

- Should any action be taken to further reduce the risk in a planned process or in an existing process?
- Which one from a number of alternatives should be adopted to reduce the risk?

- How should a company allocate resources among its facilities and process plants in order to reduce overall risk?
- Should a company undertake a new venture that has significant risk and financial uncertainty?
- When should a company get out of a business area based on risk?

These decisions are not easy and can be unintentionally affected by financial decisions made throughout an organization. The economic climate caused by globalization has increased the pressure to be the lowest cost producer. The risk assessment phase ensures sufficient protection layers are provided and focuses attention on continuous improvement in process safety.

The Health and Safety Executive (HSE) publication, *Principles and guidelines to assist HSE in its judgments that duty-holders have reduced risk as low as reasonably practicable*, states "We believe that the greater the risk, the more that should be spent in reducing it, and the greater the bias on the side of safety."

In the HSE publication *Guidance on 'as low as reasonably practicable' (ALARP) decisions in control of major accident hazards (COMAH)*, HSE states that it expects the owner/operator to have suitable controls in place to address all significant hazards and that those controls, as a minimum, must implement authoritative good practice irrespective of situation based risk estimates. HSE has established a burden of proof, allowing additional safeguards to be ruled out only if the sacrifice involved in taking them is grossly disproportionate to the risk reduction benefits.

The process industry covers a wide spectrum of process operations, making it virtually impossible to develop a single solution that meets everyone's needs. For example, refineries have different operational, quality control, and reliability goals than those of pharmaceutical manufacturers. The rate of technology evolution also demands a flexible risk management approach that allows adjustment in how resources and systems are deployed.

Thus, very few governments have specific requirements for risk reduction. Instead, most rely on the application of recognized good engineering practices (see Chapter 2.4). These practices evolve to account for technology changes, for new knowledge and lessons learned, and for new means to prevent hazardous events. Applicable recognized and generally accepted good engineering practices should be incorporated into the owner/operator's internal policies, procedures, and practices.

During the risk assessment phase, as shown in Figure 3.1, hazardous events are identified and the risk reduction strategy is developed. The process risk associated with each hazardous event is reduced below the owner/operator risk criteria using protective functions. Each protective function is allocated to an independent protection layer (IPL) that is designed and managed to achieve the required risk reduction.

Instrumented protective functions (IPFs) are identified that reduce the risk below the risk criteria. IPFs allocated to the preventive layer are protective instrumented functions (PIFs). Other IPFs may be identified, such as protective alarms and protective controls. During the design basis development (see Chapter 4), the IPFs are physically allocated to systems. When the hazardous event poses safety risk, the IPS is an instrumented safety system (ISS).

This phase relies on widely accepted methods for identifying and analyzing hazardous events, including those outlined in:

- *Guidelines for Hazard Evaluation Procedures* (CCPS/AIChE 1992),
- *Layer of Protection Analysis: A Simplified Risk Assessment Approach* (CCPS/AIChE 2001),
- *Guidelines for Chemical Process Quantitative Risk Analysis* (CCPS/AIChE 2000), and
- *Guidelines for Consequence Analysis of Chemical Release* (CCPS/AIChE 1999).

While all owner/operator's seek to minimize risk, the methods used to define the risk and to determine its acceptability are different across the process industry. The risk reduction strategy can also be quite variable. In some industry sectors, certain types of risk reduction equipment are simply expected. In others, the same equipment would be considered unnecessary.

Consequently, this chapter provides guidance concerning various risk assessment activities, but does not attempt to provide specific risk criteria or rules to use for judging the adequacy of a specific design and/or administrative control. Instead, it provides an overview of the risk assessment work process and a discussion of key management system elements.

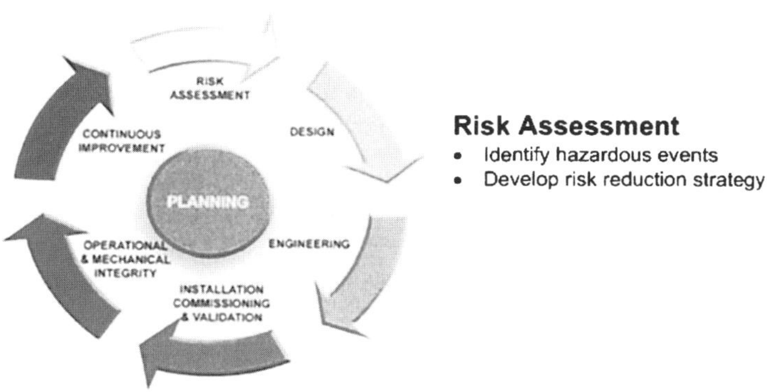

Figure 3.1. Risk Assessment Phase.

3.1 INTENDED AUDIENCE

The personnel responsible for supporting the risk assessment phase should understand the application of the specified risk criteria in the hazard and risk analysis. Each individual should have an understanding of how the process operates and have sufficient experience to assist in the identification of hazardous events and development of the risk reduction strategy employed to manage the identified events. This chapter is intended for:

- **Process safety** who need to understand how the risk criteria is used during the hazard and risk analysis, participate in and/or support the hazard and risk analysis execution, maintain the documents under revision control, and track recommendations,
- **Process** who participate in the hazard and risk analysis and respond to its findings,
- **Instrumentation & Electrical** who participate in the hazard and risk analysis and respond to its findings, and
- **Operations** who participate in the hazard and risk analysis and respond to its findings.

3.2 INPUT INFORMATION

Hazard and risk analysis is executed throughout the process unit lifecycle. Depending on when the study is performed, the information available varies widely. The choice of hazard and risk analysis method is often influenced by the available information. At the research and development stage, the information may be preliminary, such as process flow diagrams along with material and energy balances, while at the operational stage, a complete design package is available, e.g., as-built piping and instrumentation diagrams (P&IDs), design basis for the protective systems, and operating and maintenance procedures. Depending on the lifecycle phase, the following information may be available for review:

- Risk criteria,
- Applicable codes, standards, and practices,
- Process description,
- Operating modes,
- Process flow diagrams,
- Material and energy balances,
- Plot plans,

- Piping & Instrumentation Diagrams (P&IDs),
- Process and protective equipment data,
- Process simulations,
- Consequence analysis, such as mass of released materials, dispersion characteristics, and fire/explosion analysis,
- Control system documentation,
- Protective system design basis, and
- Operating and maintenance procedures.

The significant qualitative aspect of most hazard and risk analysis methods lends itself to variability in the absence of a detailed procedure that establishes ground rules for the analysis. Nothing can cause more frustration to the design and engineering team than to see significant variation in the requirements for similar hazardous events evaluated by different hazard and risk analysis teams. Without a detailed procedure clearly stating owner/operator expectations, it is possible for the hazard and risk analysis team to report highly variable findings for similar hazardous events ranging from no special requirements to significant risk reduction requirements.

Consequently, procedures should be systematic, produce similar results when executed by different people, take application information into account, be auditable, be under revision control, and be consistent with good engineering practice. Each procedure should include some common elements, such as:

- Scope,
- Applicability of the method,
- Identification of required resources and responsibility,
- Definition of risk criteria,
- Required information for the study,
- Assumptions,
- Guidance on how to assess the consequence,
- Guidance for determining any required frequency (or likelihood) assessments,
- Work process for identifying hazardous events and independent protection layers,
- Implementation of a tracking system to ensure that any recommendations are addressed with an action plan,

- Documentation requirements, and
- Documentation version management.

Sometimes, a failure can cause hazardous events in several units simultaneously. For example, loss of cooling water and loss of power are often controlling cases in the design and sizing of flare relief systems that receive streams from several process units.

Global hazardous events include events, such as:

- Multiple relief scenarios,
- Significant spills or releases of flammable, combustible, toxic or airborne materials,
- Overpressure of steam header from source utility,
- Loss of steam,
- External fires,
- Loss of cooling water,
- Loss of power, and
- Loss of other utilities.

Global hazardous events may be assessed as part of a process hazard analysis, but are more frequently assessed in a focused study (e.g., atmospheric dispersion modeling, quantitative risk assessment, flare design). The following information is typically required for the analysis:

- Process equipment layout (plot plan),
- Utility distribution drawings, showing various supply parameters for steam, condensate, nitrogen, electrical supply, etc,
- Hazardous events with associated initiating causes, consequence severity, protective functions, allocation to protection layers, and required risk reduction, and
- Relief system design specifications.

Sometimes, a global hazardous event presents unique risks, requiring the implementation of special classes of IPLs. For example, atmospheric release modeling may identify areas where a fire suppression system should be used to initiate evacuation, deluge, and water curtains. Plant and community emergency response procedures should also consider suitable response strategies for hazardous events posing plant-wide or community impacts.

The evaluation of risk and the development of a risk reduction strategy is an iterative process. Over the life of a process, personnel gain a deeper understanding of the process operation, potential hazardous events, and protective systems. Operating history may demonstrate the events that were predicted in the analysis or may reveal unidentified events.

Lessons learned should be brought into the hazard and risk analysis to ensure that assumptions made in the analysis agree with the real world. Continuous improvement also seeks to identify means to reduce risk further, so the hazard and risk analysis should be periodically reviewed to see if there are new opportunities for risk reduction. Therefore, the analysis should include an examination of the following, if applicable:

- Previous experience, including historical performance and collected data/information,
- Review of other hazard and risk analyses,
- Incident reports and near-miss investigations,
- Review of trip reports (i.e., process demands on IPSs),
- Review of failure reports for IPS equipment,
- Review of operating and maintenance procedures, including examination of human factors, and
- Review of approved changes since last hazard and risk analysis.

3.3 BASIC WORK PROCESS

The hazard and risk analysis should be started as early as feasible within the project lifecycle to minimize the cost of implementation. When projects are in scoping and front-end loading phases, the cost of making changes to the design or to the protection layers is relatively low, and numerous options for risk reduction can be evaluated on a cost/benefit basis. As the project proceeds toward full implementation, the cost associated with making changes increases dramatically and the options for change decrease sharply. Consequently, it is important to begin hazard identification early in a project life to identify options that lower the cost of safe operation.

During process design, the risk associated with potential hazardous events should be reduced, when practical, through inherently safer design. The process designer should consider inherently safer design strategies, such as minimize, substitute, moderate, and simplify, to reduce the risk or to eliminate the potential event. These design strategies are best employed during the earliest stages of process design, where the opportunities for process modification are greatest and the cost of design change is least. However, inherently safer design can be employed during any lifecycle stage.

Inherently safer design provides substantial benefits by reducing long-term operational and maintenance cost. Further, if the process is inherently safer, it is generally more reliable, because the process is more tolerant to operational upsets. Refer to *Inherently safer chemical processes: A lifecycle approach* (CCPS/AIChE

1996) for more information. Inherently safer design has economic and engineering limitations. For example, it is sometimes necessary to use a specific process chemistry or technology to produce a product.

If the is higher than the specified risk criteria, protection layers are implemented to control the risk in accordance with good engineering practice. *Guidelines for Safe Automation of Chemical Processes* (CCPS/AIChE 1993) introduced the concept that protection layers can be used to avoid the occurrence of or to reduce the effect of an identified hazardous event. When significant hazards cannot be avoided by inherently safer design, protection layers are used to reduce the frequency of the event. The identification, design and management of protection layers is, consequently, very important.

The protection layers (Figure 3.2) are as follows:

- Inherently safer design--reduction or elimination of hazardous events through effective use of process technology, design methods, and/or operational techniques;
- Control--standard operating procedures, basic process control systems, and process alarms--this layer is generally focused on maintaining the process within the normal operating limits;
- Supervisory--protective alarms, operator monitoring and supervision, and process actions--this layer is designed to achieve or maintain a safe state of the process to reduce the frequency of the hazardous event;
- Preventive--protective instrumented systems, such as safety instrumented systems, environmental protective systems and asset protective systems--this layer is designed to achieve or maintain a safe state of the process to reduce the frequency of the hazardous event;
- Mitigative--mechanical equipment, such as pressure relief devices, and instrumented systems, such as life safety systems, high integrity protective systems, and reactor kill systems--this layer is designed to reduce the frequency and/or consequence severity of the hazardous event;
- Barriers--dikes, berms, bunds, and explosion barriers--this layer minimizes the consequence severity due to its physical design;
- Limitation--fire & gas, fire protection, water/steam curtain, deluge, emergency dump, and emergency shutdown systems--this layer acts to reduce the consequence severity of the hazardous event; and
- Response--emergency response systems--this layer notifies personnel and/or the community to shelter-in-place or to evacuate to safe zones and musters emergency response personnel.

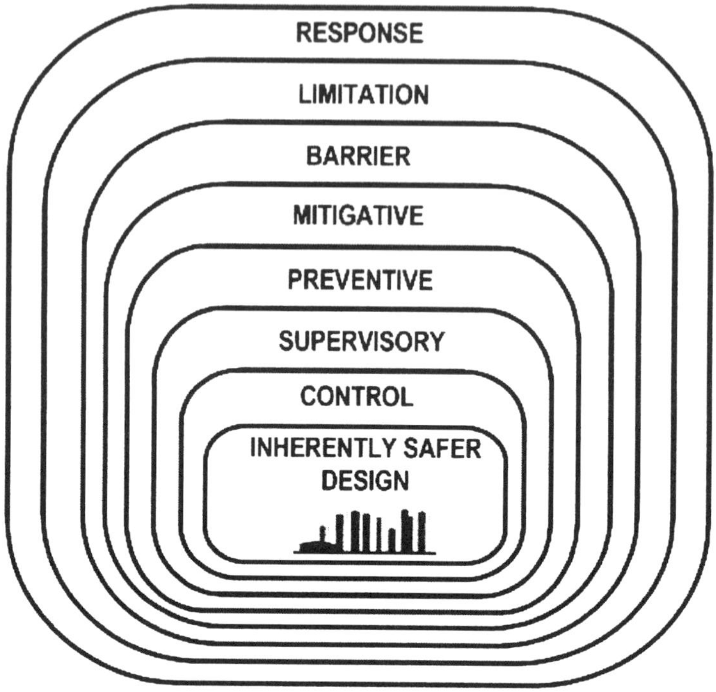

Figure 3.2. Protection Layers.

Some protection layers are more effective than others in reducing the risk. A protection layer may only partially mitigate the event, resulting in an undesirable, though reduced severity, event. Some layers are not sufficiently effective in detecting and responding to the particular types of events. Other layers can achieve predictable risk reduction based on proven design and management practices.

Protection layers are known as IPLs when they are designed and managed to achieve the following seven core attributes (see Appendix C):

- Independence--the performance of a protection layer is not affected by the initiating cause of a hazardous event or by the failure of other protection layers;
- Functionality--the required operation of the protection layer in response to a hazardous event;
- Integrity--related to the risk reduction that can reasonably be expected given the protection layer's design and management;
- Reliability--the probability that a protection layer will operate as intended under stated conditions for a specified time period;

- Auditability--ability to inspect information, documents and procedures, which demonstrate the adequacy of and adherence to the design, inspection, maintenance, testing, and operation practices used to achieve the other core attributes;
- Access Security--use of administrative controls and physical means to reduce the potential for unintentional or unauthorized changes; and
- Management of Change--formal process used to review, document, and approve modifications to equipment, procedures, raw materials, processing conditions, etc., other than "replacement in kind," prior to implementation.

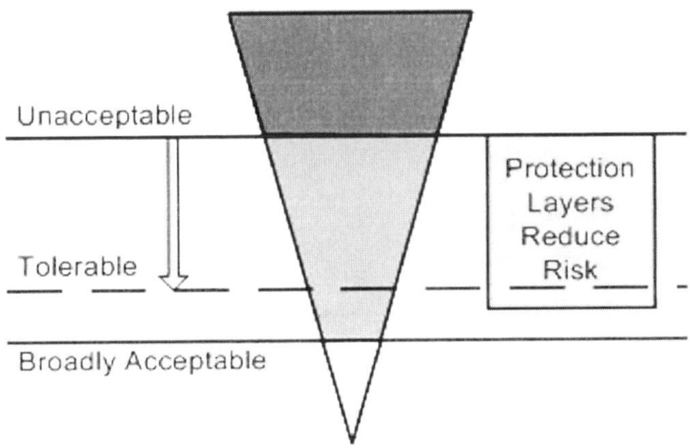

Figure 3.3. Risk Reduction Triangle.

As shown in Figure 3.3, the objective is to reduce the risk from an unacceptable risk to one that is considered tolerable as defined by the owner/operator's risk criteria. When the risk is unacceptable, action must be taken to eliminate (e.g., inherently safer design) or reduce the risk (e.g., apply protection layers). When the risk is considered broadly acceptable, further action (or risk reduction) is generally taken only when required by applicable good engineering practices. In the middle region, the risk must be systematically analyzed, assessed, and controlled using protection layers until the cost of further reduction is grossly disproportionate to the benefit gained.

Many owner/operators provide specific risk criteria for use in the risk assessment. The risk criteria establish the minimum performance to be achieved within the tolerable region. The risk criteria are often presented in different forms as necessary to support the chosen hazard and risk analysis method, the nature of potential consequences, and the level of rigor employed during the assessment. Personnel responsible for risk assessment activities should be trained in the intended use of the risk criteria.

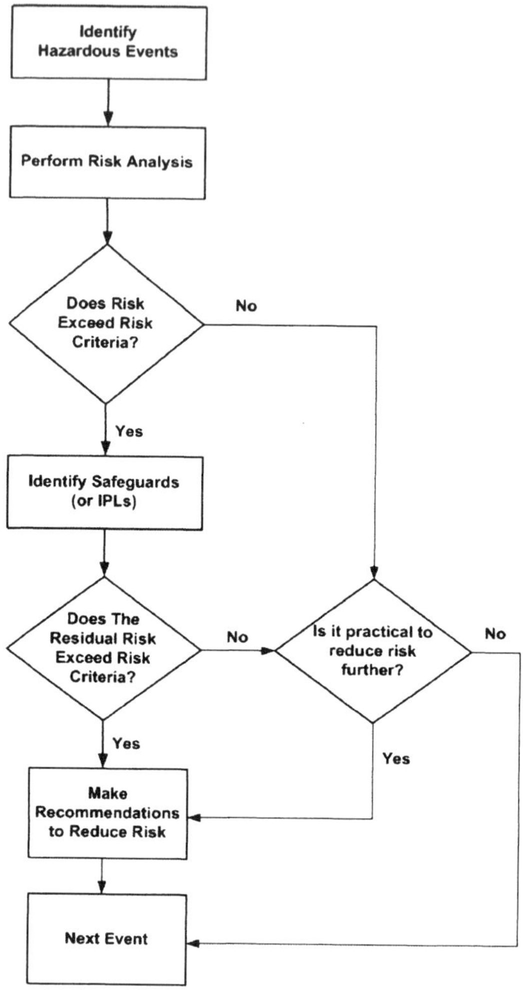

Figure 3.4. Hazard and Risk Analysis Work Process.

The work process, shown in Figure 3.4, relies on a team to identify hazardous events and to develop and verify the risk reduction strategy. To accomplish this, the team must include personnel that have expertise in the operation of the process unit and its protection layers. Hazardous events may be caused by events internal or external to the process. This guidelines book only addresses those hazardous events that are the result of a failure to maintain the process within the safe operating limits. For more detailed guidance on hazard evaluation, refer to *Guidelines for Hazard Evaluation Procedures* (CCPS/AIChE 1992).

Hazardous events are identified using an analysis technique appropriate for the event under consideration and the process lifecycle stage. Events may be screened using one or more criteria throughout the process life. Screening is used in different ways throughout the lifecycle (see Section 3.5.2). It may trigger the requirement for more rigorous analysis, verification, or approval. Common screening criteria are consequence severity, initiating event frequency, process risk, and residual risk.

An initiating cause starts the chain of events that propagates into the hazardous event if something does not act to stop it. For example, a basic process control loop may fail such that the control valve is driven open, allowing full supply pressure into the vessel. The control loop failure could be caused by failure of the sensor, controller, positioner, valve, or other relevant equipment. The initiating event frequency is estimated with different levels of rigor, depending on the risk analysis method. See Section 3.5.4 for further discussion of initiating events and frequency.

The consequence is the logical conclusion if the hazardous event is allowed to propagate unchecked. For example, can the vessel be over-pressured? Will the vessel fail? If so, will it fail at a flange, seal or weld?

Consequences are the undesirable result of the hazardous event, usually measured in health and safety effects, environmental impacts, loss of property and business interruption costs. The consequence severity is a measure of the expected impact when the consequence occurs.

The consequence severity should be evaluated based on the worst credible scenario, which is the most severe incident of all identified outcomes and their consequences that are considered plausible or reasonably believable. The consequence severity is often used to determine the level of rigor employed in the risk analysis, which estimates the event frequency and consequence severity. The process risk should be reduced below the owner/operator risk criteria using IPLs. Refer to Section 3.5.3 for discussion related to the evaluation of the consequence severity.

The initiating event frequency and the consequence severity define the process risk, which may be compared to specified risk criteria. When the process risk is less than the risk criteria, the team is not required to make recommendations for additional risk reduction. However, if the team can identify practical means to reduce risk further or to improve IPL performance, these recommendations should be collected to support continuous improvement in safety and reliability.

Additionally, some good engineering practices and owner/operator practices may require specific IPLs or safeguards in certain applications, regardless of the risk reduction requirements. Safeguards may also be required to address other issues, such as housekeeping, minor personnel protection, quality, operability, production, maintenance, or compensating measures.

When the process risk exceeds the risk criteria, the team identifies IPLs that reduce the risk as required. As shown in Figure 3.5, the risk gap between the process risk and the risk criteria can be managed by a single protection layer or by multiple layers. Each identified layer implements a specific protective function. The team identifies functions that achieve or maintain a safe state; thereby preventing the hazardous event. These functions are allocated to IPLs designed and managed with sufficient rigor focused on the core attributes to achieve the required risk reduction.

The appropriate identification of IPLs is extremely important, because the successful execution of the remaining lifecycle phases is dependent on it. Potential common cause failure between the initiating causes and the protective functions should be considered for each hazardous event. Common cause occurs when protection layers share management systems, equipment, controllers, support systems, utilities, interfaces, etc.

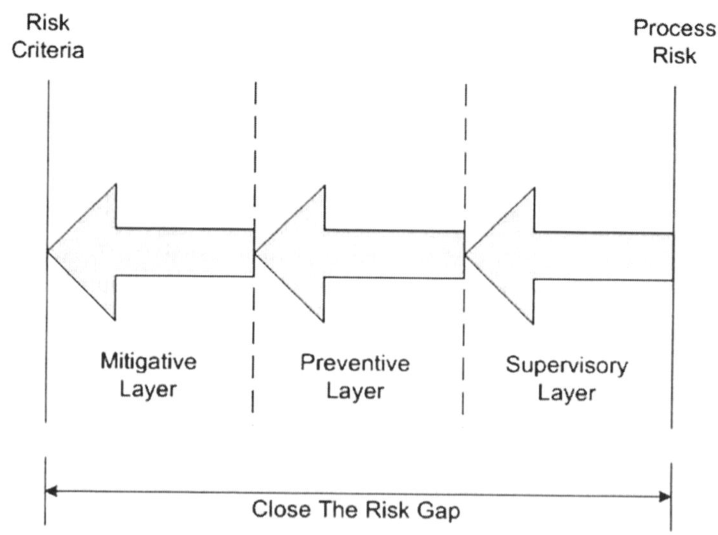

Figure 3.5. Using IPLs to Close Risk Gap.

To identify common cause, the initiating cause is examined to determine the layer in which it resides. If the protective functions are allocated to the same layer, they are not independent of the initiating event. When multiple protective functions are used to address the risk associated with an initiating cause, the independence between the protective functions should also be evaluated. If the protective functions are allocated to the same layer, they are often not independent and should be considered one IPL, unless a detailed analysis confirms independence.

Example--IPF with diverse inputs

A runaway reaction is identified, which results in a significant loss of containment event with the potential for serious injuries or fatalities within the process area. The target risk reduction for this protective function is 1,000. A runaway reaction is detected using temperature and pressure.

The analysis lists two protective functions: 1) high temperature instrumented protective function (IPF) closing the inlet reactant valves with a target risk reduction factor of 10; and 2) high pressure IPF, which closes the inlet reactant valves with a target risk reduction of 100. Both protective functions are implemented in the same instrumented protective system (IPS) using the same logic solver and final elements.

These protective functions are not independent; there is really only one IPF with diverse inputs: high temperature or high pressure initiates the closure of the inlet reactant valves. The IPF would need to meet the overall risk reduction requirements, 10 x 100 = 1000.

The identification of protective functions and their allocation to IPLs continues until the risk gap is closed and the team is satisfied that the risk has been adequately addressed. When there is insufficient risk reduction available from existing or planned IPLs, the team makes recommendations to address the residual risk. The remaining risk can be reduced by various means, such as adding or modifying an IPL, such as new process design limits, adding an independent verification step to an operating procedure, or using a higher integrity IPL. Recommendations are generally prioritized based on the magnitude of the consequence severity and/or residual risk level.

Figure 3.6 summarizes a risk assessment process used to evaluate hazardous events, to compare the risk to the owner/operator's risk criteria, and to question the economic viability of dealing with the identified risk. If it is technically and economically feasible to reduce the risk, the business activity is considered viable and protective functions are implemented.

3.4 OUTPUT DOCUMENTATION

The risk assessment phase identifies hazardous events and the IPLs that reduce the hazardous event risk below the owner/operator risk criteria. A hazard and risk analysis team defines the required protective functions and allocates them to IPLs designed and managed to achieve the required risk reduction.

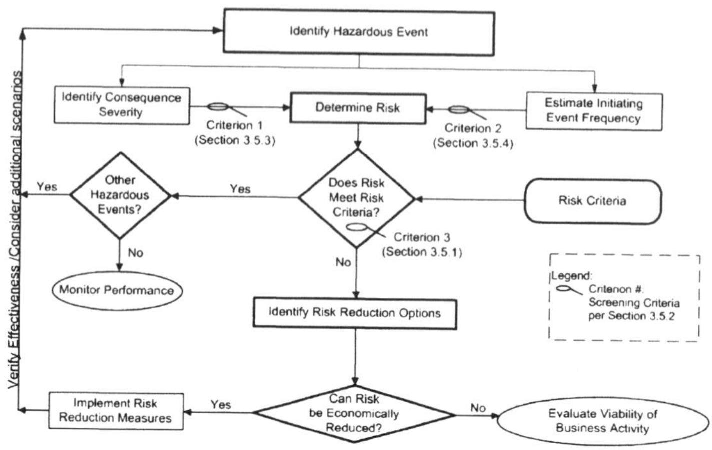

Figure 3.6. Risk Assessment Process.

The hazard and risk analysis is the first step toward the protective function specification, so improvements in hazard and risk analysis content and quality result in better-informed decisions during later lifecycle phases. Expectations with regard to the hazard and risk analysis findings and documentation should be clearly stated, so essential information is collected as needed. These expectations will typically vary with project phase and with the chosen hazard and risk analysis method.

At a minimum, the risk assessment phase should demonstrate that the following have been completed:

- Identification and description of hazardous events,
- Estimation of the initiating event frequency,
- Evaluation of the consequence severity, and
- Identification of IPLs with allocated protective functions and risk reduction.

A list of protective functions should be issued for review and approval. This list can be used to determine the resources necessary to complete the design phase for each protective function. Any safety functions must have a documented design basis to serve as process safety information. For example, design capacity calculations should be provided for each identified pressure relief valve clearly documenting the overpressure cases it was specified to protect against.

3.5 KEY MANAGEMENT SYSTEM ELEMENTS

The protective management system establishes the policies, procedures and practices used to execute the risk assessment phase. Government authorities, insurers, or other parties may require specific activities during this phase. Any applicable requirements should be incorporated into the overall work process to improve efficiency and to ensure that relevant requirements are considered at the appropriate lifecycle phase.

The identification of IPLs considers the core attributes necessary to achieve specified levels of risk reduction. As discussed in Chapter 2, the management system defines how these attributes are achieved by the various protection layers, so these attributes can be assessed, verified, validated, and audited, as necessary. The rigor used to administratively and technically control the attributes determines the risk reduction reasonably achievable by each IPL.

Periodically, the installed equipment should be compared to current good engineering practices. If the installed equipment does not agree with current practices, the variance should be assessed through engineering processes to determine whether the current design provides acceptable risk reduction. See Chapter 2 for a discussion of good engineering practices, as well as a discussion of the "grandfather clause" for safety instrumented systems.

The risk assessment phase cannot be successfully completed without a definition of the risk criteria. Many methods are available for identifying the hazardous events and developing the risk reduction strategy. The following are considered key management system elements:

- Risk Criteria,
- Screening hazardous events,
- Evaluation of consequence severity,
- Estimation of initiating event frequency,
- Identification of protective functions, and
- Conceptual allocation of functions to IPLs.

3.5.1 Risk Criteria

Risk criteria define the owner/operator's risk tolerance and enables prioritization of investments related to IPLs. To ensure consistent application of the risk criteria, internal practices should explain how the criteria are used at different stages of the process unit life. The intent is to reduce the risk below the risk criteria, unless a deviation from the risk criteria is justified and formally approved by management.

Risk criteria can be established for a site, business unit, or corporation. It is seldom site specific, unless there are unique risks associated with the process, such as toxic substances not present at other owner/operator locations. The risk criteria should be stated in a manner that is clear and understandable to personnel assigned responsibility for risk assessment activities. Assigned personnel should also receive training on how the frequency and consequence severity are evaluated and how the risk criteria are used to define the risk reduction requirements.

Risk criteria should be established for business or asset losses. Protective functions are often implemented to prevent equipment damage and process downtime. This is especially true for process-critical pumps, boilers, furnaces, and rotating equipment (e.g., compressors, centrifuges, etc.). In a sold out market, downtime is production that cannot be made up. How much an owner/operator is willing to spend to reduce asset risk is dependent on the downtime duration, the process criticality, and the estimated cost of the event.

The two most common methods for defining the risk criteria are 1) the risk matrix and 2) the quantitative target. These two methods are briefly introduced below.

3.5.1.1 Risk Matrix

The risk matrix is widely used in the process industry (Figure 3.7A,B,C). The risk matrix presents the risk criteria as a two-dimensional plot with frequency on one axis and consequence severity on the other. Each X-Y coordinate in the risk matrix indicates whether additional risk reduction is required.

Consequence Severity					
	H	D	B	A	A
	M-H	E	B	A	A
	L-M	E	D	B	A
	L	E	E	D	C
		L	L-M	M-H	H
		Likelihood			

(A)

Consequence Severity						
	5	100,000	10,000	1,000	100	10
	4	10,000	1,000	100	10	TR
	3	1,000	100	10	TR	TR
	2	100	10	TR	TR	TR
	1	10	TR	TR	TR	TR
		1	10	100	1,000	10,000
		Frequency (1 in x years)				

(B)

Consequence Severity						
	1000+	100,000	10,000	1,000	100	10
	100 to 1000	10,000	1,000	100	10	10
	10 to 100	1,000	100	10	10	10
	1 to 10	100	10	TR	TR	TR
	0.1 to 1	10	TR	TR	TR	TR
	#lbs release	1	10	100	1000	10,000
		Frequency (1 in x years)				

(C)

Figure 3.7. Example Showing Risk Matrices Using Qualitative (A), Semi-Quantitative Frequency (B), and Semi-Quantitative Frequency and Severity (C).

Figures 3.7A through 3.7C provide examples of risk matrices. Figure 3.7A shows an example of a risk matrix that might be used in an early hazard and risk analysis. In this case, the likelihood and consequence severity are defined by qualitative categories. Figure 3.7B provides an example where the frequency is defined by order of magnitude numerical ranges and the consequence severity is assessed by qualitative rankings. Finally, Figure 3.7C shows a risk matrix where both frequency and consequence severity are evaluated using order of magnitude numerical ranges. In this latter case, the consequence severity is defined by the size of the gas release. Figure 3.7C is an example matrix created as an illustration and is not based on any actual chemical.

The risk matrix can be different for each risk type, i.e., a different matrix for safety, environmental, or asset protection. Or, the risk matrix can be the same, but the definitions for each column in Figure 3.7A through 3.7C are aligned to achieve the desired risk prioritization or risk reduction for the consequence under consideration.

The values associated with specific X-Y coordinates in these risk matrices vary with the hazard and risk analysis method. Hazard and Operability (HAZOP) studies, what-if/checklists, and similar qualitative methods often use the risk matrix to establish when analysis of greater rigor should be performed or to prioritize team recommendations. When the IPL and risk reduction are analyzed, the X-Y coordinates typically represent the required number of protection layers or required risk reduction.

3.5.1.2 Quantitative Target

Quantitative criteria can be used in conjunction with semi-quantitative analysis, such as LOPA, failure mode, effects and criticality analysis, or quantitative risk analysis, such as fault tree or event tree analysis. It can be defined in many different ways, focusing on worker, individual, public, societal, and geographic risks. Efforts are underway at CCPS to provide more detailed guidance on this complex topic. Table 3.1 provides two examples of quantitative risk criteria for illustrative purposes.

Typical Hazardous Event[(1)] Criteria (per year)	Personnel		Community	
All hazardous events	10^{-3}	10^{-5}	10^{-4}	10^{-6}
Each hazardous event	10^{-4}	10^{-6}	10^{-5}	10^{-7}

Note 1: Worst credible consequence involves a personnel or community fatality as indicated.

Table 3.1. Examples of Quantitative Targets.

Quantitative risk assessment is used when qualitative or semi-quantitative techniques are not yielding the required risk discrimination. Quantitative assessment is especially helpful when evaluating high risk or complex events having a high degree of system interrelationship where independence is not provided between layers. These events are difficult for the team to assess qualitatively, so further analysis by more rigorous methods can be beneficial.

3.5.2 Screen Hazardous Events

In initial project stages, potential hazardous events are identified and documented. A risk reduction strategy is developed that addresses the identified hazardous event. As the project progresses, increasingly rigorous analysis is performed on some, if not all, of the identified hazardous events. This refines the risk reduction strategy and ensures that the frequency of significant consequence events is less than the owner/operator risk criteria. As studies are performed throughout the lifecycle, the focus may also expand to include evaluation of asset and business losses, as well as harm to people.

The hazard identification can be accomplished using any accepted hazard analysis method, such as those shown in Table 3.2. The choice of hazard analysis method is influenced by many factors, including:

- Government requirements,
- Lifecycle phase,
- Information available,
- Process complexity, and
- Previous experience with process.

Each method guides a team experienced in the process design and operation in determining how causes lead to undesirable consequences. Brainstorming is the least structured method, but it is often used during process development to define the process design basis, identifying target areas for the application of inherently safer design. HAZOPs and FMEAs are more structured and are used in later lifecycle phases to assess the adequacy of the process design basis in addressing potential hazardous events.

Methodologies
Brainstorming
Preliminary Process Hazards Analysis
Checklists
What-if/Checklists
Hazard and Operability (HAZOP) Study
Failure Mode and Effects Analysis (FMEA)

Table 3.2. Hazard Analysis Methods.

Risk analysis is used to estimate the risk of each hazardous event based on its potential frequency of occurrence and consequence severity. Safeguards are then identified which reduce the risk to the owner/operator risk criteria.

Table 3.3 provides a list of methods used for the risk analysis. The method selection is affected by similar factors as the hazard analysis selection, as well as risk aversion, lack of independence between layers, and safety margin reduction.

Methodologies
Layer of Protection Analysis (LOPA)
Failure Mode, Effects and Criticality Analysis (FMEA)
Fault Tree Analysis (FTA)
Event Tree Analysis (ETA)
Cause-Consequence Analysis (CCA)
Quantitative Risk Assessment (QRA)

Table 3.3. Risk Analysis Methods.

Simple criteria can be used to screen hazardous events and to prioritize events for analysis with more complex techniques. The screening criteria may also trigger other forms of analysis, such as human reliability analysis, consequence modeling, or job safety analysis. Greater analysis rigor is often employed when the hazardous event meets one or more of the following criteria:

- Criterion 1: high consequence severity,
- Criterion 2: high frequency, and
- Criterion 3: process risk falls above a specified risk level.

The intent is to estimate the process risk using the most cost effective method, while considering the degree of risk discrimination that can be achieved by each method. In general, the more complicated the method, the higher the cost (dollars and time) to perform the analysis. Complex techniques may yield additional information better defining the risk reduction strategy.

All methods produce results with some degree of uncertainty. This may be due to many factors, including the method assumptions and rigor, the team's experience with the process and its risk, the team's ability to assess the various factors considered in the method, and the team's experience with the method. Although quantitative methods may be useful in prioritizing or benchmarking hazards, qualitative judgments are often adequate given the information available at the time of the analysis.

Figure 3.8 provides an illustration of a risk screening process. Qualitative techniques are used to identify hazardous events that exceed specified severity, frequency, or process risk thresholds. Catastrophic or significant consequence events are typically assessed using of more rigorous methods. Hazardous events that do not exceed these thresholds proceed through the qualitative analysis. A large percentage of the total hazardous events are typically reviewed using only qualitative techniques.

Selected hazardous events are then screened a second time using semi-quantitative assessment. As shown in Figure 3.8, the team assesses the consequence severity and frequency according to the defined semi-quantitative ranges. If the process risk is very high, as defined by a semi-quantitative threshold or by specified cells (i.e., X-Y coordinates) in the risk matrix, additional study using consequence modeling or quantitative risk assessment should be considered, subject to management review. Such hazardous events are typically noted as high risk and the team continues with the analysis of these events, as well as the remaining ones.

For high risk hazardous events, management review is conducted to determine the level of confidence in the frequency and consequence rankings. If confidence is high, or it is not expected that further study will yield significantly different results, the findings from the semi-quantitative study are used for the development of the design and operating basis. If confidence in the consequence severity is low, consequence modeling is often performed to improve the understanding of the event consequence. Likewise, low confidence in the frequency estimation indicates the need for a quantitative risk assessment, such as event tree or fault tree analysis.

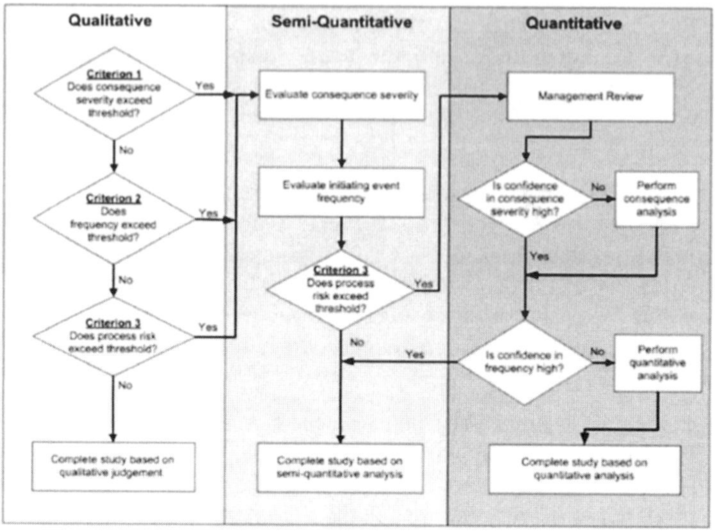

Figure 3.8. Example Risk Screening Process.

3.5.3 Evaluate Consequence Severity

The consequence is the undesired outcome of the identified hazardous event. The following serves as the basis for the consequence evaluation:

- Process equipment is designed and constructed in accordance with appropriate codes, standards, and practices,
- Protective equipment is installed, operated, inspected, maintained, and proof tested to meet the core attributes,
- Process safety information is available for review and is up-to-date,
- Personnel assigned responsibility for safe operation are trained on the hazardous events and the protective systems used to prevent their occurrence,
- Staffing levels are typical for the process operating mode under consideration,
- Personnel wear personnel protection equipment (PPE) in accordance with owner/operator practice (e.g., eye protection, flame retardant clothing, etc.), and
- Facility has an emergency response plan to minimize potential harm to people and to the environment when these consequences are possible.

The consequence severity represents the magnitude of the undesired impact of the hazardous event. The severity should be assessed based on the worst credible scenario considering conditions specific to the hazardous event:

- Process conditions at loss of containment,
- Ambient conditions, including wind speed, direction, temperature, and humidity,
- Release point, size, and direction of release,
- Flammable, explosive, or toxic nature of the release,
- Dispersion potential,
- Equipment density or congestion,
- Containment (dikes, bunds, etc.) works,
- Event is easily recognized, e.g., noise, smell, process indications, alarms, etc.,
- Speed of event escalation, i.e., time from event recognition to hazardous event, is sufficient for personnel escape from area,
- Escape routes are clearly marked and accessible during event, and
- Personnel are trained on emergency response plans, hazard communications, safe area location, etc.

The hazard and risk analysis procedure should define the consequence basis and the conditions to be considered. Some owner/operators use frequency modifiers to estimate the frequency at which harm is caused by the hazardous event. Other owner/operators find that these factors add complexity to the method, while not yielding significant improvements in risk discrimination. Also, since the goal of most owner/operators is no incidents, no harm to people and no damage to the environment, many prefer to focus on preventing the hazardous event and loss of containment.

The consequence analysis rigor and the consequences assessed, e.g., safety, environmental, asset protection, and business interruption, often change as risk is screened and evaluated using increasingly complex methods. Consequence analysis techniques are listed here in order of increasing complexity:

- Plant history and process knowledge,
- Look-up tables developed based on the assessment of the chemical and physical properties, and
- Consequence assessment, such as release rate, limiting oxygen concentration, dispersion modeling, and fire and explosion modeling.

When the consequence severity is highly negative (e.g., catastrophic event with multiple fatalities), case specific analysis should be considered. For example, when the community is potentially affected, a consequence analysis can be performed to better define the impact zone. The evaluation chosen by an owner/operator depends on many factors, including the lifecycle phase, consequence severity, and whether additional expenditure is necessary to gain confidence in the consequence assessment. More information on consequence analysis can be found in *Guidelines for Consequence Analysis of Chemical Releases* (CCPS/AIChE 1999).

3.5.4 Evaluate Initiating Event Frequency

The initiating causes are evaluated for each hazardous event. Each process operating mode (e.g., start-up, shutdown, maintenance, normal operation, etc.) should be assessed separately, so that risk unique to a particular mode is adequately addressed by the risk reduction strategy. Initiating causes generally fall into one of five types:

- Equipment failures--piping failure and malfunction (e.g., improper action or inaction) of fixed equipment, such as pumps or compressors,
- Instrumentation failures--action or inaction of instrumentation and control systems, such as the controller, operator interface, communication system, or field devices,
- Utility failures--loss of instrument air, cooling water, hydraulic oil, or power,
- Human errors--any incorrect action or inaction, such as the failure to execute all steps of a procedure in the correct order or the failure to take appropriate action in response to an abnormal process condition or an alarm, and
- External events--natural disasters, plane crashes, major incidents in surrounding areas, sabotage or terrorism.

Management system failures, such as poor supervision or inadequate training, are not typically listed as initiating causes. The presence of an effective management system is a basic underlying principle of risk management.

An initiating event is the minimum combination of failures (or errors) necessary to start the propagation of the hazardous event under consideration. It can be comprised of a single initiating cause, multiple causes, or initiating cause(s) in the presence of enabling conditions. The initiating event frequency may be estimated based on historical performance, published industrial data, previous experience, or defined criteria. Internal practices are often developed to make the evaluation of the initiating event frequency easier and more consistent.

Table 3.4 provides examples of qualitative, semi-quantitative, or quantitative evaluation criteria. Rules, means, or techniques for frequency estimation should be provided in the procedure. The perceived level of rigor often increases as the method becomes more quantitative. However, achieving quality results from more complex methods often requires higher skills and expertise than simple methods. This adds cost to the risk assessment process.

For this reason, screening (see Section 3.5.2) is often used to identify the hazardous events that would benefit from more detailed analysis. As the analysis becomes more complex, resources will require more training and tools. The considerations for estimating the frequency should be defined in the hazard and risk analysis procedure.

Method	Basis	Example Definition
Qualitative	Experience	Once in process life
Semi-quantitative	Benchmark/Experience	Once in 100 years
Quantitative	Numerical analysis	Once in 425 years

Table 3.4. Example of Frequency (or Likelihood) Rankings.

3.5.5 Identify Protective Functions

The risk assessment phase identifies hazardous events with undesirable consequences and documents the risk reduction strategy. The team analyzes how various causes lead to hazardous events and identifies means to control these hazardous events. A safeguard is an engineered system or administrative control that likely interrupts the chain of events following the initiating cause or mitigates the event consequence. Safeguards include compensating measures, and IPLs. An IPL is a device, system, or human action which meets the core attributes to the necessary level of rigor and is capable of preventing an initiating cause from propagating into an identified hazardous event.

Safeguards are documented, which provide some measure of protection against the hazardous event under review. Not all safeguards satisfy the core attributes necessary to provide sufficient risk reduction and thus be considered IPLs. The more rigorously the core attributes are managed throughout its lifecycle, the greater the level of confidence that the IPL can achieve its allocated risk reduction and stop the propagation of the initiating cause.

Each identified safeguard is evaluated against the appropriate design and management criteria. The assessment considers relevant factors, such as independence, documentation, proof testing, training, procedures, access security, and management of change. Typically, rules are provided in the procedure for assessing the safeguards against the core attributes. Some safeguards will simply not meet the required attributes. Other will, and these are identified as protective functions.

In general, risk reduction is accomplished by implementing one or more protective functions. The required functionality should be documented, so that the effectiveness of each protective function in reducing the risk can be assessed. For example, the protective function detects high vessel pressure and responds by isolating the feed supply to the vessel.

A key requirement is that each protective function is designed and managed independently of other protective functions used to address the same hazardous event. In this way, a single failure does not cause the failure of multiple functions. Independence is generally achieved by implementing each protective function in a different protection layer, i.e., an IPL.

3.5.6 Conceptual Allocation

Protective functions lower the process risk by reducing the hazardous event frequency. The risk reduction is based on the use of an appropriate management system, the core attributes of the IPL, historical performance, and maintenance practices. A list of typical protective functions may be provided with estimated risk reduction based on a set of screening criteria. The rules should focus on attributes that can be assessed by the team using the process safety information.

Protective functions are typically allocated to the control, supervisory, preventive, and mitigative layers, depending on the source of the initiating cause, the type of function to be executed, and the required risk reduction. Figure 3.9 provides a conceptual illustration of the four IPLs being challenged by an initiating cause external of the process, which leads to a hazardous event. Each IPL has a management system as its foundation and a physical system that implements the IPF stopping the propagation of the initiating cause to the hazardous event.

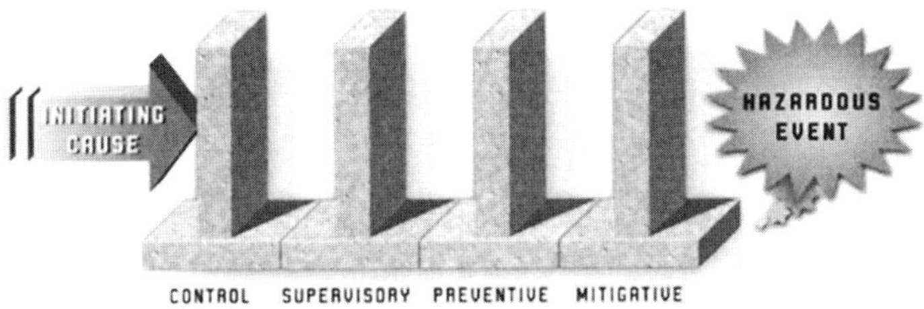

Figure 3.9. Initiating Cause Challenging Four IPLs.

IPLs are not perfect. This fact is generally acknowledged in the risk assessment by restricting the risk reduction that can be allocated to an IPL. Figure 3.10 illustrates the IPL imperfections as cracks in the management system and in the physical system used to implement the protective function. These cracks represent a loss of integrity due to errors or failures within these systems. The risk reduction provided by each IPL determines the hazardous event frequency. The capability of an IPL to provide the specified risk reduction is limited by its management system and the design and operating basis.

For example, the equipment inspection frequency is reduced due to a management decision regarding the workforce. The reduced inspection means that minor problems begin to accumulate in the system, making the IPLs increasingly vulnerable to failure.

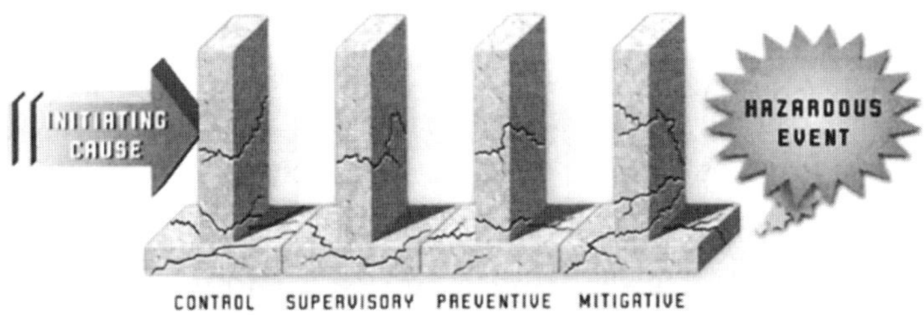

Figure 3.10. Initiating Cause Challenging Four IPLs With Vulnerabilities.

The initiating cause may be internal to the process, such as when a control function fails initiating the event as shown in Figure 3.11. A risk reduction factor can be used as a measure of how much the protective function reduces the frequency of the hazardous event. For example, a risk reduction factor of 10 reduces the frequency of the event by 10. Appendix B provides examples of risk reduction factors for protective functions implemented in various IPLs.

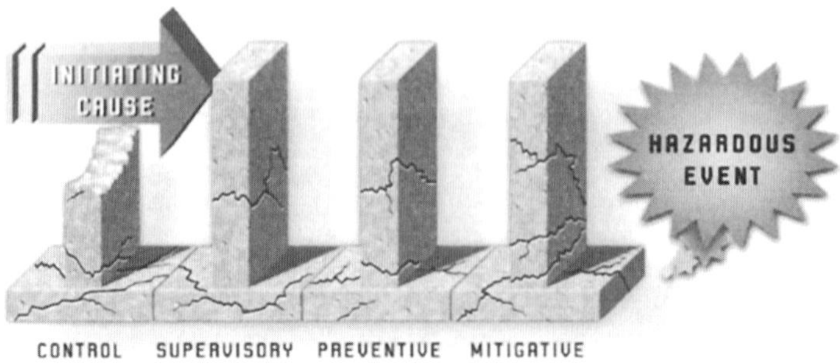

Figure 3.11. Control Function Fails Leading to Challenge on IPLs.

The failure of one layer leads to a challenge on the next layer. If the control system has a dangerous failure rate of one in 10 (1/10) years, the control system failure demands the supervisory layer to take action once every 10 years. If the supervisory layer is designed and managed to meet a risk reduction factor of 10, the demand rate on the preventive layer is 1 in 100 years (1/10 years x 1/10).

The estimated demand rate on each layer can be determined by the initiating cause and the risk reduction provided by layers designed to act earlier. If the layer is used to address more than one hazardous event, the demand rate for the layer is the culmination of the individual demand rates. The protective function's operating mode is related to the demand rate on it. In this case, the demand rate for the preventive layer, which executes IPS, is 1 in 100 years.

The control functions are typically implemented in the basic process control system (BPCS). The BPCS responds to input signals from the process and its associated equipment, other programmable systems, and/or from an operator and generates output signals causing the process and its associated equipment to operate in the desired manner. The BPCS could be a distributed control system, a programmable logic controller, or a single station digital or analog controller.

Supervisory functions may be implemented in the BPCS, with appropriate consideration for separation and the core attributes. The amount of risk reduction achievable by the BPCS is limited by its design and management. In Figure 3.12, the control function and supervisory function are implemented in the BPCS. The functions are supported by separate management systems.

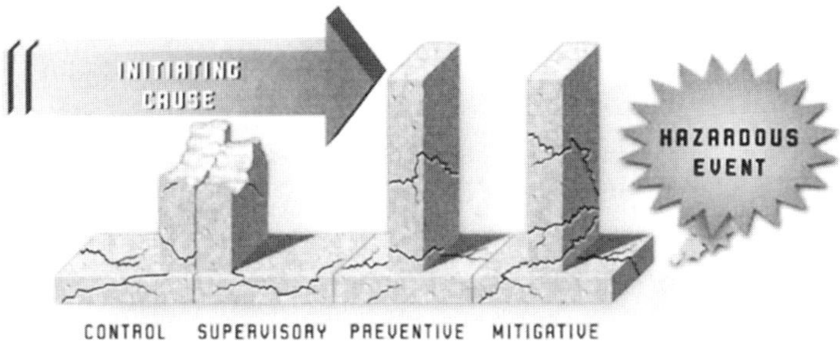

Figure 3.12. Initiating Cause Due to Failure Within BPCS.

The control and supervisory function are sufficiently independent that the failure of the control function does not cause the failure of the supervisory function. Any shared component must be designed such that its failure cannot cause concurrent failure of both functions or its failure rate is sufficiently low compared to the overall target hazard rate.

For successful management of the BPCS, the management system should be rigorous enough to reduce the potential of human error causing failure of the control and supervisory functions to a sufficiently low level. For a more comprehensive discussion of human error, refer to *Guidelines for Preventing Human Error in Process Safety* (CCPS/AIChE 1994). See Appendix B for more discussion on the control layer, supervisory layer, and the integration of these into the BPCS.

If all IPLs fail, the initiating cause propagates to the hazardous event (Figure 3.13). Again, assume that an order of magnitude method is used. If the initiating cause frequency associated with failure of the control function is 1/10 years and each IPL provides a risk reduction factor of 10, the hazardous event frequency for the sequence shown in Figure 3.13 is as follows:

Hazardous event frequency = 1/10 years x 1/10 x 1/10 x 1/10 = 1/10,000 yrs.

The hazardous event frequency is compared to the risk criteria to determine whether there are sufficient IPLs.

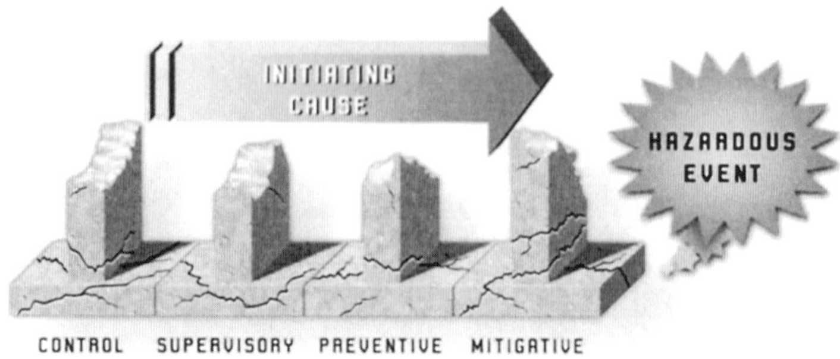

Figure 3.13. Initiating Cause Leads to Hazardous Event due to Multiple IPL Failure.

During the hazard and risk analysis, protective functions are identified, allocated to IPLs, and assigned risk reduction necessary to reduce the process risk below the risk criteria. Potential common cause failure should be evaluated between the following:

- The initiating cause and its protective functions, and
- The various protective functions, especially those implemented in the same IPL or same IPS.

Common cause failures are discussed further in Appendix D. The work process associated with defining the how the protective functions are physically allocated to IPS is discussed in Chapter 4.

3.6 SPECIAL TOPICS

3.6.1 Develop Risk Reduction Strategy

The hazard and risk analysis findings may include recommendations for further analysis, modification of existing functions, or addition of new functions. These recommendations should be reviewed and reconciled prior to the initiation of the IPS design basis.

The risk reduction strategy should be finalized early in the project, because there are many activities that occur in sequence after the strategy is defined. The project budget and schedule can be significantly impacted when the risk reduction strategy is changed late in a project, due to the vast number of engineering documents that are potentially affected.

The risk reduction strategy should address inherent safety, preventive measures, such IPS, and post-incident measures, such as limitation and emergency response measures. The risk reduction strategy related to IPS should be finalized by examining the hazardous events to:

- Gain better understanding of the initiating causes of the hazardous events,
- Determine whether the identified protective functions are the best approach to reduce the risk of the hazardous event,
- Reduce the number and complexity of the protective functions, and
- Identify where a single function is mitigating multiple hazardous events. The function may need to meet a higher risk reduction.

The latter two can be very important depending on the method chosen for identifying the hazardous events. Single cause-consequence methods tend to produce a risk reduction strategy based on preventing deviations rather than hazardous events. A single hazardous event may be caused by multiple process deviations (e.g., tank overpressure may be caused by high temperature, high flow, and high level). Conversely, a single process deviation may cause multiple hazardous events.

Example--Group hazardous events

Consider the protection of a process furnace from a potential fire or explosion within the firebox. This event could be caused when low fuel gas pressure leads to flame-out. If the fuel gas flow continues to the furnace, it may ignite uncontrollably, leading to furnace damage. An IPF is typically used to detect low fuel gas pressure and to isolate the furnace fuel gas feed.

The hazard and risk analysis may also identify another initiating event leading to flame-out caused by inerts in the fuel gas. The team should consider the possibility of improving the fuel gas quality to achieve an inherently safer design. Otherwise, an IPF may be specified that uses a flame scanner to detect flame-out and to initiate closure of the fuel gas valves.

The process engineer realizes that the flame scanner would also detect the flame-out related to low fuel gas pressure. The fuel gas pressure IPF is consequently modified to include the flame scanner as an input. This provides diverse process measurements with detection of low fuel gas pressure and loss of flame initiating closure of the fuel gas block valves.

Many owner/operators examine how each cause leads to various hazardous events. When the hazardous event is derived from several initiating causes, the independence of the initiating causes is considered and the team uses their judgment to qualitatively determine whether the potential hazardous event frequency is higher than that predicted by any single event. This method is more subjective than the cumulative risk method, but is simple, straightforward, and conservative. It can be implemented as part of any hazard and risk analysis without interruption of the typical work process.

Some companies use cumulative risk criteria, in addition to individual event criteria. Once the hazardous events are identified, the cumulative risk is compared to societal or individual risk criteria. The benefit of using cumulative risk criteria is that it potentially yields the least number of protective functions, while ensuring that adequate protection is provided for hazardous events resulting from many different causes. This method is complicated, is time consuming, and requires substantial expertise to yield meaningful results. There is also the potential that additional risk is assumed due to the reduction in protection layers.

The process impact of successful protective function operation should be considered. Protective functions act in response to process operation outside the normal operating limits. The successful action of a protective function may result in a safe state with regard to the identified hazardous event, but cause another problem, e.g., a block valve closes and blocks a pump's discharge. This "secondary event" may not be acceptable.

Finally, protective functions are subject to spurious failure, where some or all of the protective function's safe state actions take place. These actions often have significant process impact, such as stopping flow through equipment, injecting reaction kill materials, shutting down pumps, opening of vent valves, or opening of relief valves. Any hazardous event caused by the protective function's action should be addressed in the overall risk reduction strategy.

3.6.2 Mitigated Versus Unmitigated Risk

In a qualitative risk analysis, the team judges the likelihood of the hazardous event given the presence of the safeguards. This mitigated event frequency along with the consequence severity defines the residual, or mitigated, risk. If the residual risk is too high, additional risk reduction should be provided. The residual risk establishes the priority level for any action plans. Risk ranking is commonly practiced and is discussed in *Guidelines for Hazard Evaluation Procedures* (CCPS/AIChE 1992) and *Layer of Protection Analysis: A Simplified Risk Assessment Approach* (CCPS/AIChE 2001).

The work process presented in Section 3.3 uses the process (or unmitigated) risk to initiate more detailed evaluation of the identified safeguards. The initiating event frequency and the consequence severity define the process risk. The gap between the process risk and the risk criteria establishes the required risk reduction.

Based on the risk gap, protective functions are identified that achieve or maintain a safe state of the process to prevent the identified hazardous event. Each protective function is allocated risk reduction as necessary to close the risk gap. Recommendations are made when there is insufficient risk reduction. The magnitude of the risk gap is often used to prioritize response to the recommendation. For example, a risk gap of 1,000 would trigger a higher priority response compared to a risk gap of 10.

3.6.3 Identify Non-Protective Actions

When a protective function action is initiated, the process is taken to a state where the hazardous event can no longer occur. This state is called the "safe state." In addition to the safe state action, other actions are often performed as part of the protective function. Process actions may address many different issues, including operability, facilitation of unit re-start, acceleration of shutdown, or sequencing of shutdown. When the hazard and risk analysis team identifies a protective function, they may also identify and document these non-protective actions. These actions are specified in the process requirements specification as discussed in Chapter 4.

> Example--Reactor with three inlets
>
> Consider a reactor that receives Chemical A and Chemical B. Chemical A is a reactive material that is converted within the reactor. Chemical B is a stabilizer that causes the reaction to yield the right polymer properties. Chemical B is used for quality purposes only and does not react with Chemical A.
>
> A potential runaway scenario is reviewed by the team, who recommend a protective function that senses high pressure and closes two block valves on the Chemical A feed line. This stops the reaction and fully mitigates the hazardous event. In addition, Chemical B's control valve is closed to prevent wastage of Chemical B and to facilitate restart. The documentation clearly shows that Chemical B's control valve is not safety-related.

4

DESIGN

The argument that "It can't happen to me" or "It can't happen here" is wrong. Most consider their homes to be quite safe. Yet, according to the United States National Safety Council's Injury Facts® 2004 Edition, 33,100 fatalities and 7,900,000 disabling injuries occurred in homes in the United States in 2002. This represents 11.4 fatalities per 100,000 people. There was a fatal injury every 16 minutes and a disabling injury every 4 seconds. Homes are not inherently safe; it is the actions taken in response to known hazards that determine an individual's chance of injury.

Human beings learned to build safely by trial and error, learning from mistakes and passing down to others lessons learned the hard way. As science and engineering knowledge matured, designers were able to predict potential risk and build safety margins into the process to provide an additional measure of safety.

Safety margins alone have proven to be insufficient. Human error, poor inspection and maintenance, unrecognized interactivity of processes, equipment wear and tear, forces of nature, and other failure can cause the process to behave in an unintended manner with potentially catastrophic consequences. Effective risk management acknowledges potential hazardous events and establishes an administrative priority for their analysis and control.

Years of process operation provide valuable insight into probable hazards and risk associated with process operation. Historically, owner/operators implemented prescriptive risk reduction strategies based on experience and good engineering practice. They did not use performance criteria to determine how to manage risk. Instead, they relied upon internal practices that detailed the required architecture, including redundancy and voting, diagnostics, and installation details, and prescribed a maximum proof test interval. Many owner/operators mandated a 6 to 12 month proof test interval for safety instrumented system (SIS) equipment. These practices were further supported by an approved equipment list, which provided the model/version of specific equipment demonstrated to work in the operating environment.

Due to successful mechanical reliability and preventive maintenance programs, many owner/operators began to extend the maintenance interval (or maintenance turnaround) for the process unit. These programs yielded significant economic returns through improved production and enhanced product quality.

While extension of the equipment maintenance interval conflicted with the documented proof test requirements, evidence mounted that an effective mechanical integrity program yielded performance benefits.

Unfortunately, prescriptive practices were not flexible enough to adjust for improved equipment integrity, the use of redundancy, or the implementation of on-line diagnostics. The desire to tailor the equipment design and management to meet specific risk reduction requirements caused a shift from prescriptive to performance-based practices starting in the late 1980s and continuing through today.

As early as 1993, the Center for Chemical Process Safety (AIChE/CCPS) published *Guidelines for Safe Automation of Chemical Processes*, which addressed the importance of designing safety systems to meet performance expectations established by a hazard and risk analysis. It also introduced the concept of safety integrity level (SIL) to benchmark performance of safety instrumented systems (SIS), implemented to reduce the risk of hazardous events resulting in catastrophic consequences.

The SIL was defined by three discrete values related to the probability that the SIS fails to perform as required when needed. As the risk reduction requirement increased from SIL 1 to SIL 3, the performance expectation also increased, i.e., the tolerable probability of the SIS failure was reduced. Achieving a higher SIL requires more rigorous design, operating, inspection, and maintenance practices.

SISs are covered by the international standard, IEC 61511, which has been adopted in many countries, including the United States, where it is known as ANSI/ISA 84.00.01-2004 (IEC 61511 mod). The standard is referred to as ISA 84.01/IEC 61511 in this guidelines book. ISA 84.01/IEC 61511 uses the SIL concept established by previous practices as a platform for various requirements affecting all aspects of the SIS lifecycle. ISA TR84.00.04, Guidelines on the implementation of ANSI/ISA 84.00.01-2004 (IEC 61511 mod), provides guidance related to specific topics and requirements of ISA 84.01/IEC 61511.

Hazardous events can also result in risk associated with business loss, downtime, equipment damage and lost production. Consequently, this chapter extends the work process to cover the design basis development for all IPS, including the SIS. This chapter also provides guidance on achieving the seven core attributes; namely, independence, functionality, integrity, reliability, auditability, access security and management of change.

The requirements and guidance contained in ISA 84.01/IEC 61511 and ISA TR84.00.04 are embedded within this guidelines book. This is not intended to indicate that all IPS should be designed according to ISA 84.01/IEC 61511. Rather, it is to acknowledge that the use of the ISA 84.01/IEC 61511 management system and work processes provide substantial benefit to IPS design and management.

While performance-oriented standards provide flexibility in IPS design and management, they do not eliminate the need for internal practices. Flexibility can lead to inconsistencies in installation and configuration practices, fault detection and response practices, and proof test facilities. This inconsistency increases the potential systematic error during long-term operation, when personnel are responsible for several process units designed according to different practices.

Prescriptive internal practices ensure consistency in the IPS design and implementation across a process facility. Internal practices can also express the work processes and expectations using local language and terminology. For SIS, many owner/operators require that internal practices be followed, unless the deviation is justified and approved.

Some governments have specific regulations for the design of certain classes of IPS, especially those intended to reduce risk associated with catastrophic events, such as IPSs. Regulatory authorities, good engineering practices, and insurers may require specific protective systems, such as burner management systems (see Appendix B.4) and high integrity protective systems (see Appendix B.5).

To support auditing, it is highly recommended that the design basis clearly distinguish the IPS classification, such as safety, environmental, or asset protection, and IPS type, such as protective alarm, BMS, SIS, HIPS, etc.

The SIL can also be adapted as a metric for protective instrumented systems (PIS). This performance metric will be referred to as the integrity level (IL) to acknowledge that PIS may be used to reduce risk associated with non-safety consequences. Internal practices may further expand on "IL" to define specific consequence categories, such as SIL for safety, EIL for environmental, or APL for asset protection. For IPS, the metric is typically stated in terms of risk reduction.

Performance-based processes are only as good as the data and information fed into it. Operating experience and an understanding of historical performance is more essential than ever to the development and implementation of an effective risk reduction strategy. The expected risk reduction is based on operating experience, which evolves over time. The understanding of how equipment functions within the operating environment improves as equipment performance is monitored and tracked.

Operating history affects two key areas: 1) the hazard and risk analysis that identified potential hazardous events and developed the risk reduction strategy and 2) the design basis that reflects the known effectiveness of protective equipment in the operating environment.

Performance-based design provides flexibility but also complexity, because there is a wide range of options that can achieve any required risk reduction. These options present different capital, installation, and long-term operational costs. Additional capital investment in the form of equipment redundancy, diagnostic capability, and test/bypass facilities often yields lower long term operational costs.

However, some owner/operators prefer to implement systems with lower capital cost and absorb the higher operational cost in the product cost. Others strive to be the lowest cost producer and are willing to expend the capital necessary to achieve that goal.

The protective management system shown in Figure 4.1 covers the complete IPS lifecycle. This chapter focuses on the design phase, where an IPS design basis is developed to implement the risk reduction strategy established by the risk assessment phase. The design basis documents required operation of each IPF and the IPS. The complexity and size of today's processes, compounded by rapid technology evolution, makes it important that everyone who has responsibility for the IPS design understand how various design choices impact the IPS performance and its capability to meet the core attributes.

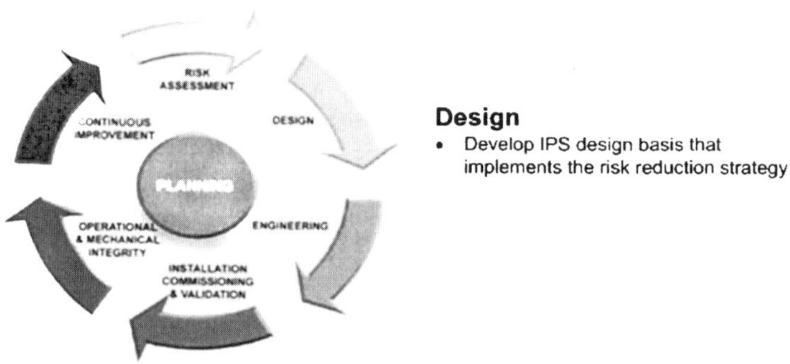

Figure 4.1. Design Phase.

4.1 INTENDED AUDIENCE

Achieving safe and reliable operation is not the work of a single individual; it is a matter of teamwork. Design basis development is a collaborative effort involving many parties across an organization. The owner/operator should be involved in the earliest stages of design to ensure that the IPS achieves the desired operability, functionality, maintainability, and reliability requirements, along with meeting the required risk reduction.

Reliability cannot be assumed as an inherent outcome of safe design practices; targets should be established and verified during the design phase. Operability and maintainability objectives should also be stated to ensure consistency with site operating and maintenance philosophy.

Information must be transferred between disciplines and departments to create a design that meets the intent of the risk assessment recommendations and the needs of the operating facility. This chapter is intended for:

- **Process** who develop and maintain the process requirements specification, establish constraints on the IPS operating procedures, determine when it is safe to bypass IPSs, and define compensating measures,
- **Instrumentation and Electrical** who develop and maintain the I&E requirements, establish constraints on the maintenance procedures, and develop and maintain the Approved Equipment List, and
- **Manufacturers** who wish to develop and provide products that achieve user approval.

4.2 INPUT INFORMATION

The design phase defines how the IPS achieves the risk reduction strategy developed in the risk assessment phase. Project scope can be highly variable, such as a new IPS, a major upgrade to an existing system, or a single new function. Project goals related to operability, maintainability, and reliability should be defined as early as possible, since these affect many choices made during the design phase. For efficient project execution, a schedule should be created showing when major tasks must be completed, identifying requirements for specialized resources, and highlighting critical path items that impact the project schedule.

Performance-based design encourages focus on individual functions and how the design of each function achieves the desired performance. This can lead to an IPS where every function is implemented differently even when these differences do not achieve significant performance or economic benefits. Instead, the focus should be placed on long-term operation and maintenance where consistency in the design and installation reduces errors.

Internal practices should be sufficiently flexible to account for specific process applications and sufficiently prescriptive to ensure similar results when the design is executed by different people. Internal practices should be consistent with accepted good engineering practice and should provide a mechanism for assessing, verifying, and auditing compliance.

The design basis evolves over the process equipment life as personnel gain a deeper understanding of how equipment functions in the operating environment and how installation, inspection, and maintenance practices affect equipment performance. Operating history may demonstrate that specific design and installation methods are required to achieve good performance. Lessons learned may result in the need to modify the design basis, as continuous improvement activities recommend the upgrade or replacement of part or all of the IPS.

The design phase should include an examination of the following:

- Process information, including process description, equipment and process design limitations, and operational modes (e.g., start-up, normal, reduced rates, maintenance, and shutdown),
- Hazardous events, describing initiating causes, consequence severity, and frequency of occurrence,
- Independent protection layers including allocated protective functions and risk reduction requirements,
- Internal practices, and
- Cost benefit criteria.

4.3 BASIC WORK PROCESS

The design basis is a collection of documents and information clearly defining how the IPS is designed to achieve the risk reduction strategy and the core attributes. The design basis consists of two parts: the Process Requirements (refer to Section 4.4) and the Instrumentation and Electrical (I&E) Requirements (refer to Section 4.5). The design basis is an evergreen set of information that evolves throughout the process unit life. During early front-end loading, only high level philosophical decisions are typically made (e.g., online testing is/is not necessary or a maintenance bypass will/will not be used). As the design progresses through detailed engineering, the design basis becomes equally detailed.

To reduce the design phase cost, many owner/operators develop prescriptive approaches to the IPS design basis. Internal practices may include an approved equipment list, typical architectures, approved equipment configurations, recommended installation practices, etc. Internal practices reduce the tendency to "re-invent the wheel" on every project, substantially lowering engineering cost. The design basis should reference the good engineering practices to be considered during design and detailed engineering.

If the design team sees alternatives to the risk assessment findings that appear to be more attractive, such as lower lifecycle costs, better technical solution, easier to implement, or easier to operate and maintain, these alternatives should be identified and evaluated early in the design process. Changes made during early design work occur with relative ease and little project budget or schedule impact. Additional upfront capital expenditures can often be justified to achieve desired reliability, operability, and maintainability. Lifecycle costs for inspection, test, maintenance, and operation support should be considered in the economic analysis.

Engineering and cost benefit analysis can be used to select the best option for long-term risk reduction. Where safety risk is involved, the analysis should err on the side of increased safety. Inherently safer design should be considered when practical, because it almost always represents the most effective risk reduction and often the lowest long-term cost of ownership. The addition of protective function increases operating and maintenance complexity, raising the likelihood of human error.

Deviation from the risk assessment findings should be reviewed using hazard and risk analysis to ensure that the change meets the original intent. When changes are made, assigned personnel should be notified of the change and documentation should be updated as necessary to ensure that other work process activities keep pace with the change.

> Example--Changing design pressure specification
>
> Consider a pressure vessel containing flammable hydrocarbons. During abnormal situations, the pressure in the vessel can reach 180 psig. For non-fire cases, the pressure vessel could be designed to be inherently safe by specifying a 200 psig design pressure. This ensures that the maximum allowable working pressure (MAWP) of the manufactured vessel will be 200 psig or more. A lower rated vessel presents an overpressure scenario which may require an IPS. The choice between the higher design pressure and the IPS is driven by risk management philosophy and/or by cost benefit analysis.

During the risk assessment phase, various hazard and risk analysis teams identify independent protection layers (IPL) that reduce the process risk below the owner/operator risk criteria. As shown in Figure 4.2, the process requirements and I&E requirements are developed based on the risk reduction strategy defined in these studies.

Process, I&E, and operations representatives should review the hazard and risk analysis findings to gain an understanding of the initiating events and consequence severity for the identified hazardous events requiring IPS. A key aspect to safe and reliable design is ensuring that the IPS is independent of the initiating events and any other protection layers used to reduce the risk of the hazardous event. With operations input, process engineering personnel finalize the process requirements, which document the operability, functionality, maintainability, and reliability needs.

The I&E requirements define the instrumentation and controls necessary to achieve the required risk reduction. It should define how the functions are physically allocated to the system(s) and how the individual functions are configured to achieve the required integrity and reliability.

The fault detection strategies are defined, including diagnostic expectations, proof test interval, and mean time to repair. Finally, the human interface requirements are established by specifying manual shutdown and reset requirements.

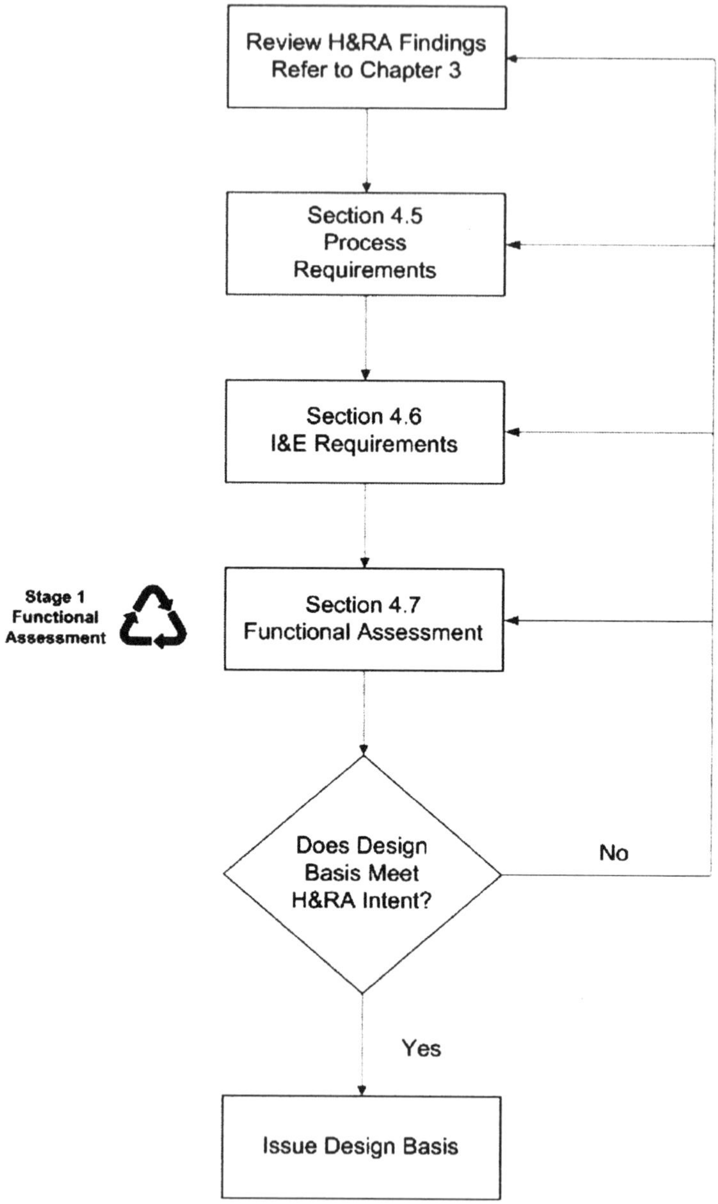

Figure 4.2. Overall Work Process.

In Chapter 5, the process and I&E requirements are further developed into detailed specifications for the IPS equipment. As the design proceeds, project team members should verify the design decisions against the intent of the hazard and risk analysis and internal practices. Systematic errors in the design basis may result in the IPS not operating as intended.

A functional assessment should be conducted prior to the design basis completion to identify and correct errors. Depending on the project size and complexity, more or less review may be required. At a minimum, a formal review of deviations from the hazard and risk analysis or internal practices should be conducted prior to finalizing the design basis. If the deviation is not approved, the design basis should be modified, as necessary.

4.4 OUTPUT DOCUMENTATION

The design basis is a collection of information that evolves throughout the project lifecycle and the IPS life. It establishes the requirements for the operating basis, which is discussed further in Chapter 6. The design basis is required for SIS. The work process provided in this chapter addresses the information that should be collected to fulfill the intent of the safety requirements specification in ISA 84.01/IEC 61511. A design basis should be developed for all IPS, because there are many benefits to this work process:

Enhanced understanding of hazardous events and how the IPS acts to stop their propagation,

- Increased focus on achieving the seven core attributes,
- Better rationale for expenditures on the IPS,
- Improved IPS design with fewer unplanned shutdowns,
- Improved overall design quality,
- Better justification for the required diagnostic and proof test requirements, and
- Improved consistency during the design and implementation of risk reduction strategies across the owner/operator organization.

There are no specific requirements for how the documents should look or how they should be assembled. There is tremendous flexibility to fit any project size. The documentation evolves as the project progresses and should be sufficiently detailed to clearly communicate the design requirements. In early project stages, the design basis may consist of simplified diagrams, logic narratives, and input/output (I/O) lists. Early project scoping is generally based on rough estimates of instrumentation requirements. Later project phases involve more detailed documentation, such as an instrument index, logic flow charts, and electrical system diagrams.

The design basis covers for each process operating mode:

- Good engineering practices to be followed,
- Physical allocation of functions to systems,
- Integrity and reliability requirements,
- Operability and maintainability requirements,
- Functional requirements, including the process measurements, safe state actions, process actions, required sequencing, and process safety time,
- Architecture, including specification of redundancy and configuration requirements for selected equipment technologies,
- Fault detection and response requirements, including diagnostic expectations, proof test interval, and mean time to repair, and
- Human interface requirements, including alarm, bypass, manual shutdown and reset.

The ability to prove performance is a core attribute of any independent protection layer. The design basis serves as the control document for the verification, assessment or audit of the design, maintenance, inspection, proof test, and operating practices. Actual performance of the installed and operating systems should be periodically compared with the design basis and its assumptions. If deviations are found, they should be flagged and routed to the appropriate parties for review, documentation and justification. For cases where the field installation or current operating/maintenance procedures do not meet the design basis, a hazard and risk analysis should be conducted. The risk introduced by the design basis deviation may necessitate immediate action.

The approved design basis is the process safety information for any SIS and is part of the safety case for the process equipment. It should be maintained sufficiently up-to-date to support any hazard and risk analysis, verification, functional assessment, or audit.

4.5 PROCESS REQUIREMENTS

Input from process engineering and operation personnel is essential to the correct specification of IPSs. The process requirements specification is a collection of information covering the operability, functionality, reliability and maintainability requirements for the process. This work process finalizes the risk reduction strategy and is used to develop the I&E requirements. As shown in Figure 4.3, the process safety information and hazard and risk analysis findings are used along with input from operations to define the following:

- Operability requirements, including operating modes, process composition, operating conditions, external conditions and criticality,
- Functionality requirements for each process operating mode, including a definition of the process measurements, safe state actions, process actions and process safety time,
- Reliability requirements, including sensitivity to spurious trips, and
- Maintainability requirements, including proof test opportunities and availability of compensating measures.

The process requirements evolve throughout the process and IPS life. During the design phase, verification is performed as part of the project execution to ensure that the process requirements specification continues to achieve the hazard and risk analysis requirements. Over the IPS life, management of change (MOC) should be followed when changes to the process requirements specification are proposed. The review should ensure that the intent of the hazard and risk analysis is met by any change and should examine how the change impacts other functions within the system.

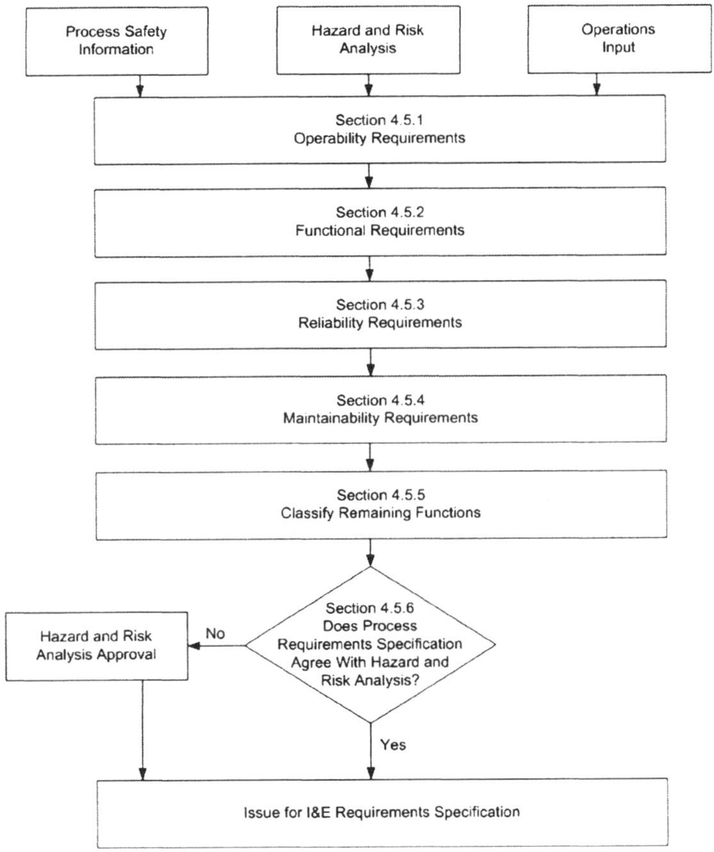

Figure 4.3. Process Requirements Work Process.

4.5.1 Operability Requirements

All planned or expected process operating modes should be examined for each piece of equipment or set of equipment (e.g. a process unit or a process). These modes include many planned activities such as start-up, normal production rates, reduced production rates, scheduled maintenance outages, cleaning (e.g., regeneration and decoking), safe state, and shutdown.

Each operating mode may have different IPF requirements. Specific protective functions or additional logic may be necessary to achieve or maintain a safe state for certain modes. The risk reduction requirements may also be higher in some modes due to increased risk. For example, when the operating mode involves manual operator actions, such as burner light-off or valve line-up, operators are likely to be present in the process unit during the hazardous event, increasing personnel risk.

Many catastrophic incidents have occurred during start-up when operator or maintenance mistakes resulted in catastrophic releases with large numbers of personnel present due to intense work activity. Each operating mode should be assessed during the hazard and risk analysis to determine whether special safeguards and/or protection layers are required.

Operations personnel should participate in defining the required operating modes. Operational flexibility is established during this lifecycle stage. Special operating mode requirements should be identified early in the design process to minimize project rework. Implementing major changes during detailed engineering to accommodate an unforeseen operating mode can result in significant schedule and budget impact, as these decisions typically affect many design aspects. A small change in how an operating mode is addressed can propagate into a major change in the IPF design.

Some operating modes may require specific IPF logic, including bypasses or permissives. Start-up may involve the increased exposure of a field operator to potential hazardous events, due to required field activities. The increased exposure may result in higher risk reduction requirements for some functions than what is required for normal operation.

Start-up bypasses may be required to facilitate safe unit start-up, such as the use of a time delay to enable an automated bypass of a low discharge flow IPF for pump start-up. Permissives may be needed to allow some modes to occur, such meeting a specified process operating state or condition prior to allowing the next step in the sequence. Permissives are often required to prevent unacceptable operating states, such as regeneration cannot be started until the feed valves are confirmed closed. Permissives are often implemented to reduce the risk associated with manual operating steps.

Batch operation often involves multiple operating modes. This complicates the IPF design, due to the frequent interaction required between the basic process control system and the IPF. Batch control relies on the use of defined sequences that often require operator input to initiate specific processing steps.

Batch operation may also include the use of multiple "recipes," which represent different combinations of reactants to produce multiple products from the same equipment. Each recipe with its associated process steps represents a set of operating modes that should be subjected to analysis. For reactive processes, extensive permissive logic is often required to ensure proper batch sequencing. For example, 5000 gallons of solvent must be metered into the reactor before the SIS block valves are reset.

Operability should also examine any potential process or application issues impacting the IPF's ability to operate as intended. The application and operating environment should be examined to identify special or unusual conditions that should be considered when selecting IPS equipment, such as:

- Process composition (e.g., solids, salts, corrosives, interferences),
- Process operating conditions (e.g., extremes in temperature or pressure),
- External conditions (e.g., need for winterization, potentially corrosive environments and hazardous area classification), and
- Criticality (e.g., accuracy, precision, repeatability, drift, fire survivability, and leak tightness).

  Example--Criticality of valve closure

  A hydrocarbon source must be tightly isolated from oxygen sources requiring the specification of a special valve technology, the installation of double block and bleed arrangements, or implementation of other means of positive isolation.

  In other applications, partial failure may not significantly impair the IPF's ability to respond to the unacceptable process condition. A partial failure of a cooling water valve, such as the valve failing stuck in a partially open position rather than going full open, may be acceptable as long as the feed is isolated.

4.5.2 Functionality Requirements

The functional requirements for each instrumented protective function (IPF) are developed by answering three key questions:

- What is measured?
- What decision is made based on the measurement?
- What action is taken?

The answers to these questions identify the chain of events that must occur to detect and respond to unacceptable process conditions. In most cases, the measurement is a process variable, but it can also be other input variable types, such as a trip signal from another IPF or protection layer, the operational status (run/stop) of mechanical equipment, or the position of final elements. Any approved equipment technology can be specified, such as differential pressure or radar for a level application.

The process requirements define the functional relationship of the measurement, logical decision, and action, e.g., de-energizing a solenoid operated valve to close a block valve when high flow is detected. Taking action may require one or more process variables to be present. Time delays are sometimes incorporated in the logic to minimize false trips caused by transient process conditions that do not present a hazard. These time delays are often finalized during start-up and operation when such conditions are better understood.

The IPF takes protective and non-protective actions. The protective actions include any output needed to achieve or maintain a safe state. Non-protective actions, which are referred to as process actions, are generally specified to support process recovery after IPF action. Process actions may taken to address cascade trips, to prepare the facility for access following shutdown, and to facilitate process equipment start-up after the cause of the IPF action has been investigated. Process actions often include taking action on control system equipment to prevent automatic equipment re-start when process conditions return to the normal operating range. Process actions should be clearly distinguished from the protective, or safety, actions.

4.5.2.1 Process Measurement

The process measurement should detect the process condition of concern with minimal detection lag. For example, thermocouple or RTD installation should consider the potential for stagnant fluid around the thermowell, potentially causing erroneous readings. Each hazardous event must be detected early enough that the required safe state of the process can be achieved within one-half the process safety time. Often a single location is chosen, such as the measurement of pressure at the top of a tower. In other cases, multiple locations may be required to detect the event, such as the detection of hot spots, requiring multiple thermocouples to be distributed throughout a reactor bed.

> Example--Reactor hot spot detection
>
> Hot spots can develop within a plug flow reactor. A hot spot is a localized area of high temperature which can result in a variety of highly undesirable events, such as decomposition, runaway reaction, and de-rating of vessel MAWP. To detect hot spot formation, multiple thermocouples are often used within the reactor. Sometimes, the reactor is divided into zones and thermocouples are distributed horizontally across each zone.
>
> Thermocouples are prone to burn-out, which causes an open circuit. For SIS applications, failed equipment should be configured to take the specified safe state. In a multiple thermocouple application, this could result in a high spurious trip rate for the reactor. Fault tolerance against spurious trips should be considered for the SIS design.
>
> Depending on the reactor, multiple thermocouples may be capable of detecting the hot spot as required. It is quite common to use three thermocouples within each reactor zone. If all three thermocouples are able to detect high temperature within the zone, the thermocouples could be voted 2oo3, reducing the spurious trip rate. If each thermocouple location represents a distinct location, multiple thermocouples should be located in each location voting 2oo2 or 2oo3.

For reactors with multiple zones, hot spot detection typically relies on 2oo2 or 2oo3 voting in each location or zone. This voting is often simplified in later I&E implementation as 2ooN for the reactor. For instance, if there were 3 zones with 3 thermocouples per zone capable of detecting the event, the voting per zone is 2oo3. The voting within the SIS is typically implemented as 2oo9, which is more conservative and accounts for potential channeling and localized contamination. The SIS description should clearly state that the event is detected by any 2oo3 thermocouples within a zone and the verification should be based on 2oo3 voting.

Whenever possible, process variables should be directly measured, rather than inferred or calculated from other process variables. Verification, testing and validation of calculated process variables is more difficult. Calculations require extensive testing to ensure that the algorithm is correct.

When calculations are used, consideration should be given to establishing an acceptable range for the calculated variable and restricting system response to the acceptable range. Simulation techniques should also be used to thoroughly test the stability of the calculation. The potential impact of calculation errors or unexpected values should be evaluated and their impact limited by the logic to ensure a predictable response.

Example--Function complexity

The Research and Development (R&D) group at a major chemical company wanted to shutdown a reactor based on a calculated variable to determine the optimum time to begin additive feed and increase the mixing rate. The equation eventually had 21 inputs, some of which were inferred calculated values based on analyzer inputs. This degree of complexity requires thorough verification and validation to ensure that the logic behavior is understood and managed during normal and abnormal conditions, including expected ranges of each variable and bad or out-of-range data.

4.5.2.2 Safe State Actions

An IPF takes action to achieve or maintain a safe state, thus stopping the hazardous event propagation. The safe state is a state of the process operation where a hazardous event cannot occur. The safe state is initiated automatically by the IPF when predetermined conditions are violated. Equipment is said to be specified as “fail-safe” when its dominant failure mode is toward the safe state.

Incorrect specification of the safe state is a serious systematic error that can result in the IPF failing to function as required. The safe state definition should be considered a critical path item for the project schedule. Changing how the IPF functions becomes more costly as design progresses, so it is imperative to define the safe state early in the project and not change it without justification.

Sometimes, several actions are taken in response to an identified hazardous event. The need to sequence a process can be identified by analyzing the affect of the process if some output actions occur and others do not. Some processes require a sequenced shutdown to achieve the safe state with minimum impact. Process chemistry, design, and dynamics may determine the best groupings of related equipment to be shutdown and the correct sequence to execute.

> Example--Dangerous combination of outputs--sulfur recovery unit
>
> A sulfur recovery unit is shutdown when high thermal reactor chamber temperature is detected to prevent refractory damage. The shutdown isolates the acid gas feeds, combustion air, induction air, and recycle gas from the sulfur pit to halt combustion.
>
> If the acid gas streams are successfully isolated but the combustion air valve fails to close, a temperature excursion will occur due to oxygen breakthrough into the catalytic reactor. Another example would be isolating the combustor from the sulfur pit without stopping induction air flow. This allows the induction air to reverse flow back to the sulfur pit and sweep H2S to the atmosphere.

4.5.2.3 Process Actions

Process actions are work in conjunction with safety or protective actions to facilitate a complete and orderly process shutdown. Process actions may include those taken to:

- Prevent cascade trips, e.g., if PIF #1 closes a valve on the discharge of pump 1, the pump experiences a block-in condition, potentially resulting in pump damage. PIF#2 acts on detecting high pump discharge pressure and shuts down the pump. Consider having PIF #1 also trip the pump,
- Prepare facility for access, inspection, and personnel response after shutdown is successfully completed, e.g., fail open cooling water valves to reactor jacket or turn on snuffing steam to burner,
- Prepare the facility for start-up after information has been gathered for investigation, and
- Prevent automatic restart of the process after clearing of the unacceptable process condition.

Process actions are often used to prevent the control system from immediately resuming operation upon clearing of the unacceptable process condition. The three most common ways to take control function equipment to the safe state are:

- Trip signal is sent from the IPS to the BPCS, causing the BPCS to take action on its final elements,
- The IPF takes direct action on the control function final elements, and

- The trip function is mirrored in the BPCS so the process condition is directly measured by the BPCS, which takes action on the control function final elements.

4.5.2.4 Process Safety Time

The IPS must be able achieve or maintain a safe state of the process within the process safety time with a very high degree of certainty. The IPS response time is the time required to fully execute the required sequence of events from detecting the process condition to achieving the safe state of the process. The equipment specification establishes the process design limits. The process safety time is constrained by the design limits and the set points chosen for each IPF.

Each function should be capable of acting within one-half the process safety time allocated to it by the selected set point. The process safety time is the difference between the time at which the unacceptable process condition occurs ($T_{CONDITION}$) and the time where the hazardous event occurs (T_{EVENT}). The process safety time can be long (increments of seconds to minutes) or short (milliseconds).

$$\text{Process Safety Time} = T_{EVENT} - T_{CONDITION} \qquad 4.1$$

The operation of the first protection layer (or IPF) establishes the boundary for the normal operating range and the beginning of abnormal operation. The equipment specified must be able to act fast enough to prevent the incident. In some cases, dynamic analysis may be required to determine whether the hazardous event can be avoided by the IPF within the process safety time.

The process dynamics determine how quickly a process deviation propagates into a hazardous event. The process dynamics are influenced by the process design, operation, chemistry, mass transfer, fluid flow, heat transfer, thermodynamics, kinetics, etc. In many cases the process responds within milliseconds, e.g., a pump is shutdown, immediately stopping feed into a tank. In other cases, the response might take seconds, e.g., closing fuel gas valves. The response could also take minutes, e.g., the fuel gas is isolated, but the retained heat provides energy for some time after the burners are shutdown.

For any process excursion, multiple protective functions may be used to detect and respond to the process deviation. As illustrated in Figure 4.4, a process alarm (non-protective) indicates a process condition resulting in a quality control excursion. The operator takes action to maintain the process within the normal operating range.

If the control action is ineffective, a protective alarm indicates that the process condition has propagated outside the normal operating range. In this illustration, the operator takes immediate action on the process to bring the process to a safe state. Again, if the operator action is ineffective, the process condition worsens, leading to a challenge on a protective instrumented function (PIF), which takes the process to the safe state. Finally, if the PIF fails to operate correctly, the process condition exceeds the design limits and vessel failure may occur, leading to a hazardous event.

Process lag time affects the choice of set point. Longer lag times require lower (or higher) set points to prevent the process condition from exceeding the design limit or to achieve a particular protection layer sequence. Process lag is illustrated in Figure 4.4 by the continued progression of the process variable for a time period after each set point. Set points should be selected to provide sufficient time for each IPF to complete its action in the proper sequence.

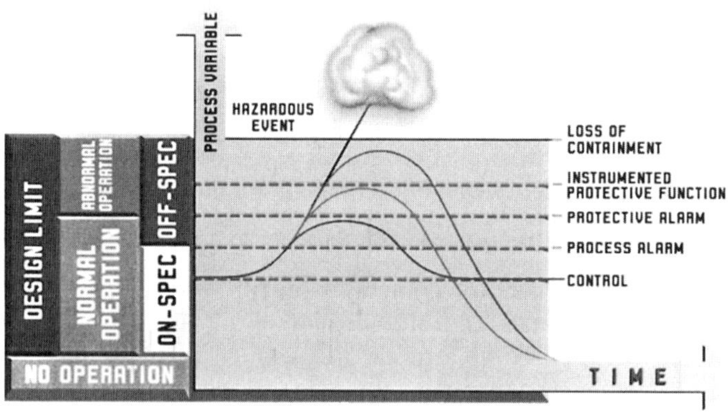

Figure 4.4. Process Condition Changes With Time.

For the protective alarm, the allocated process safety time is the time between the alarm occurrence T_{ALARM} and the hazardous event occurrence T_{EVENT}. For this alarm to be allocated risk reduction, the operator should be able to complete the required actions in less than one-half this process safety time. The operator response time begins with alarm initiation and stops when the operator actions have resulted in a safe state of the process.

The time available for the operator to prevent an automated trip is limited by how close the process alarm set point is to the protective alarm and/or the PIF set point. For protective actions, trained operators should receive a clear and unambiguous alarm and the intended response should be covered by written procedures. Operators should be trained and tested on procedures related to IPF indications and alarms.

The more time available for the operator's overall response, the more likely the operator will be able to bring the process under control and prevent the need to take the process to a safe state. But at some point, the operator should take a specified action to achieve or maintain a safe state based on "never exceed never deviate" process conditions.

Example--Slow dynamic response

A knock-out drum for a compressor is large enough that the drum level is manually controlled by a field operator by monitoring the level each shift and draining the drum when level is detected. The drum is provided with a high level alarm to notify the control room operator when a specified drum level exists. The available response time is based on the quantity of water collected by the drum during normal and upset conditions and the volume of the drum and the interconnecting piping design. In this application, more than 40 minutes are available for the operator to take action in response to the alarm prior to the overflow of liquid from the knock-out drum to the compressor.

Example--Fast dynamic response

Consider an oxygen mixing station, where hydrocarbons and oxygen are mixed. If the ratio of hydrocarbon and oxygen are not controlled within an established range, a flammable mixture is produced that is prone to almost instantaneous deflagration and possible detonation. The extremely short process safety time requires an automated system with negligible response lag.

4.5.3 Reliability Requirements

Process reliability goals are often defined based on target service factors, on-line percentages, production targets, cost of product, profit margin, or cost of a spurious trip. IPFs are used to reduce the risk associated with identified hazardous events. However, IPF implementation can introduce reliability problems, because each piece of IPS equipment is another potential cause for an unnecessary shutdown. This is why inherently safer design should be practiced to the extent practical and should always be considered prior to adding an IPF.

The process impact of equipment spurious operation should be assessed. For example, the spurious closure of the block valve on the inlet line to a tank may result in low tank level. Or, the opening of a pressure control vent valve could result in a significant flaring event or an environmental release. The potential for hazardous events due to spurious operation should be identified during the risk assessment phase.

Processes also have different sensitivities to spurious shutdown. All process shutdowns involve lost production and downtime. Some process units are operating at 100% capacity and downtime represents lost production that cannot be recovered. Other process units are operating at less than maximum capacity and do not experience significant production losses.

Reliability requirements are often established on an individual IPF basis by estimating the cost of process disruption. The requirements are generally stated in terms of a tolerable spurious trip rate, e.g., the spurious trip rate should be less than 1 in 50 years for high reliability service. As a general rule of thumb, if the potential costs associated with a spurious trip are high, the mean time to spurious trip should be 5 to 10 times the turnaround interval, e.g., if the turnaround interval is 5 years, the tolerable spurious trip rate is once in 25 to 50 years.

4.5.4 Maintainability Requirements

The mechanical integrity of IPS equipment is maintained through inspection, preventive maintenance, proof testing, and repair. Finding failed equipment through an effective mechanical integrity program is always safer and less costly than finding it by incident. The longer one waits to verify the correct operation of IPS equipment, the greater the likelihood that the equipment is non-functional. Running IPF equipment to failure is not acceptable. More robust IPF designs, incorporating redundancy, diagnostics, and on-line repair/maintenance capability, are required to compensate for long intervals between operational checks.

Opportunities for inspection, preventive maintenance, and proof testing should be identified by process engineering and operations personnel. Production schedules should be reviewed to determine when equipment can be accessed. Short term or maintenance (or turnaround) outages are used for off-line maintenance of IPS equipment associated with continuous processes. Batch or semi-continuous processes can generally be accessed during mode transitions.

On-line testing and maintenance activities often require the use of a maintenance bypass to allow these activities to occur without causing a process disruption. When a bypass is used, compensating measures must be documented that provide protection equivalent to the loss of IPF protection. Compensating measures typically convert the risk reduction strategy from one relying on the IPF to one relying on the operator taking action. Process engineering and operations must determine whether safe operation can be sustained while on-line maintenance and testing is performed.

Compensating measures typically fall into two categories:

- Change from the normal process operating condition to one where the hazardous event cannot occur or the risk is reduced below the risk criteria. This includes potentially shutting down the process, and
- Continuous operator supervision of the specified process variables supplemented by a documented procedure that defines the "never exceed never deviate" process condition, where manual shutdown is required. Continuous monitoring may require additional staffing to adequately manage the risk at all times.

When assessing the risk reduction provided by operator response, consider the guidance provided in Appendix B under supervisory functions, in Section 4.5.1.5 concerning process safety time, and in Chapter 5.5.5 related to response time.

4.5.5 Classify Remaining Functions

Most process units, especially existing ones, have a few IPFs scattered throughout the unit that are not identified as being necessary functions during the risk assessment phase or in the process requirements specification. As instrumented systems migrate from one technology to the next, occasionally previous generations of shutdowns are left behind. These IPFs may no longer be required due to process changes or to the implementation of other IPSs.

Unnecessary and potentially outdated instrumented systems may impact operational reliability if equipment fails to operate properly. Superfluous IPFs should be addressed to obtain the best reliability from the IPS as a whole. Consequently, for existing process units, the IPFs defined in the process requirements specification should be compared against a list of existing IPFs to identify IPFs that have not been classified.

A hazard and risk analysis team, as well as operations and process engineering personnel, should review unclassified IPFs. This review should determine whether the IPF is needed or whether it can be decommissioned. If the decision is to maintain the IPF, the function should be classified, so it can be managed appropriately.

4.5.6 Verify Process Requirements

A risk reduction strategy was developed during the risk assessment based on owner/operator risk criteria. As the project evolves, the identification of various operating needs may necessitate changes to this strategy. The process requirements specification should be verified against the hazard and risk analysis to ensure its intent is met. Owner/operator procedures should define when hazard and risk analysis team and/or management level approval is required for deviation from the hazard and risk analysis. These procedures often establish different review requirements based on the type of risk, the required risk reduction, and degree of deviation.

4.6 I&E REQUIREMENTS

Before starting the I&E requirements, the hazard and risk analysis for the process unit should be reviewed to understand the initiating causes for the hazardous events and the protection layers used to reduce the process risk below the risk criteria. The I&E requirements should meet the intent of the hazard and risk analysis team, which recommended the IPS, and should fulfill the process requirements specification (see Section 4.5).

I&E personnel should review the process requirements with a process representative to ensure that operability, functionality, maintainability, and reliability constraints are understood. The risk reduction allocated to the PIFs yields an integrity level requirement for the PIF and its equipment.

Three types of evidence are required to claim that an IPF achieves a particular risk reduction. First, the equipment should be user approved for the operating environment and application. Second, the equipment subsystems should meet the requirements for fault tolerance against dangerous failure, as discussed in Section 4.4.6. Third, the overall IPF design, inspection, maintenance, testing, and operating practices achieve the required risk reduction.

The integrity level should be verified for PIFs (or SIFs) by determining and documenting that the system achieves the average probability of failure on demand (PFD_{AVG}) for a demand mode system or, if the PIF (or SIF) is operating in the continuous mode, it achieves the required hazard rate.

The overall response to a hazardous event may include safety, protective, and process actions. The IPF requirements should be traceable from the identified hazardous event to the IPF design basis. If multiple functions are used to reduce the risk of an identified hazardous event, the required independence of these functions should be understood. Any potential common cause failure impacting multiple functions within the same layer should be addressed during design and verification.

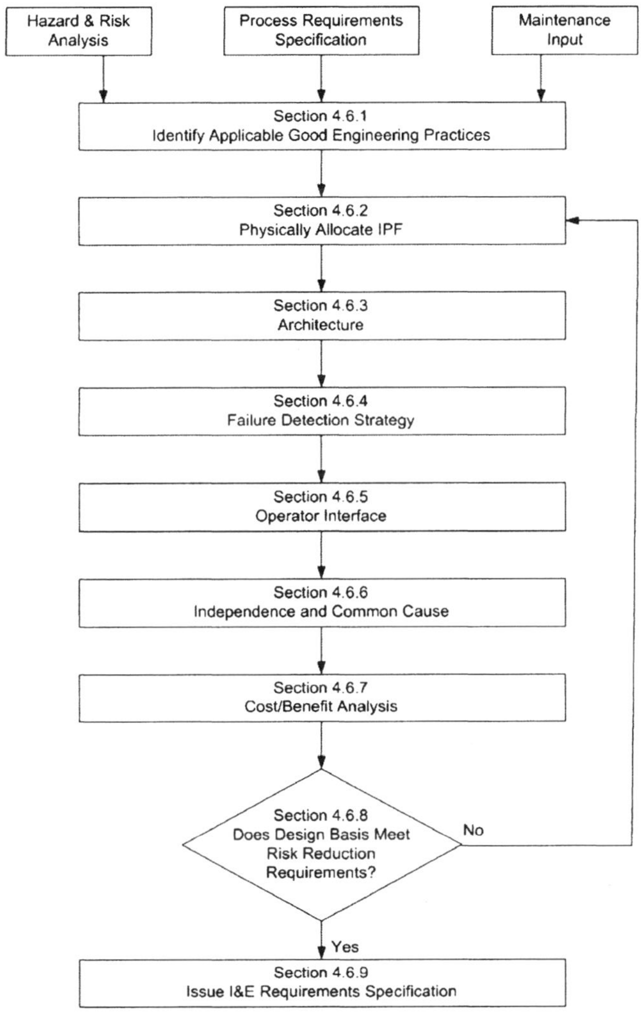

Figure 4.5. I&E Requirements Work Process.

The I&E requirements work process is illustrated in Figure 4.5. There is often quite a bit of interaction between the process requirements and the I&E requirements in the early stages of the project. For example, the ideal process measurement may not be practical in the existing installation.

At all times, the intent of the design basis is to document how the IPF will detect and respond to the unacceptable process conditions. Improvements should be encouraged if these lead to achieving process unit goals in the areas of operability, functionality, maintainability, and reliability. However, the process requirements should be completed and approved prior to the 25 percent complete point of the I&E requirements. Otherwise, the project schedule and budget may be impacted by project rework.

4.6.1 Identify Good Engineering Practices

Many owner/operators document IPF design and engineering practices in the form of revision controlled engineering guidelines and an approved equipment list. The intent of these internal practices is to improve project efficiency and consistency. Documented practices reduce the tendency to "reinvent the wheel" on every project therefore ensuring that proven design and installation practices are employed where possible. Internal practices provide reference documents for use by future projects that seek to alter the initial design. Refer to Section 4.8.1 for more discussion on technical practices.

At a minimum, relevant good engineering practices should be reflected in the design basis. When the actual design, inspection, maintenance, testing or operating practices deviate from a referenced good engineering practice, the deviation should be justified as equivalent to or exceeding the applicable practice.

4.6.2 Physically Allocate IPF

The hazard and risk analysis identifies protective functions required to reduce the risk below the owner/operator risk criteria. The identified protective functions are allocated to IPLs which are designed and managed to meet the required risk reduction. Once the risk reduction strategy is finalized, the protective functions can be allocated to physical systems.

The physical allocation process represents the transition of the protective functions from conceptual IPLs to physical systems capable of achieving the required core attributes. The allocation process often involves compromises:

- Less risk reduction for certain asset protective functions based on cost/benefit analysis,
- Reallocated risk reduction among IPLs to achieve cost constraints, and
- Physical solutions for non-safety protection that best fits into the "existing" plant maintenance capabilities.

Assumptions about the core attributes play a significant role in the physical allocation process. For example, the core attributes of independence, functionality, integrity, and reliability are generally well defined for each function as a result of the hazard and risk analysis and are relatively tangible in nature. The remaining three attributes, auditability, security, and management of change, though valid core attributes, are a bit more abstract and must be defined for each system. Consequently, physical allocation is often iterative and is a collaborative effort between I&E, process, and operations with input as needed from equipment manufacturers.

Business needs and long-term management goals should be considered during the allocation process. Blurring the lines of independence between IPLs may lower capital costs, but increases system complexity, making it more difficult to assess, verify, and audit. Open access to basic process control system (BPCS) data and software (e.g., external connections allowed) may be desirable from a business standpoint, but limits the ability to provide effective and rigorous management of change and access security for the functions within it.

Often, there is a one-to-one correspondence between the conceptual and physical allocation. In such cases, physical allocation is a straight forward process. For example, if an IPF is classified as safety, the conceptual allocation is to the preventive layer and the physical allocation is to the SIS. But other times the physical allocation may not be as simple. Consider a hazardous event caused by failure of a control function. The hazard and risk analysis identifies a supervisory function and an SIF, which take action to achieve or maintain a safe state in response to unacceptable process conditions.

Figure 4.6 illustrates three physically separate and independent systems, which do not share data, information, equipment, support systems, operator interfaces, engineering interfaces, etc. The distinction between the layers is clear and easily understood.

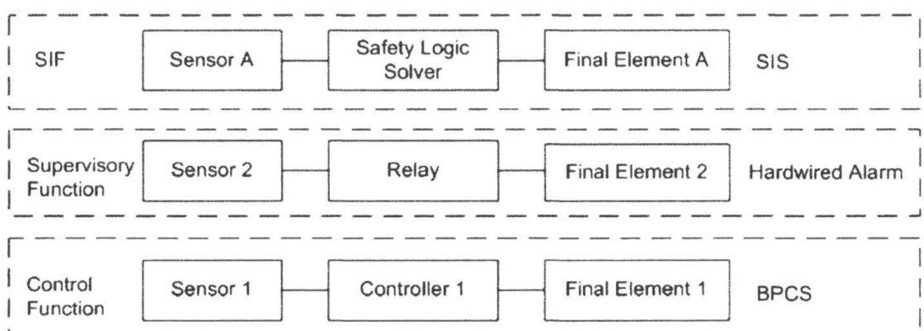

Figure 4.6. Separate and Independent Protection Layers.

As discussed in Appendix B.3, some owner/operators choose to implement the control functions and the supervisory functions within a BPCS, when the BPCS meets specific design and management requirements. Figure 4.7 illustrates a BPCS implementing control and supervisory functions. The functions are allocated to separate controllers, which communicate across a common data highway. The supervisory function is clearly separate from the control function from a hardware and software standpoint, but the common data highway increases the potential that both functions could fail simultaneously.

ISA 84.01-2004/IEC 61511 limits the performance that can be claimed for instrumented systems that are not designed and managed in accordance with the standard. The dangerous failure rate (or hazard rate) of a control function allocated to the BPCS may not be assumed to be lower than 10^{-5}/hr (approximately 1/10 years). The risk reduction allocated to a safety function in the BPCS must be less than 10. The maximum hazard rate claim for the BPCS is therefore limited to 10^{-5}/hr x 1/10 = 10^{-6}/hr (1/10 years x 1/10 = 1/100 years) even when redundant channels and tighter management systems are in place.

Access security and management of change must be carefully planned to achieve the same level of rigor of independent and separate systems. Although allocated to the BPCS, the supervisory function must be designed and managed to achieve the required core attributes. Sharing BPCS equipment between multiple functions should be justified by a detailed analysis of the equipment and their management.

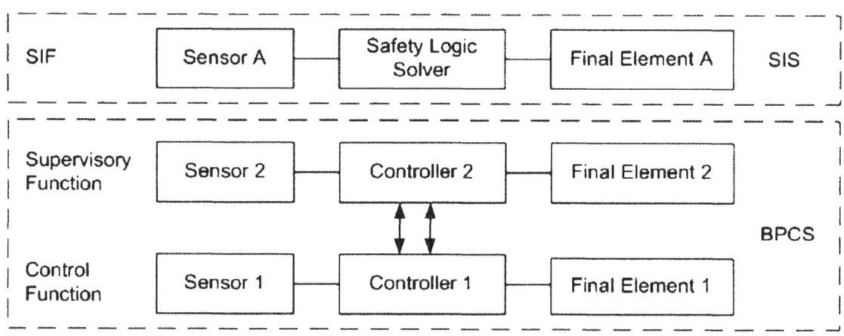

Figure 4.7. Supervisory Function Implemented in BPCS with Separate SIS.

Other owner/operators choose to implement supervisory functions in the SIS, especially if the supervisory function is used to reduce safety risk. Figure 4.8 illustrates the allocation of the supervisory function to the SIS. This allocation provides the highest level of management system rigor for the supervisory function, since it must be managed as an SIS.

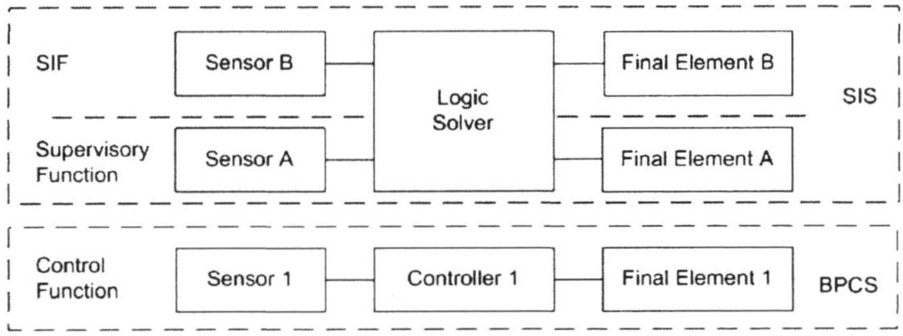

Figure 4.8. Separate BPCS with Combined Supervisory Function and SIF.

If the supervisory function and the SIF are reducing the risk of the same hazardous event, the SIS is designed and managed to the level of rigor required to achieve the risk reduction allocated to both functions. For example, if the supervisory function is providing a risk reduction of 10 and the SIF is also providing a risk reduction of 10, the overall system would need to provide a risk reduction greater than 100 (10x10).

In special applications, the control, supervisory, and SIF are allocated to a single SIS, as shown in Figure 4.9. In this case, the entire system is managed as an SIS, which provides a high degree of management system rigor for the control functions, as well as the supervisory functions and SIF. However, maintaining the core attributes for the system life can be quite challenging.

Documentation, access security, and management of change requirements apply to the system as a whole. Achieving and maintaining the required discipline increases administrative costs and reduces control system flexibility. For this reason, combined systems are typically restricted to special process applications.

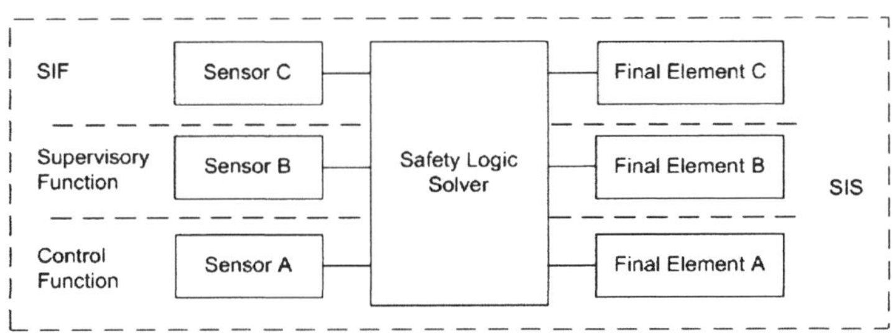

Figure 4.9. Combined Control Function, Supervisory Function and SIF.

For combined systems, the overall hazard rate should be assessed, especially if the control function can place a process demand on the supervisory function and/or SIF. For example, consider a hazardous event in which the control function is the initiating cause. Assume the control function fails at a frequency of 1/10 years and the supervisory function and SIF provide a risk reduction of 10 each. The system would need to achieve an overall hazard rate less than 1/10 years x 1/10 x 1/10 = 1/1000 years

Now, consider a different hazardous event which poses asset risk. A protective instrumented function (PIF) is identified, which needs to be allocated to a system. The tendency is to allocate the PIF to the SIS, as shown in Figure 4.10.

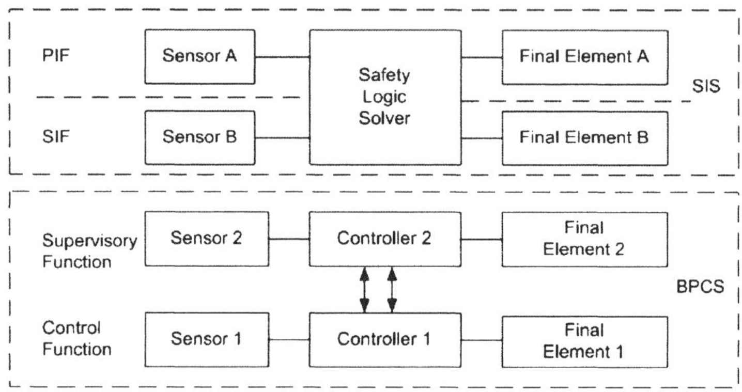

Figure 4.10. BPCS with Control and Supervisory Functions, and SIS with PIF and SIF.

Sometimes, this is done without realizing that the PIFs must be commissioned and managed to the same level of rigor as the SIFs that coexist within the same system. This can load the SIS with non-SIS functions, increasing its complexity. The larger the SIS, the greater the challenges on the documentation, validation, access security, and management of change processes.

The physical allocation to the SIS enhances the integrity and reliability of these functions, if the rigor of the management system can be maintained for the overall system. When non-safety functions begin to dominate the system, it becomes more challenging to maintain the rigor required for SIS.

Figure 4.11 shows a dedicated PIS, which is designed and managed as a separate system to meet owner/operator specified core attributes. PIS segregation is generally driven by cost-benefit analysis. The feasibility of segregating the PIFs and SIFs depends upon the quantity within a each classification. If the ratio of PIFs to SIFs is high, segregation may yield cost savings. Consider a process that requires 20 equipment PIFs and two SIFs. A cost effective solution may be a non-safety-rated programmable electronic system (PES) for the PIS and a hardwired system for the SIS. When separate, the PIS management system may be less rigorous than the SIS management system, as discussed in Chapter 2.6.1, lowering lifecycle management costs.

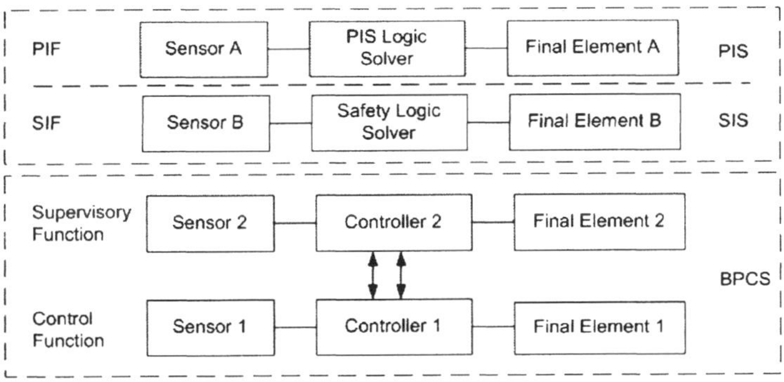

Figure 4.11. BPCS with Control and Supervisory Functions and Separate SIS and PIS.

The identified systems may be further segregated into multiple systems. This often happens naturally, since the instrumentation and control system is an integral part of the process equipment and unit design. When the decision is made to build a process unit, instrumentation and controls are also implemented, such as a BPCS and PIS.

Specialized equipment, such as boilers and compressors, are provided with dedicated BPCS and PIS (or SIS). Specialized equipment generally requires a specialized control system that operates independently of other process equipment. Boiler practices also require separate PISs.

The segregation of the functions for non-specialized equipment should consider the intended operating and maintenance schedule for the process equipment. Human error and equipment problems are significant sources of spurious operation when maintenance is conducted on PISs containing in-service PIFs protecting on-line process equipment. At a minimum, segregation should ensure that all functions within a PIS are associated with equipment that will be shutdown at the same time for maintenance. This improves reliability, reduces the potential impact of human error initiating spurious shutdowns of or hazardous events in on-line process equipment, and supports thorough PIS maintenance and proof testing.

Segregation should also consider the overall risk being managed within each system. If electrical systems are used for PIS implementation, each function operates independently within the system. The risk borne by a particular relay or trip amplifier is limited to the function that it is performing. In contrast, programmable electronic systems (PES) rely on single or multiple channel main processors and input and output (I/O) modules. The I/O can be distributed across modules to provide segregation of the field devices, but the system still relies on the main processors with embedded software, an application program, system internal communication, power supplies, etc.

When designing PES, the natural tendency is to build large systems to obtain the lowest cost per I/O. However, shared components within the system represent sources of common cause and common mode failure, potentially affecting all of the I/O contained within the system. Spurious operation of a large I/O PES may result in significant consequences, since the spurious event likely initiates the shutdown of multiple pieces of process equipment. Likewise, a large PES presents the greatest risk when a dangerous PES failure occurs. I/O segregation into multiple systems should consider the risk posed by the dangerous failure or spurious operation of the logic solver on the process equipment. Segregation should be implemented when practical to reduce the impact of systematic and common cause failure.

> Example--Multiple train process unit
>
> A process unit has 6 reactors that are configured in three trains of two reactors. The shutdown systems on all 6 reactors are identical. Each train operates the two reactors as part of an overall production plan supported by operating procedures. Each train is taken out of service for maintenance independently.
>
> If all three trains are placed in the same system, the risk posed by 6 reactors is being protected by a single PES. SIS modification, maintenance and proof testing could potentially impact all 6 reactors. A safe or dangerous PES failure would put all 6 reactors at risk. Implementing the trains in separate systems limits these impacts to two reactors. Further segregation would provide independent equipment protection with each reactor in separate systems, limiting the impact to single reactors.

4.6.3 Architecture

An IPS is comprised of many elements. Figure 4.12 provides an example of how a typical IPS is organized. Each element consists of hardware and software that is configured for specific functional and/or physical characteristics. The elements are physically organized, interconnected, or integrated into a system architecture that operates according to the design basis.

Three major subsystems are generally implemented: 1) field sensors, 2) logic solver, and 3) final elements. IPS field sensors detect the process condition. A logic solver processes the available information and executes decisions based on the detected condition. Final elements take action on the process.

Each subsystem provides the fault tolerance against dangerous failure necessary to achieve the expected risk reduction using appropriate redundancy and voting. Redundancy involves the use of two or more pieces of equipment to perform the same function, e.g., multiple sensors to detect the same process condition. Voting is typically described using the nomenclature MooN (M out of

N) where M represents the required number of success states out of N total operating states. For example, 2oo3 means that there are three pieces of equipment of which two must agree for the vote to be satisfied. For 2oo3 with equipment A, B, and C, the successful combinations are AB, AC, and BC.

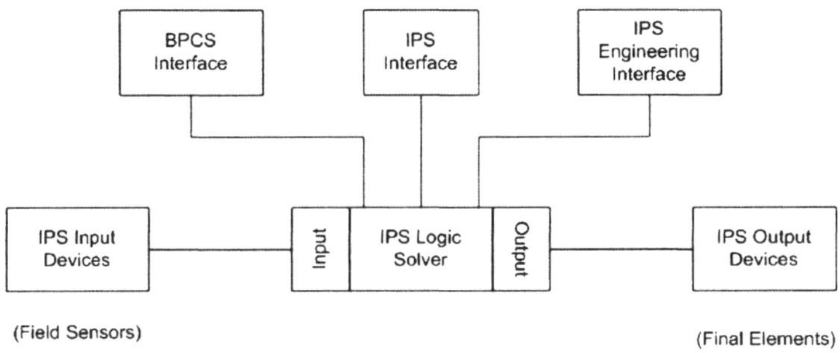

Figure 4.12. Scope of IPS.

4.6.3.1 Field devices

In the early stages of the design, equipment technology is selected to execute a specified function or set of functions. For example, a mass flow meter or a differential pressure transmitter is chosen for measuring flow. Chosen equipment should be user approved for the IPF application. It is typically not necessary to define the specific manufacturer until procurement. For example, one mass flow meter performs very similar to another, as long as both are on the approved equipment list for the application and are installed and maintained according to recommended practices.

The mean time to failure dangerous ($MTTF^D$) and mean time to failure spurious ($MTTF^{SP}$) should be estimated for the equipment in the operating environment. Ranges of $MTTF^D$ and $MTTF^{SP}$ for common IPF equipment are provided in Table 4.1.

In general, technology enhancements in field devices have improved the configuration interface, diagnostic tools, and communications. When estimating the equipment performance, all devices, components, and support systems required for its operation must be considered. Performance is limited by the weakest equipment used by the function. The risk reduction is a constraint that must be demonstrated throughout the IPF life.

In Figure 4.13, the effect of $MTTF^D$ on the average probability of failure on demand (PFD_{AVG}) is illustrated using the simplified equation (eq. D.19). A test interval (TI) of 1 year is assumed, since this benchmarks the equipment performance against past prescriptive practices as discussed in Appendix F.8.

Description	$MTTF^D$ (years)	$MTTF^{SP}$ (years)
Analyzers	0.35 - 4.00	0.35 - 4.00
Flow Switches	25 -50	10 – 50
Flow Transmitters	50 - 175	25 – 80
Level Switches	25 - 125	25 – 75
Level Transmitters	25 – 250	15 – 150
Pressure Switches	15 - 80	15 – 80
Pressure Transmitters	75 - 200	75 – 125
Temperature Switches	10 - 100	10 – 50
Temperature Transmitters	75 - 250	25 – 100
Solenoid Valves (de-energize to trip	30 - 100	10 - 30
Block Valves (failure to close)	25 – 100	50 - 200
Control Valves (failure to close)	15 - 60	30 - 100

Table 4.1. Example Ranges of $MTTF^D$ and $MTTF^{SP}$ for Field Equipment. (SIL Solver 2006)

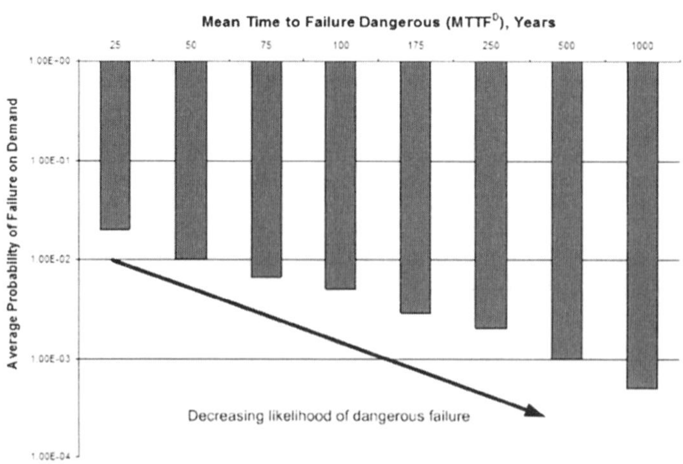

Figure 4.13. Impact of $MTTF^D$ on the PFD_{AVG} of Equipment Assuming the Simplified Equation, $\lambda_D TI/2$.

4.6.3.2 Logic Solver

The logic solver should be user approved (refer to Appendix F) for implementing IPF(s). User approval should determine the logic solver's capability to provide the required risk reduction and reliability. While the cost, ease of modification and the desire for one set of configuration tools have strong appeal, the logic solver should be selected to fulfill the design basis requirements and the long-term needs of operations and maintenance.

Any manufacturer or user restrictions related to logic solver implementation should be identified through review of the hardware and software documentation, such as the installation and maintenance manual, user manual, or safety manual. For SIS, if the manufacturer restrictions are not met, the logic solver may not achieve the specified performance and may not be compliant with its SIL claim limit.

Logic solver equipment should be analyzed to identify where "single points of failure" or "common cause failure" can result in system failure. These failures can exist anywhere within the logic solver or the utilities that support it, including the power supplies, communications, hardware, and software components. The logic solver should be designed to go to the safe state on loss of support systems and remain in the safe state until reset.

A safety-configured PES provides diagnostics for the inputs, main processor, outputs, support systems, and communications. This diagnostics is configured to take faulted component or channel to the safe state on detection of dangerous failure. Reliability is achieved through redundant channels (e.g., 2oo2, 2oo3, and 2oo4 architectures).

For PE logic solvers, the application program should be developed using a limited variability language. The program development tool, or the utility software, should use recognized language structures, such as those discussed in IEC 61131, *Programmable controllers - Part 3: Programming languages*. The utility should also support modular program development. For additional software requirements, refer to Chapter 5.7.

Description	$MTTF^{D}$ (years)	$MTTF^{SP}$ (years)
Relays	100 - 1000	100 - 500
Non-Safety Configured Single Channel PES	10 - 30	10 – 30
Safety-Configured Single Channel PES	100 - 250	5 – 15
IEC 61508 Compliant SIL 3 PES	2500 – 50,000	10 - 1000

Trip Amplifiers (programmable)	300 - 600	150 - 275
Trip Amplifiers (non-programmable)	500 - 850	150 - 250

Table 4.2. Example Ranges of $MTTF^D$ and $MTTF^{SP}$ for Logic Solvers. (SIL Solver 2006)

Once installed according to the manufacturer's recommendations, the achievable $MTTF^D$ is limited by the equipment configuration, fault tolerance, diagnostics, and proof test interval. The achievable $MTTF^{SP}$ is often dictated by how difficult it is to perform on-line maintenance, proof testing and repair/replacement of system hardware and software.

When multiple functions are allocated to the logic solver addressing the same hazardous event, the logic solver must be capable of providing the total required risk reduction. Considerations for the physical allocation of functions to systems are discussed in more detail in Section 4.6.2.

> Example--Logic solver total risk reduction
>
> For an identified hazardous event, two protective functions are identified and allocated to the same IPS logic solver: 1) a protective alarm with an allocated risk reduction of 10 and 2) an IPF with an allocated risk reduction factor of 100. The two protective functions are not independent, since they share the same logic solver.
>
> The effective risk reduction must be determined by analyzing the overall system comprised of the alarm and the IPF. Since the logic solver is the single point of failure for both functions, it must achieve a higher level of performance than indicated by the individual functions. The system must provide a total risk reduction equal to 10x100=1000.

The logic solver must be capable of executing the specified IPF or set of IPFs, while supporting the core attributes for the system. Equipment required for the IPS logic solver to operate correctly should be designed and managed to achieve the core attributes. Access security features should prevent accidental or undetected change of the IPF logic. The operator interface should not have access (or administrative rights) to modify the IPS application program when PE logic is used.

If the IPS logic solver shares equipment with other logic solvers, such as the BPCS, the impact of the failure of the shared equipment on each system should be evaluated. Shared components are managed as IPS and should be covered by IPS configuration management, access security, management of change, and validation requirements. Examples of shared components are uninterruptible power supplies (UPS), information displays, communications, and annunciator panels.

Access to the IPS engineering interface should be controlled to authorized personnel. The interface should be located in an access restricted (or controlled) area and access should be monitored and audited. The failure of the engineering interface should not affect the ability of the IPS to bring the process to the safe state.

For SIS, a dedicated engineering interface should be used, as it is inherently more secure and easier to administratively control. The SIS engineering interface should not be used for any other purpose, such as business or other unrelated engineering software. Some companies use notebook (i.e., portable) computers for the SIS engineering interface. The notebook computer is kept in a locked cabinet, requiring MOC approval to access the cabinet and the engineering interface.

4.6.3.3 Fault Tolerance Requirements

Fault tolerance against dangerous failure is provided to ensure that no single point of failure causes an IPF failure to operate. A fault tolerant subsystem continues to provide protection given a single failure in the subsystem. Fault tolerant design requires redundant equipment and appropriate voting to prevent a single failure from causing a dangerous IPF failure.

ISA 84.01/IEC 61511 provides minimum requirements for fault tolerance against dangerous failure based on the SIL. User approved non-programmable electronic (non-PE) equipment can be used in a simplex, non-fault tolerant architecture in SIL 1 and 2 applications; for SIL 3, fault tolerance against dangerous failure is required (e.g., 1oo2 final elements).

ISA 84.01/IEC 61511 allows the same minimum fault tolerance for PE field devices, such as analyzers and transmitters, when the following conditions are met:

- Equipment is selected on the basis of prior use (e.g., user approved),
- Equipment supports configuration only (e.g. range, failure direction),
- Configuration is protected, and
- Dominant failure mode is to the safe state.

The dominant failure mode is determined by the installation and equipment configuration, considering the inherent tendency of the equipment to fail to the safe state and the specified response to dangerous detected failure.

When dangerous failure is detected, the failed IPS equipment, e.g., a transmitter or a logic solver channel, should be configured to take a specified safe state. To achieve fault tolerance against spurious failure, additional equipment within a subsystem can be used, such as 2oo2 for SIL 1 and SIL 2 and 2oo3 for SIL 3. If the device is not configured to fail to the safe state, the minimum fault tolerance requirement is increased by one to create a fault tolerant system, such as 1oo2 or 2oo3 voting for SIL 1 and SIL 2 and 1oo3 voting for SIL 3.

In general, IPS equipment should fail to the safe state condition on the loss of any support system, e.g., power, instrument air, and hydraulics. Although some equipment requires a support system to operate as specified, e.g., energize-to-trip circuits or air-to-move valves, reliance on a support system should be restricted to those applications where there are no other acceptable technology choices. The support system design must achieve the desired risk reduction. In particular, the design should pay special attention to the detection of support system failure and the minimization of common cause or common mode failures. This includes examining how support systems are shared between the BPCS and SIS. Refer to Appendix F for a discussion of support systems, including energize-to-trip and instrument air design.

For SIS applications, PES must be safety configured by implementing diagnostics to detect input, output, and main processor failures. Redundant channels should be provided for SIL 3 and strongly considered for SIL 2 applications. For non-PE logic solvers, such as electrical systems, simplex equipment generally achieves SIL 1 and SIL 2 with redundant architectures providing SIL 3. For fault tolerance against spurious trips, redundant architectures should be considered at all SILs.

Safety-configured PES are available, which are designed in compliance with IEC 61508 and incorporate internal fault tolerance and extensive diagnostics. The equipment manufacturers provide a safety manual describing how the safety PE logic solver should be implemented to achieve its SIL claim limit. Manufacturer integration of diagnostics and fault tolerance can enable the safety PES to achieve higher performance than can be achieved by the safety-configuration of a general purpose PE logic solver. When SIL 3 performance is required, the safety PES should have an SIL 3 claim limit in accordance with IEC 61508.

4.6.3.4 Subsystem configuration

The process requirements provide the general functional description that is turned into an IPF architecture consisting of input subsystems (e.g., a single transmitter or voting switches), logic (e.g., discrete, calculation), and output subsystems (e.g., a pump motor control circuit or dual block valves). Redundant subsystems are often used to achieve the target performance at the desired test interval.

The most common subsystem architectures (see Figure 4.14) are simplex (1oo1), dual (1oo2, 2oo2), and triplicated (1oo3, 2oo3). Table 4.3 provides example voting architectures, a simple logic diagram for each architecture, and the voting subsystem fault tolerance against dangerous and spurious failure.

Voting Logic	Logic Diagram	Dangerous Fault Tolerance	Spurious Fault Tolerance
1oo1		0	0
1oo2		1	0
2oo2	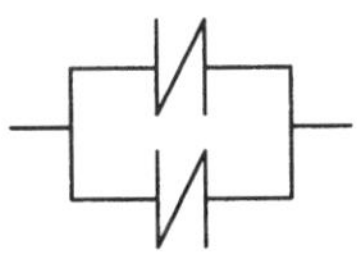	0	1
2oo3	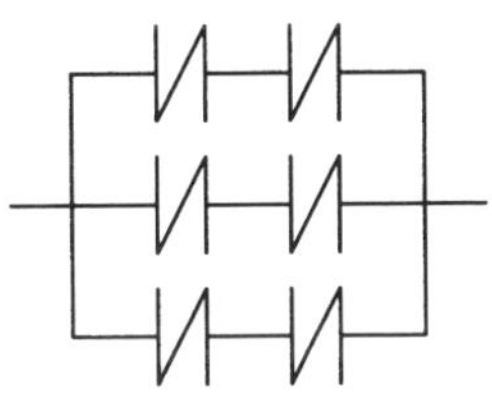	1	1
2oo4	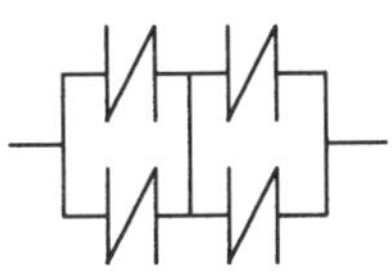	2	2

Table 4.3. Voting Considerations.

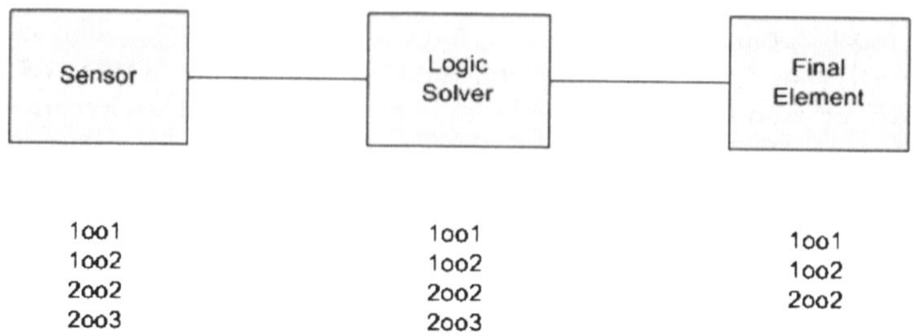

Figure 4.14. Common Voting Architectures.

Some process requirements specifications will include voting architectures that are 2ooN, where N can be a large number, e.g., 2oo9, 2oo16, and 2oo50. While the IPS logic may be implemented in this manner, these architectures are misleading from a verification standpoint, because they often indicate more fault tolerance than is actually provided by the installation. For example, a 2oo9 architecture requires that 8 devices fail dangerously before a hazardous event occurs; 2oo16 requires 15 failures, and 2oo50 requires 49 failures. It is unlikely that 9, 16 or 50 devices would be installed if all were measuring the same process condition and addressing the same process hazard.

2ooN architectures with large N values typically indicate applications where the unacceptable process condition can occur in multiple distinct locations, e.g., a hot spot on the vessel wall. What appears to be a redundant architecture is actually a simplification of what could be a more complex voting scheme. Section 4.5.1.2 discusses process measurement and provides an example of the use of redundant equipment in multiple locations. For example, in the case of a plugged flow reactor, it may be that 2oo3 voting sensors are used in each of three reactor zones, leading to an overall 2oo9 voting scheme. The verification of the risk reduction should be based on the voting architecture that detects the hazardous event. In the case of 2ooN architectures, this is usually 2oo2 or 2oo3 at best.

Some manufacturers advertise other architectures, such as 1oo1D, 1oo2D, 2oo2D or 2oo3D. The "D" architectures generally indicate that diagnostics are provided for the equipment subsystem that is configured in a specific manner to achieve a claimed performance. 1oo2D and 2oo2D have been used by different manufacturers to describe the same architecture and configuration. Product literature should be reviewed to determine how the equipment is implemented and configured to obtain the claimed performance.

Redundancy is sometimes required to meet the requirements for fault tolerance against dangerous failure. Redundancy is often necessary to meet the PFD_{AVG}. It may be provided using identical equipment (e.g., pressure measurement using two pressure transmitters) or diverse equipment (e.g., flow measurement with a vortex meter and a differential pressure (dP) transmitter).

Redundant final elements may also be identical or diverse, such as flow isolation using a block valve versus a pump shutdown.

Figure 4.15 provides a plot of the PFD_{AVG} as a function of test interval for four architectures: 1oo1, 2oo2, 1oo2, and 2oo3. The calculations were conducted using the simplified equation (eq. D.19) with an assumed 100 year $MTTF^D$ and proof test intervals (TI) ranging from 1 year to 7 years. Non-fault tolerant architectures (1oo1 and 2oo2) have higher PFD_{AVG} than fault tolerant ones (1oo2 and 2oo3) at the same proof test interval.

The process requirements also provide the target spurious trip rate (STR) for each IPF. The target STR is established based on process unit production goals (refer to Section 4.5.3). The spurious trip rate is affected by the $MTTF^{SP}$ and, for redundant architectures, on the ability to detect equipment faults. Section 4.6.4 provides further discussion on fault detection.

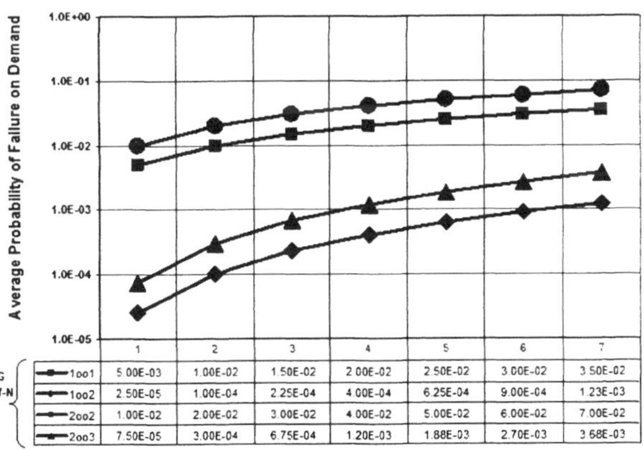

VOTING M-out-of-N	1	2	3	4	5	6	7
1oo1	5.00E-03	1.00E-02	1.50E-02	2.00E-02	2.50E-02	3.00E-02	3.50E-02
1oo2	2.50E-05	1.00E-04	2.25E-04	4.00E-04	6.25E-04	9.00E-04	1.23E-03
2oo2	1.00E-02	2.00E-02	3.00E-02	4.00E-02	5.00E-02	6.00E-02	7.00E-02
2oo3	7.50E-05	3.00E-04	6.75E-04	1.20E-03	1.88E-03	2.70E-03	3.68E-03

Proof Test Interval (Years)

Figure 4.15. Effect of Architecture on the PFD_{AVG} for Proof Test Intervals Between 1 and 7 Years.

As illustrated in Figure 4.16, when two sensors are used, the voting architecture is either fault tolerant against dangerous or spurious failure. Voting establishes the number of devices required to execute the specified action. 1oo2 is fault tolerant against dangerous failure, because the system trips when either signal reaches the trip condition. It is not fault tolerant against spurious failure; a single safe or dangerous detected failure results in spurious operation. 2oo2 is fault tolerant against spurious failure, because the system does not trip unless both signals are in agreement. It is not fault tolerant against dangerous failure; a single dangerous failure results dangerous subsystem failure.

When a dangerous failure is detected in the sensor, the sensor output is configured to fail toward a specified safe state condition. This will result in a step change in the sensor signal rather than the drifting slope shown in Figure 4.16. In 1oo2 voting, the detected failure results in a safe state. If 2oo2 voting is selected, the process remains on-line, but the sensor voting degrades to 1oo1, since the failed sensor has provided a vote to trip.

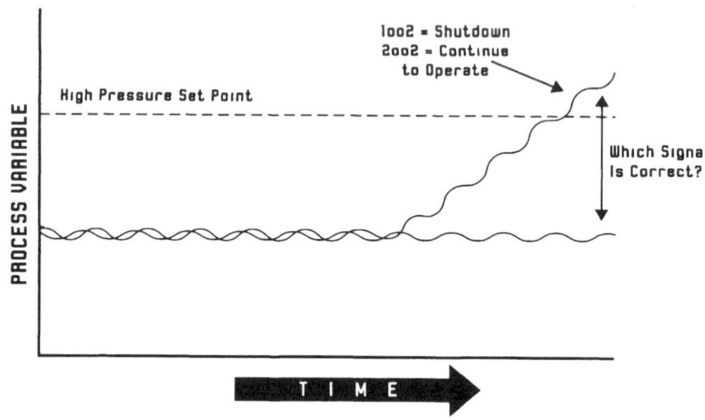

Figure 4.16. Dual Voting Architecture.

The primary benefit of using three sensors in a 2oo3 voting architecture is the protection it affords in preventing a single point of failure from causing spurious operation or dangerous IPF failure. A 2oo3 voting architecture provides fault tolerance against dangerous and spurious failure. Figure 4.17 illustrates the situation where the sensor signal fails dangerously within the normal operating range. The 2oo3 architecture is fault tolerant against dangerous failure, since the subsystem is still capable of operating based on the agreement of the remaining two (2oo2).

Figure 4.18 illustrates the situation where the signal drifts toward the trip condition. The 2oo3 architecture is fault tolerant against spurious failure, because a single sensor does not result in the subsystem taking its safe state action. When a dangerous detected failure is identified, the system remains on-line with the remaining sensors voting 1oo2.

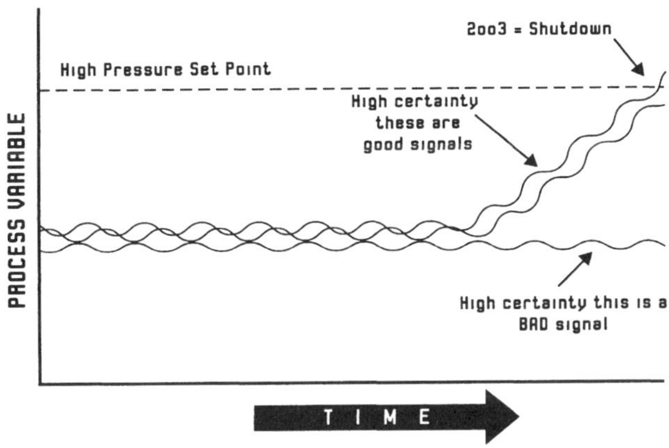

Figure 4.17. 2oo3 Voting Architecture with Single Failure in the Normal Range.

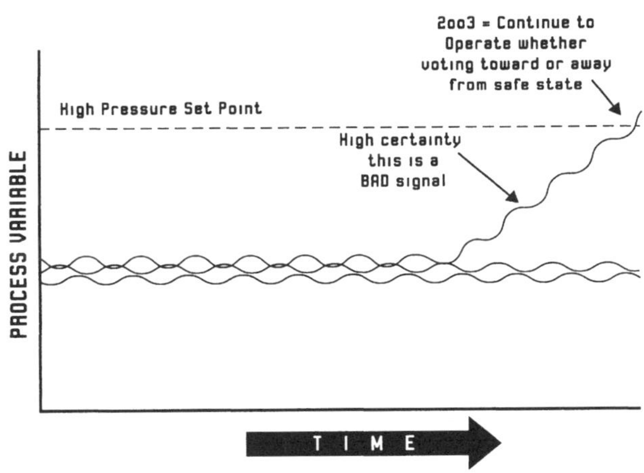

Figure 4.18. 2oo3 Voting Architecture With Single Failure Toward the Trip State.

Figure 4.19 illustrates how voting architecture affects the achievable risk reduction and reliability. The x-axis ranges from low reliability applications where downtime is not critical or costly to high reliability where outages result in significant economic penalties. This is referred to as process sensitivity. The y-axis ranges from a low risk reduction to a high risk reduction requirement. On both axes, there is a transition between low and high that cannot be explicitly represented, because this region is highly affected by the choice of the specific equipment and test interval.

For applications involving low risk reduction and reliability requirements, simplex equipment (i.e., 1oo1 voting) is generally used. The plant strategy is to keep the IPF as simple as possible, while meeting the risk reduction. Simplex systems have no fault tolerance against either spurious or dangerous failure. A single safe failure results in a spurious trip, and a single dangerous failure results in the loss of the function. With simplex equipment, there is little ability for diagnostics, so this architecture has the shortest, or most frequent, test interval requirements to achieve the risk reduction.

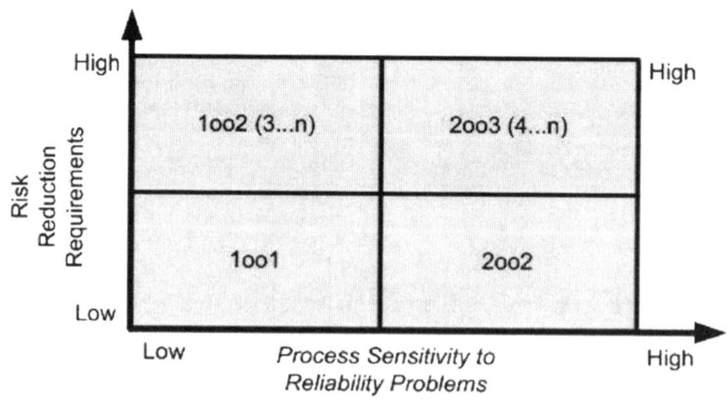

Figure 4.19. Voting Considerations. (adapted from Englund and Grinwis 1992)

Sometimes, reliability concerns are great, but the risk reduction requirements are not. For low risk reduction, dual (2oo2) equipment can be used to achieve high reliability when diagnostics and/or testing of the two channels is sufficient to achieve the required risk reduction. When 2oo2 voting is utilized, a single dangerous failure results in a failure of the subsystem. Since the subsystem is not fault tolerant for dangerous failure, it is generally implemented with additional diagnostics, or it is inspected and proof tested more frequently. This architecture can be tested on-line without bypass.

Initiating cause equipment should not be used in a 2oo2 subsystem, since equipment failure would result in the inability of the IPF to achieve or maintain the safe state. For example, the control transmitter should not be used in a 2oo2 subsystem, if it is a potential initiating cause.

The 2oo2 architecture is fault tolerant against spurious failure and provides the best immunity to spurious trips. The intent is to meet the risk reduction with an architecture that maximizes on-line performance.

Many specialty chemical processes fall into the next category: high risk reduction but low reliability requirements. To protect these processes, IPF are often expected to provide a substantial amount of risk reduction. However, the impact of spurious trips is minimal. This is especially true in reactive processes where the IPF is expected to detect and prevent a runaway reaction. In reactive processes, a shutdown may only cause a batch interruption with rapid recovery and minimal production impact. Consequently, the IPF reliability may not be a significant concern for batch operations.

For low reliability but high risk reduction requirements, 1ooN (where N is the quantity of equipment) is used. With this architecture, there is no single point of dangerous failure; multiple dangerous failures are required for IPF failure. The 1ooN architecture is fault tolerant against dangerous failures, because the subsystem continues to function even if there is a dangerous failure of a piece of equipment. This architecture is not fault tolerant against spurious failure and is prone to spurious trips, because a single safe failure results in the subsystem taking its safe state action.

Fault tolerant against dangerous failure architectures require less diagnostics and proof testing to achieve the same risk reduction as compared to other architectures. A single safe failure does result in a spurious trip, so the plant strategy is focused on the high risk reduction requirements and not on spurious trips.

The final category involves high risk reduction and high reliability requirements. This is the most equipment intensive region. Achieving a high degree of risk reduction with high reliability requires redundant subsystems throughout the IPF. The voting configuration should provide fault tolerance against both spurious and dangerous failures. For field sensors and logic solvers,

2ooN voting (2-out-of-N, where N≥2) is typically used. Final elements are often 2ooN solenoid operated valves on series (1oo2) block valves.

As with other fault tolerant against spurious failure architectures, diagnostics and proof testing are used to detect equipment failures, allowing the subsystems to achieve very high levels of risk reduction. On-line testing can be performed without bypass and with minimal loss of protection. The plant strategy is focused on maintaining safe and on-line operation.

4.6.4 Fault Detection Strategy

All equipment eventually fails. To achieve safe and reliable operation, these failures must be detected before they propagate into a plant shutdown or to an incident. Failures can be detected through operator observation, diagnostics or proof tests.

4.6.4.1 Diagnostic Expectations

Diagnostic techniques typically address only a portion of the equipment failures, especially for complex equipment. Five items should be examined when considering the diagnostics:

- What failure modes are detected by each diagnostic,
- What is the effect (failure classification) of the detected failure modes,
- What percentage of the dangerous failures can be detected,
- How and where the diagnostic is executed,
- Where, when, and what priority of fault alarm is displayed, and
- What action is taken when equipment faults are detected?

Automated diagnostics can be specified using hardware and/or software to detect specific failure modes in the sensors, logic solver, final elements, communications and utilities. Diagnostic requirements should be specified for each IPF. Identified faults should initiate alarms on an operator interface, notifying the operator to take a documented and trained response. With effective diagnostics, faults can be detected and the operator notified within minutes of fault occurrence.

Operator response to fault detection should be covered by procedure. Continued process operation with a degraded IPS requires the implementation of compensating measures, which provide equivalent risk reduction to any loss of protection from the degraded IPS. See Chapter 6 for further discussion of fault alarms and compensating measures.

Equipment selection has a significant impact on the degree of diagnostic capability that can be incorporated into the IPS. For example, discrete switches are digital devices that have only two states: "on" or "off." Since the signal only changes when the set point is reached, diagnostics are not possible with discrete switches. For this reason, many owner/operators use transmitters rather than discrete switches, because the transmitter generates an analog signal suitable for diagnostics.

When making the choice between switches and transmitters, the normal operating range of the process variable should be considered. If the process variable does not vary with time, such as in a normally empty vessel, the use of a transmitter does not improve diagnostics. Refer to Appendix F.4.2 for more discussion concerning switches and transmitters.

Various equipment failure modes can be detected by diagnostics external to the equipment, such as signal loss, drifted analog signal, out of range, or final element is in wrong position. These diagnostics may be executed in any user approved equipment, such as the BPCS or IPS. When executed in the BPCS, the IPS signals are shared with the BPCS using a user approved technique. Sharing of IPS signals should not prevent the IPS from being able to detect unacceptable process conditions or to respond by taking its designated action. When executing the diagnostics in the IPS, it is important to consider the additional code complexity and potential for increased processing (or scan) time.

The hardware and software used to implement the diagnostics should be covered by IPS management system and should be subjected to proof tests, access security, and management of change. For example, if two analog transmitter signals are compared and a 5% deviation for more than 3 minutes is alarmed, the proof test procedure for the transmitters, logic solver, and operator interface should incorporate steps to validate the deviation alarm.

The diagnostic coverage is the fractional decrease in the probability of dangerous failure resulting from diagnostics that report faults to the operator and take a specified action on fault detection. The diagnostic coverage is often limited to the detection of specific failure modes, e.g., partial stroke testing can only detect actuator failures and a limited number of stem problems. The diagnostic coverage is calculated by dividing the number of dangerous detected failures by the total dangerous failures.

The assumed diagnostic coverage should be limited to what is reasonably achievable given the design and management system. Detecting a fault generally requires multiple pieces of equipment, including diagnostic equipment, communication equipment, and an operator interface. Fault detection is only the beginning of a process, which requires prompt and correct response from multiple personnel, including operators, maintenance, and management.

As shown in Figure 4.17, the diagnostic coverage has a significant impact on the calculated average probability of failure on demand (PFD_{AVG}). For this illustration, the $MTTF^D$ is assumed to be 100 years and the proof test interval is 1 year. When 90% diagnostic coverage is assumed, the PFD_{AVG} is reduced by a factor of 10. This assumption is a significant reduction to be realized in the actual field application, so the design basis should clearly specify diagnostic expectations and the operating procedure should define the desired expected response should be clearly defined.

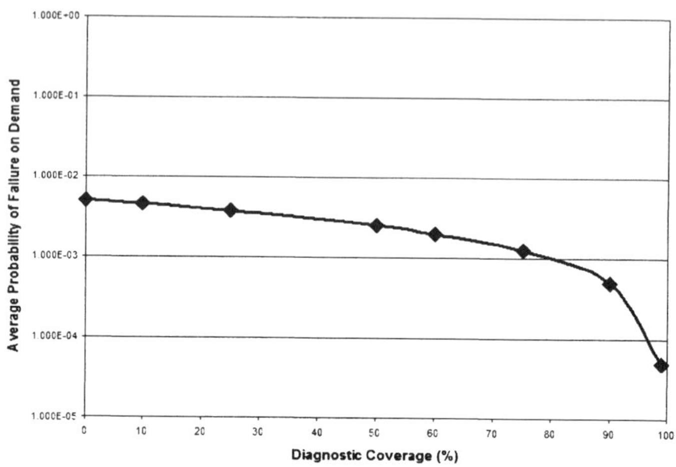

Figure 4.20. Effect of Diagnostic Coverage on the PFD_{AVG}.

Factors greater than 90% should not be assumed without a failure mode and effects analysis, a human factors analysis covering the operator response, and a substantial volume of operating experience. For SIS, the allowable diagnostic coverage factor is limited to 90% by ISA 84.01/IEC 61511 when the diagnostic or its alarm involves the BPCS.

Diagnostics improve maintainability by providing a means to identify degraded operation, so faults are corrected before they become critical failures. When dangerous failure is detected, the failed equipment should be configured to take a specified safe state action. Otherwise, the fault alarm is a protective alarm, which should meet the appropriate practices.

Example—Deviation alarm

When IPS designs use analog signals, external diagnostics are often implemented using a median select algorithm, where the median of the three signals is selected and the other two are compared to the median. The median is generally used, because it is less sensitive to gross errors than the mean or average.

A deviation from the median indicates degraded operation, so when the deviation exceeds a specified percentage, a deviation or fault alarm is initiated. The operator issues a work order for maintenance to troubleshoot, repair, and test the equipment. The operator also implements any required compensating measures until the equipment is placed back in service in its "as good as new" condition.

Operational reliability can be severely impacted when IPS equipment fails frequently even when detected and safe. Fault tolerant against spurious failure architectures can be specified, as well as provisions for on-line maintenance and testing. However, detected faults still increase operation and maintenance cost. Equipment demonstrating higher than expected failure rates should be benchmarked against other manufacturers of similar equipment to determine if better equipment is available.

4.6.4.2 Proof test interval

An effective mechanical integrity program assures the IPS equipment is designed, inspected, maintained, tested and operated in a safe manner. IPS equipment should not be run to failure, but should be maintained in an "as good as new" condition by focusing on preventive rather than corrective maintenance.

Testing and on-line diagnostics can be employed to identify equipment degradation, such as incipient and degraded failures, making it possible to correct problems before they propagate into failed equipment and possibly to process incidents. Section 4.6.4.1 discusses automated diagnostics. This section focuses on proof testing.

The IPS design should incorporate means of proof testing each IPF from the sensor to the final element. The proof test should cover all equipment required for IPF operation, including support systems such as batteries, power supplies, or instrument air. The proof test should involve physical inspection and functional testing to demonstrate that the equipment operates according to the design basis and is maintained in the "as good as new" condition.

Proof tests document the current equipment condition related to its specification. Proof test records are reviewed during functional assessments and continuous improvement efforts. Proof tests also include preventive maintenance to ensure that the equipment is maintained in its "as good as new" condition until the next proof test. Proof testing must be sufficient to ensure the IPF provides the required risk reduction. The maximum interval is related to the equipment failure rate and IPF architecture. The process requirements specification identifies opportunities for off-line proof tests. Given the process limitations, additional equipment and facilities for on-line testing are specified.

Test interval selection impacts the requirements for bypass, test, maintenance, and repair facilities. Interval selection should take into account the equipment failure rate in the operating environment, manufacturer's recommendation, and operating experience.

In general, the lower the equipment failure rate, the less proof testing is required to be reasonably certain that the IPF is functional. Decreasing the time interval between tests increases the ability to detect hidden failures. However, shorter test intervals increase operation and maintenance costs and the potential for maintenance errors that could possibly disable IPS equipment or cause spurious operation.

Figure 4.21 illustrates the effect of test interval on the average probability of failure on demand (PFD_{AVG}) for various values of $MTTF^D$. Theoretically, equipment with a 25 year $MTTF^D$ proof tested annually can provide equivalent PFD_{AVG} performance to equipment with a 100 year failure rate proof tested every four years. This "linear" effect is seen in MooN (M-out-of-N) configurations where M=N, such as 1oo1 and 2oo2.

The effect is non-linear for configurations where $m \neq n$, such as 1oo2 and 2oo3. The non-linearity makes a significant difference in the calculated PFD_{AVG}. For double block valves with $MTTF^D$ of 50 years (includes single solenoid-operated valve for valve actuation), a change in the test interval from 1 year to 6 years results in the PFD_{AVG} changing from 1.00E-4 to 3.60E-3. The same double block valve configuration is at least 36 times more likely to be failed when tested every 6 years rather than annually.

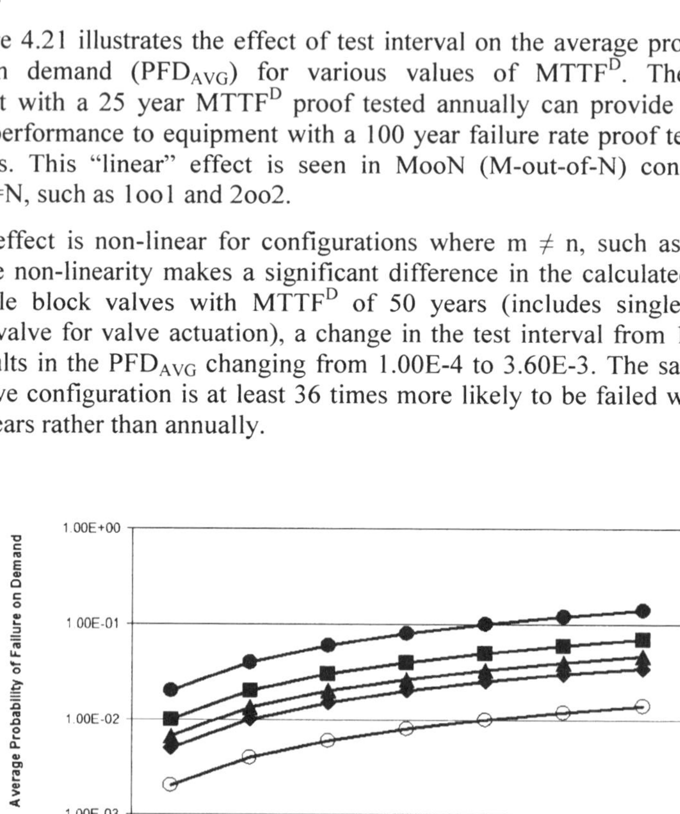

	1	2	3	4	5	6	7
25 years	0.020	0.040	0.060	0.080	0.100	0.120	0.140
50 years	0.010	0.020	0.030	0.040	0.050	0.060	0.070
75 years	0.007	0.013	0.020	0.027	0.033	0.040	0.047
100 years	0.005	0.010	0.015	0.020	0.025	0.030	0.035
250 years	0.002	0.004	0.006	0.008	0.010	0.012	0.014

Mean Time to Failure Dangerous

Proof Test Interval (Years)

Figure 4.21. Effect of Test Interval on Average Probability of Failure on Demand.

The reader is cautioned that an important assumption in this analysis is that the failure rate remains constant whether the block valve is tested annually or every 6 years. In practice, this may not be true. The equipment failure rate is impacted by the frequency of inspection, preventive maintenance and proof testing activities. This affect is especially pronounced in equipment with mechanical components. Block valve failure may be significantly increased by the decreased proof test interval, depending on the technology, service conditions, and frequency of valve operation.

Aspects to be considered in selecting the test interval are as follows:

- Equipment capability: internal diagnostics, automated test capability, and on-line repair,
- Human factors: operator displays, ergonomics, complexity of task, and potential for human error,
- Operating environment factors: external environment, process environment (i.e., deposition, pluggage, corrosion, etc.), electrical environment, and support system quality,
- Manufacturer's recommendations (refer to installation and maintenance or safety manual), and
- Applicable regulations/standards or insurance practice.

Equipment required for proof tests should be specified in the design basis and software requirement specification. The hardware and software used to support proof tests should be assessed to ensure that its failure cannot cause a dangerous IPF failure or a process disruption. The test equipment should be included in the verification of the risk reduction if it causes a failure to operate. Test equipment reliability should also be evaluated to ensure that the equipment design minimizes the likelihood of spurious operation. Under no circumstances should final element testing cause a hazardous event or present unacceptable risk.

4.6.4.3 Bypass

Bypass facilities should provide alarming, recording and reporting of bypass initiation. IPF bypass should not result in the loss of the process condition detection or indication. Use of bypasses should be covered by procedure and should be monitored and audited. Bypasses should not be used to allow process operation outside the normal operating limits.

Bypasses should only be used during on-line process operation when compensating measures are applied to ensure the risk is adequately managed during the bypass period. The availability of compensating measures often limits the use or constrains the design of bypasses.

Manual start-up bypasses should be minimized due to the likelihood of systematic errors. Whenever practical, start-up bypasses should be automated so that they can be manually applied, but automatically reset when the specified process condition is satisfied. As with any bypass, the operator procedure should define the acceptable conditions for start-up bypass initiation, bypass removal, and status confirmation.

Maintenance bypasses are typically required to allow on-line proof testing and maintenance, especially where the proof test interval is shorter than the off-line maintenance interval. Maintenance bypasses are implemented using hardwired or software switches, which may be under operator or maintenance control. Bypass use should be covered by procedure, including permit requirements, bypass initiation, bypass removal, and status confirmation.

Maintenance bypasses should bypass individual pieces of equipment within a redundant subsystem. This provides continued protection during equipment maintenance, test and repair. Bypasses modify the voting logic of redundant configurations to allow proof test, preventive maintenance, and repair without process interruption. For example, in a 2oo3 configuration, the logic may degrade from 2oo3 to 1oo2 or 2oo2. For 2oo2 or 1oo2, the logic degrades to 1oo1.

Bypasses may be implemented using manual and automated features. Bypass equipment include many different types of hardware, such as full flow bypasses to allow complete testing of the final elements, key switches to bypass logic, and manual isolation valves. Bypasses relying on software should be covered in the design basis and software requirements specification. Logic should be implemented to allow bypass without use of software forces.

Bypass facilities should provide a clear indication of status. A bypass alarm should be activated when the bypass is put in place and should require operator acknowledgement. Consideration should be given to repeating the bypass alarm if the bypass remains active across operator shifts.

4.6.4.4 Mean time to repair

The mean time to repair (MTTR) is the average time required to detect failure occurrence and to return the equipment to service in the "as good as new" condition. The achievable MTTR is affected by maintenance personnel availability, travel time, equipment location, safe access, spares, service contracts, environmental constraints, and permit requirements. These factors determine how quickly maintenance can safely plan, execute and validate work activities.

The equipment is considered *non-repairable*, when the IPF equipment cannot be repaired on-line. For the process, the MTTR represents the minimum process downtime. Depending on the time required to achieve the shutdown condition and the time required to re-start the process, the total process downtime can be substantially longer.

The equipment is considered repairable, when the IPS equipment is repairable on-line. In this case, the MTTR typically represents a time interval where the process is operated with a degraded IPF. Additional risk is not acceptable even for a short time. The loss of IPF protection should be off-set by a compensating measure, which provides equivalent protection to the failed equipment for the repair period. Compensating measures are often implemented by the operator to safeguard the process during repair.

Compensating measures may restrict production rates, limit operating modes, and/or require additional operator staffing. Some compensating measures may be difficult to maintain for long time periods. Limitations on operational activities should be considered during the process requirements specification and operating procedure development. Compensating measures should be capable of acting within one-half the process safety time.

Equation 4.2 for the PFD_{AVG} assumes that dangerous failure and repair occur once in the $MTTF^D$. Many calculations assume that the contribution of the MTTR is negligible, since the MTTR is much smaller than the TI. This assumption was made in the derivation of Equation 4.2. If safe failures are detected and repaired on-line, they also contribute to a loss of IPF availability. However, most calculations only account for the safe failures as part of the spurious evaluation.

While it is true that the MTTR is typically much smaller than the TI, the MTTR does represent a period of unavailability. For example, assume the equipment is out of service for repair once per year (8760 hrs) rather than once in the $MTTF^D$ and the repair takes 72 hours. The fractional dead time is 72 hrs/8760 hrs = 0.008. The equipment is available only 99.2% due to repair and test. If the repair time was reduced to 8 hrs, the fractional dead time is reduced to 0.0009. The equipment availability is now 99.91%. Multiple on-line repairs can significantly decrease the equipment availability, because each repair is another increment of fractional dead time.

The significance of MTTR to safe and reliable operation is not adequately addressed by conventional calculations, which focus on the random hardware failures. Frequent or extended repair is considered a systematic problem that should be addressed through management system work processes. User approval should give preference to reliable equipment which experiences few failures. Priority should be placed on IPF equipment maintenance, repair, and testing to minimize the out of service periods.

The allowable repair time should balance reasonable maintenance expectations for MTTR with operations expectations for the compensating measures. The equipment safety manual may specify a maximum MTTR, which should be met, unless deviation is justified by the design basis. When repair cannot be completed within the allowable repair time, additional review should be conducted and higher management approval should be obtained. This review and approval is typically a management of change activity.

4.6.5 Operator Interface

The operator interface design directly affects the likelihood that the operator will correctly detect the event, diagnose the cause, and take specified action. The greatest challenge in interface design is creating situation awareness, where an operator in one area, such as a control room, has the ability to perceive an event happening in the process equipment. Situation awareness requires a variety of tools, such as control system data, field operator communication, historization systems and trends, cameras, and other equipment.

Operator response to a hazardous event is often time critical. The operator must recognize the condition, take action and have the process return to the safe state within one-half the available process safety time. The hazardous event begins when the process deviates from normal operation. The operator response cannot begin until the operator recognizes this deviation. An effective interface design seeks to minimize the recognition time and maximize the likelihood of correct response.

In the control room, the operator is faced with a wide variety of equipment which competes for attention, including CRTs, panel annunciators, video monitors, flashing beacons, etc. IPS alarms and information are presented to the operator in a number of ways:

- Graphic displays on the BPCS operator interface,
- Dedicated graphic displays on stand-alone video display units,
- Panel mounted graphic displays, and
- Panel mounted annunciators.

The operator also takes action on the process using various means, such as keystrokes, touch screens, switches, and pushbuttons.

Figure 4.22 illustrates a well organized control room with a high density of operator displays. When the control room is not well planned, the equipment placement can become a function of where there is space rather than what is best for information prioritization. During hazardous events, information can simply get lost in the dizzying display before the operator. The operator can become so overloaded with information that they simply “freeze up” and do not take any action. They also can get overwhelmed by the complexity and refuse to take action without confirmation from others.

Human factors influence the ability of the operator to respond promptly and correctly to alarms and abnormal situations. Current technology provides the ability to alarm enormous amounts of information at a relatively low cost. Good engineering practices steeped with common sense are needed to ensure that information is provided such that a consistent operator response is obtained.

While the predominant pitfall with most control rooms is information congestion, the situation is typically the opposite in the field. The lack of reliable information is often the limiting factor for effective response. The field operator does not have a cockpit display of information, but instead uses local indications, such as gauges and displays, for process measurements. This often puts the field operator within close proximity of the potential hazardous event.

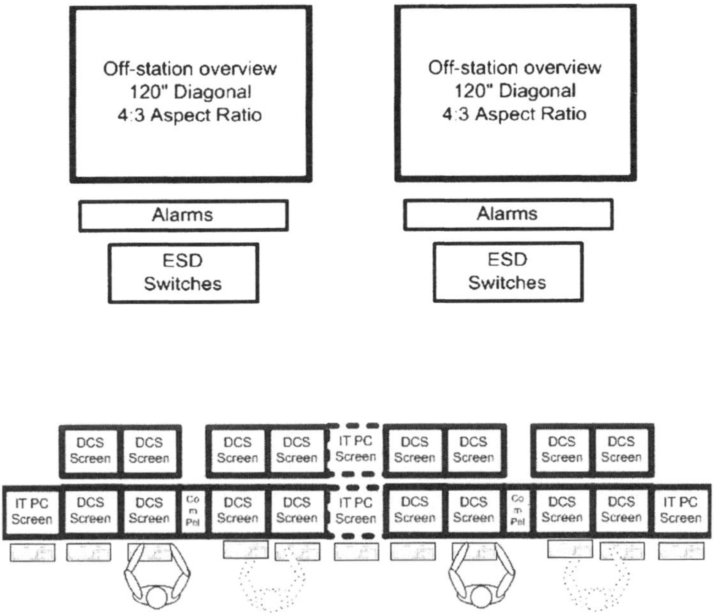

Figure 4.22. Illustration of Control Room Display. (Nimmo 2006)

Situation awareness is greater for the field operator due to hazard proximity. Some hazards can be detected through normal human senses, e.g., the operator may hear audible sounds, smell chemical odors, feel skin irritation, or see a release. The field operator communicates identified problems to the control room operator and other relevant personnel. When required, the field operator takes direct action on the process using field devices or other means.

Whether in the control room or in the field, the operator interface should present the required information as clearly and simply as possible to minimize confusion. Protective alarms should be highly visible, audible and distinguishable from less critical alarms. The optimum design of the operator interface is highly dependent on the number of hazardous events for which the operator has responsibility and required risk reduction expected from the operator's response.

One practice that has gained wide acceptance is alarm management, which addresses the various aspects of how alarms are prioritized, designed and implemented. An effective alarm management program seeks to direct the operator's attention to the most critical condition by controlling what and how information is displayed. Refer to guidance from ISO 11064 and EEMUA Publications 191 and 201.

To maximize the integrity and reliability of any operator response, the interface design should consider the following:

- Prioritizing alarms, and
- Graphic and/or audible display and annunciation of alarms and information.

If the control system does not have alarm prioritization capability, it is recommended that a separate alarm interface be considered for safety alarms requiring operator action to achieve or maintain a safe state.

4.6.5.1 Display

Alarms should be displayed in an area that is highly visible to the operator. New alarms should be easy to distinguish from previously acknowledged alarms. The alarm should be quickly identifiable and should provide a clear and unambiguous indication of what caused the alarm. When possible, protective alarms should be displayed on a dedicated interface, which typically provides the tag name, equipment symbol, short descriptor, and a symbol indicating priority and acknowledgement status.

Alarm interface design should consider the following:

- Clear and unambiguous indication to take action,
- Alarm priority for alarm displayed during the event,
- Alarm priority within an operating mode (e.g., critical during start-up, but low priority during normal operation),
- Assistance on what corrective action is appropriate or recommended,
- Operator alarm acknowledgement, and
- Indication of return to normal conditions.

Consideration should be given to varying the audible tone with each category of alarm. A common practice is to assign the highest priority alarms the most urgent tones, while the low priority alarms may be lower in volume and intensity. In all cases, the volume should accommodate the operator's hearing ability and should be at least 10dBA above the control room background noise.

When graphic displays are used in conjunction with audible alarms, the ability to silence the audible feature should be considered. During times of unit upset, when a high volume of alarm traffic occurs and communications are often strained, many operators complain of being distracted by the noise created by the seemingly endless horns and buzzers and the need to constantly press the "acknowledge" key. The operator can be provided with the ability to silence certain sets of alarms during major disturbances with a single acknowledgement. The alarm set is silenced for a prescribed period, which is generally limited to 10 to 15 minutes. Graphic display and priority level would remain unaffected.

Alarms displayed on graphic interfaces often use flashing icons and color to indicate priority. The use of color should follow a consistent practice. For example, green means safe, yellow means caution, and red means danger. It should be noted that the generally accepted colors are different across the various market sectors.

Color alone should not be the only indication of a change of state, as statistics show that many people suffer some form of color blindness. Color should be used in combination with another graphic change, such as placement, flashing, solid versus hollow, etc.

The information display should be understandable and should match the equipment orientation in the field. Preferably, the field layout and labeling should also conform to human expectations and the display should agree with this layout. If there are three transmitters, PT101A, PT101B, and PT101C, the transmitters should be installed in that order in the field. The transmitter display should present the information from left to right (or top to bottom). A hierarchy for graphics should also be developed, which allows the operator to navigate from the alarm interface to process-specific graphics to provide process details as efficiently as possible.

Lightbox annunciators or panel alarms are recommended when the BPCS does not provide alarm prioritization, access security, and management of change features. Lightbox annunciators typically cost more per alarm point and require more room for installation than the BPCS interfaces. Their use is typically restricted to safety or emergency alarms, because large annunciator displays can become confusing to the operator.

Lightboxes provide an independent alarm display, which is potentially visible by multiple operators within the control room. Lightbox alarms should be replicated, when practical, in the BPCS interface for acknowledgement and logging purposes. This allows the lightbox alarm to be subjected to the same review and tracking as other alarms.

Lightbox annunciators should be located close to the operator's work station or work areas. It should be located so that it is viewable from all locations where its information might be considered important. Properly located near the related process controls used to take action, the dedicated display can direct attention to the appropriate location more efficiently than alarm messages in a PE-based system.

Incident investigations have found installations where the operator had to walk several meters around the console to observe and acknowledge a panel mounted alarm. The operator was then expected to walk back to the normal work station to take corrective action. The extra steps delayed operator response and added stress to the decision making process.

The alarm reset actions should be specified during the design basis. Consideration should be given to ringback capability for long-standing alarms, especially bypass and protective alarms.

4.6.5.2 Manual shutdown

Manual shutdown can be initiated by the operator using the BPCS, remote or local pushbuttons and switches, or directly with process equipment, e.g., manual closure of valves. Software shutdown switches should require confirmation prior to action to prevent unintentional process shutdown.

Operators are expected to initiate trained responses to a variety of shutdown indicators, such as process indications, process alarms, observed events, protective alarms, emergency alarms, or procedures. The operator procedure should define the "never exceed never deviate" process condition requiring process equipment shutdown. Manual shutdown capability should be provided for any IPS where maintenance bypasses are used to support on-line equipment repair, maintenance and proof test.

The ability to shutdown the process independent of the BPCS controller and IPS logic solver may be required for some applications. The current editions of NFPA 85 and NFPA 86 require independent means for equipment covered by these practices. Equipment safety manuals may specifically require independence, especially for SIL 3 applications involving programmable electronics. When non-fail-safe design is used, e.g., energize to trip, independent shutdown facilities should be provided that do not require the IPF support system to be operational.

The process requirements specification defines how manual shutdown should be executed from the process perspective. Operators are often provided with manual pushbuttons to initiate equipment shutdown.

Manual shutdown facilities should be defined including the physical means of initiating the shutdown through the final elements that take action on the process. The manual shutdown should be designed to support the IPS core attributes. Physical means may be located as required, such as in the field and/or control room. Field mounted shutdowns are often provided for mechanical equipment, such as pumps and compressors.

Manual shutdown switches should clearly indicate normal (or run) and shutdown positions. Pilot lights can be used to indicate output signals. Manual shutdown activation should initiate an alarm.

The manual shutdown equipment location should consider the potential events that might necessitate their use. For example, consider locating the manual shutdown so that a seal failure does not result in the pushbutton being within the area covered by the released materials. The installation should minimize risk to personnel to the extent possible.

Remote shutdowns are provided for many types of process and mechanical equipment. The remote interface may be through the control system interface, through an IPS interface, or through a completely independent means, such as a switch or pushbutton, as required.

4.6.5.3 Reset

When an IPF takes action, the final elements should be taken to the safe state and should remain in the safe state until the IPF is reset. The reset may be executed manually or automatically based on process conditions. Manual reset ensures that equipment restart is under the operator control, providing operators with adequate time to physically inspect relevant equipment and maintenance personnel with time to investigate the trip cause prior to process start-up. Use of automatic reset should be reviewed and approved by a hazard and risk analysis team.

Internal practices should define expectations regarding reset design to facilitate engineering and operator procedure consistency. Any deviation from internal practice should be documented and clearly identified in the operator procedures to prevent confusion during process start-up.

Process constraints should be defined in the process requirements specification, including the process conditions under which reset is considered safe. Reset may be provided on a local panel or in the control room. Reset should not occur until the process conditions that caused the shutdown has been corrected. Final element reset is generally covered by the process equipment start-up procedure.

Resets and start-up bypasses are often interrelated. For example, if a pump is shutdown, start-up will not occur if reset is applied without a conditional bypass around the low flow IPF. While conditional logic is sometimes required, it can increase the likelihood of systematic error.

Reset logic should be validated during software and hardware tests. All resets and bypasses should be tested for identified errant input/output combinations, e.g., attempting to bypass when in trip condition or attempting to reset without satisfying the necessary process conditions (e.g., permissives).

4.6.6 Independence and Common Cause

Common cause failure occurs when a single failure results in system failure. Common cause can be the result of random and systematic events. A review and approval process should be used to reduce the occurrence of systematic errors. The potential for random hardware failure should be formally assessed during the design basis, so common cause failures should be identified and steps taken to minimize their impact. Appendix D provides further discussion of random, systematic, and common cause failure.

Throughout this book and ISA 84.01/IEC 61511, emphasis is placed on assuring the independence of equipment operation in each subsystem and between protection layers. Dependences should be minimized during design and implementation. In general, design errors are reduced through the use of approved equipment and architectures, while implementation errors are reduced using internal technical practices and standard installation details.

4.6.7 Cost-Benefit Analysis

While most owner/operators have fairly rigorous and definitive requirements regarding the management of safety and environmental risks, more leeway is often granted to those systems reducing asset and business interruption risks. Management practices should define how economic risks are managed. What may constitute a catastrophic business interruption or asset loss to a niche player may have little impact on a multinational, vertically integrated operation. A structured methodology, such as cost-benefit analysis, may be used to justify the investment in IPFs to address asset and business interruption risks.

Cost-benefit analysis assesses the cost of the hazardous event, including various costs such as investigation, liability, legal, fines, clean-up, repair, and start-up. It evaluates the cost of implementing the IPF against the cost associated with the risk. By approaching the economic requirements this way, owner/operators can ensure a consistent investment philosophy is used throughout the company, while providing a means to account for any site-specific peculiarities.

> Example--Process outage affects design choices
>
> Consider a site that has three parallel production trains. While the loss of a single train due to a compressor failure does pose a monetary loss until the compressor is repaired, two-thirds of the facility is still producing product, and the co-product units are still functioning as well.
>
> Now replace the three parallel compressors with a single compressor sized at three times that of the first facility. Loss of the compressor results in a total plant outage. All co-product units are down during the outage time, so business losses are substantially higher.
>
> In the first scenario, the incremental cost to reduce down time of one train out of three may be greater than the production losses. In the second scenario, the incremental cost to reduce compressor damage and subsequent production losses may justify the additional expenditure.

4.6.8 Risk Reduction and STR Verification

The IPF integrity is determined by the IPS design and the rigor of the management system used to identify and correct random failures and systematic errors. The protection afforded by the IPF is dependent on a series of subsystems that must operate like links in a chain. The measure, decision, action, and time elements of the design establish the chain of events that must occur to achieve or maintain the safe state. The equipment design, inspection, maintenance, proof test, and operation practices limit the achievable chain strength, or the claimed equipment integrity.

Operational reliability is affected by the IPF spurious trip rate. The impact of unnecessary IPF operation can be significant as discussed in Chapter 6.10. The likelihood of IPF spurious operation is reduced using fault tolerant against spurious failure architectures, good quality equipment, and a comprehensive inspection and preventive maintenance program.

The risk assessment phase allocates risk reduction to each identified protective function. The process requirements specification integrates the identified IPFs into an overall risk reduction strategy that reduces the process risk below the owner/operator risk criteria. The risk reduction claim is justified by determining and documenting that approved equipment operates as required in the operating environment and that equipment subsystems provide the required fault tolerance for dangerous failure. Process actions (refer to Section 4.5.1.4) and provisions for bypass, proof test repair, and reset should be evaluated to ensure the equipment does not contribute significantly to IPF failure.

In general, risk reduction is achieved by implementing:

- Independent and separate IPF that reduce susceptibility to common cause or common mode failures,
- Voting subsystems that provide the required fault tolerance against dangerous failure,
- Diagnostics to detect and respond to failure, and
- More frequent inspection, preventive maintenance, and proof tests.

For PIFs or SIFs, the risk reduction is also verified quantitatively by the PFD_{AVG} (or hazard rate). Reliability block diagrams, fault tree analysis, Markov models, and failure mode, effects, and criticality analysis are the most common quantitative methods for verification. ISA TR84.00.02 provides an overview of the various approaches used in the process industry to calculate the average probability of failure on demand (PFD_{AVG}), the hazard rate, and the spurious trip rate (STR). Appendix D also provides an overview of the analysis considerations.

When choosing a verification method, the following should be considered:

- Complexity of method: affects training and skills required to correctly implement the method,
- Data precision: lack of suitable data may make quantitative methods meaningless, and
- Resources: training and skills to correctly implement the chosen method.

Inherent uncertainties (e.g., data, engineering judgment) often result in diminishing returns when considering more complex methods. The relevance of the predictive calculation is highly dependent on how well the chosen failure rate data matches the actual equipment performance in the operating environment.

While Tables 4.1 and 4.2 provide failure rate data for equipment installed in general process industry applications, equipment in a particular application may experience more or less failures than indicated. Due to the uncertainty related to the data selection, the design should allow for some margin of error in the verification of the risk reduction and the spurious trip rate.

In many cases, highly complex methods do not yield more meaningful answers than empirical ones. The calculation's primary value is demonstrating how changes made to the design and maintenance strategy affect the integrity and reliability performance. The design verification determines the:

- Expected IPF performance,
- Weak links in the IPF design, and
- Trade-offs between redundancy, proof test intervals, and reliability necessary to achieve the required integrity.

4.6.9 Verify the I&E Requirements

The design basis should provide the information and documentation necessary to ensure that IPFs achieve the core attributes throughout their life. The design basis should define how the IPFs are allocated to systems. The allocation should ensure that the IPFs are sufficiently separate and independent from the initiating cause and any other protection layers reducing the risk of an identified hazardous event. The IPF design should reduce the occurrence of common cause failure to a sufficiently low level.

Functional, integrity and reliability requirements are addressed using equipment that is proven to work in the operating environment, redundancy and voting to achieve the necessary fault tolerance, and diagnostics and proof testing to identify equipment faults. The dominant failure mode of IPF equipment should be to the safe state. Loss of support systems should result in the safe state. Further, when dangerous failures are detected in field devices or within the logic solver, these failures should result in the failed device or channel taking its specified safe state action.

Design choices affect the operability and maintainability ease. The IPF can only achieve the required risk reduction when the IPS equipment is maintained in the "as good as new" condition. The design establishes the minimum requirements for the mechanical integrity program that ensure this condition. The implementation of bypasses, resets, and manual shutdown should consider the conditions of their use and how the procedures and equipment will be maintained and controlled. Access security provisions should be provided to administratively and physically control and monitor access to equipment.

The operator interface requirements are also defined to ensure that the operator can troubleshoot system failures and effectively respond to the hazardous events. The requirements for diagnostic, protective, pre-trip, trip and status alarms should be specified, as well as bypass, manual shutdown and reset facilities.

4.7 FUNCTIONAL ASSESSMENT

The intent of the functional assessment is to verify that the IPF design basis:

- Follows approved internal practices and industry good engineering practices,
- Uses equipment that is user approved for the operating environment,
- Achieves the risk reduction strategy established in the risk assessment phase,
- Employs the necessary fault tolerance against dangerous failure for the claimed risk reduction,
- Fulfills the operability, functionality, maintainability, and reliability requirements, and
- Supports the management system rigor required for the IPS classification.

The assessment should be a formal activity based on a specific revision of the design basis documents. Chapter 2.5.3 provides a general discussion on functional assessments. Attention should be paid to the clarity of the documents in communicating the design requirements, focusing on consistency in approach, achievement of risk reduction and agreement with the good engineering practices.

Criteria should be established to define the review and approval process when the final risk reduction strategy differs from the recommendations of the risk assessment phase. The review typically involves the project team and representatives from the owner/operator of the process equipment. These representatives may include various personnel, such as environmental, health and safety, operations, maintenance, and engineering. Hazard and risk analysis team review and approval may be required based on owner/operator management of change practice.

4.8 KEY MANAGEMENT SYSTEM ELEMENTS

4.8.1 Technical Practices

Owner/operators must adhere to government regulations concerning design and management of the process and protective equipment. While recognized and generally accepted good engineering practices should be applied throughout the IPS lifecycle, the sheer number of potential practices can sometimes be overwhelming. Many national and international good engineering practices potentially affect the IPS design basis.

Compounding this situation is the specialized jargon often used in industry practices and the tendency of committees to write at its level of understanding. Weaving through the jargon and reading between the lines often makes practices highly subject to interpretation, depending on the technical background of the reader. The goal of compliance is to achieve the intent of the good engineering practice, but without the proper background (and sometimes direct participation) it may not be possible to understand what is required.

Internal practices should be developed incorporating the intent of applicable good engineering practices and utilizing terminology familiar to plant and technical personnel. Site needs can be directly addressed, including specific reliability, operability, and maintainability requirements. The practices should be sufficiently prescriptive to ensure that the IPS is designed to meet the risk reduction.

Internal practices can be structured in many different ways, but should include the following elements:

- Scope,
- Applicability of the practice,
- Identification of required resources and responsibility,
- Required input information,
- Work process for designing the IPS,
- Specific design requirements,
- Management of change,
- Documentation requirements, and

- Documentation revision management.

Work processes can be outlined to highlight specific tasks, to emphasize significant compliance issues, and to drive consistency in design. Prescriptive practices can be developed to yield a design "cookbook" providing detailed requirements. Practices often cover issues such as expected risk reduction, architectures, fault tolerance, and test intervals for specific IPF applications. Common applications include heaters, furnaces, boilers, rotating equipment, tanks, high integrity protective systems, and reactor kill systems. Even when detailed practices are used, it is still necessary to perform the hazard and risk analysis to ensure that the practice provides the required functionality and sufficient risk reduction given site specific hazards.

Various disciplines and interested parties should be consulted so the internal practices meet the needs of stakeholders. The earlier design issues and concerns are brought forward, the more flexibility the project team has in resolving them without undue stress to the schedule or budget. Internal practices are often developed on various subjects, such as:

- Wiring system philosophy,
- Power and grounding,
- Lightning protection,
- Separation philosophy,
- Utility requirements,
- Building constraints,
- Minimum design requirements,
- Bypass and reset requirements,
- Typical sensor / final element architectures,
- How the design basis is maintained under revision control for the IPS life, and
- How the equipment configuration is maintained for the IPS life.

4.8.2 Approved Equipment List

Owner/operators should establish and maintain a list of equipment that is user approved for various applications and operating environments. The user approval process should determine and document that specific equipment is capable of operating as required in the operating environment. Equipment approval should consider the hardware and software version and should determine the equipment's fitness for service.

The approval process should yield an equipment manual, describing the installation, configuration, and mechanical integrity requirements. The equipment manual is a collection of documentation including:

- Installation guidelines or drawings,
- Operation and maintenance manuals,
- Configuration requirements, and
- Recommended inspection, preventive maintenance and proof test intervals and procedures.

In ISA 84.01-2004/IEC 61511, the equipment manual is known as the "safety manual." The manufacturer often provides recommendations related to installation, configuration, inspection, preventive maintenance, and proof testing. All of these manufacturer recommendations should be considered when developing the design basis.

Equipment should have a demonstrated operating history of at least two years in a similar operating environment prior to approval in IPS or SIS applications. New equipment is often installed in control system applications first. This allows operations and maintenance to gain experience in essentially continuous duty service. If the equipment does not perform, it is typically evident in a control system service.

It takes time to understand how equipment fails, how to detect its failure, and how to repair it to the "as good as new" condition. Before equipment can be used in IPS, operating experience is necessary to gain sufficient confidence in the equipment that it can be installed in what is typically a demand mode service.

Mistakes, such as installation and commissioning issues, and equipment problems, such as early failures due to lack of robustness, can be identified and corrected in control system applications. To ensure that the process operates safely, independent and proven IPS equipment is used when necessary to detect unacceptable process conditions to achieve or maintain a safe state of the process.

Some equipment never achieves user approval for IPS, because it fails to demonstrate the required performance. Safe operation depends on the use of equipment with known failure characteristics. Some equipment works well in some operating environments and fails miserably in others. Also, in a time of rapid change in programmable electronics, some technologies never progress to the point where they are considered sufficiently safe or reliable for IPS applications.

When manufacturers obsolete equipment, they often recommend a substitute. The manufacturer can assist in the user approval process by providing detailed information about the differences between the products. The recommended substitute should still be reviewed and approved through a formal process which should involve some period of demonstrated use in a similar operating environment.

The performance of approved equipment should be reviewed periodically as part of the continuous improvement process. The review should determine whether the equipment is still appropriate for continued use. Product quality, manufacturer support, expected service life, and advisory notices should be reviewed.

Historical maintenance records, alarm records, trip investigation reports, and failure investigation reports should be reviewed to identify issues associated with the equipment in specific or general services. These issues can include a wide variety of application problems, such as excessive maintenance requirements, installation problems, unanticipated failure modes, and shorter than anticipated equipment useful life. The equipment should be removed from the approved equipment list, if the performance cannot be improved to an acceptable level.

4.9 SPECIAL TOPICS

4.9.1 Independence Evaluation

Many owner/operators mandate independence and separation between protection layers, because it is the simplest approach to implement and maintain. It follows a defense-in-depth strategy to reduce the potential that a systematic error can disable more than one protection layer.

The owner/operator should document internal practices to ensure a consistent approach for assessing and achieving independence. The approach should be clear and unambiguous, so everyone assigned responsibility for the IPS operation can easily recognize the IPS from other systems.

Practices should be established that are easy to implement, rely on common sense, and are simple to verify. For example, internal practice may allocate a particular controller in an overall system for IPFs and require that this particular controller be managed under stricter requirements than the overall system. However, this segregation relies on personnel discipline, so training and administrative controls are more important. Personnel must understand the reason for and importance of the dedicated controller. Any deviations from approved practices should require justification and approval prior to implementation.

Most hazard and risk analysis assume the layers are independent. Simple qualitative and semi-quantitative analysis may not be sufficient to evaluate the residual risk if the layers are not independent. This is particularly true if the IPF and initiating cause are in the same layer. In this latter case, the hazard rate for the overall system should be determined and compared to requirements.

The decision to combine layers should consider many factors including:

- Hazard rate claim limit when random failures and systematic errors are considered,

- Restrictions or requirements related to the protection layer defined by applicable good engineering practice,
- The equipment failure rate based on the operating environment,
- Restrictions identified during user approval process, and
- Required fault tolerance against dangerous failure to achieve claimed hazard rate.

When other layers are combined with the IPS, the associated systems must be managed under the IPS management system. This means that any shared components including hardware and software must be traceable to information, documents and procedures, which demonstrate that the design, maintenance, inspection, testing, and operation practices are sufficient to meet the IPS requirements. Further, when the IPF and initiating cause are allocated to the same layer, the IPF may be operating in a continuous mode. Continuous mode systems have special design and operating basis requirements.

4.9.2 Continuous and Demand Mode Operation

IPF operate in one of two modes of operation: continuous or demand. The mode of operation impacts the fault detection and response strategy. Some IPFs are designed to operate in demand mode, but operate very frequently due to the process demand rate. Some literature refers to these IPFs as operating in a "high demand mode." However, rather than being another distinct operating mode, the frequency of the IPF operation is typically indicative of either excessive control layer failures or a misclassified control function. Often, the IPF is acting as a control function rather than a protective function under certain operating conditions or modes.

High demand mode IPF should be identified and means should be implemented to reduce the demand rate. Chapter 6.10 of this guidelines book discusses the monitoring and tracking of process demands so that their frequency can be minimized. For example, a high tank level IPF shutdown of a pump may be specified to reduce the risk of the operator failing to stop the pump at the specified control set point. In practice, the operator may be allowing the IPF to trip the pump due to other distractions or tasks. By recording and investigating the pump trips, the procedure deviation can be identified and its root causes can be assessed and corrected.

A continuous mode system is one whose dangerous failure leads to a hazardous event without further failure. This operational mode is best exemplified by the inherently safer and control layers. These layers are continuously active and when compromised, tend to be initiating causes for hazardous events.

If IPS equipment failure can initiate the hazardous event and cause the IPF to fail to operate, the shared IPS equipment is operating in a continuous mode. The design and user approval process should specify how dangerous undetected, systematic, and common cause failures are reduced to a sufficiently low likelihood. When dangerous failures are detected in continuous mode equipment, the affected equipment must be taken to the specified safe state.

The vast majority of IPFs operate in the demand mode, where the IPF only acts when another layer has failed to maintain the process within the normal operating range. For example, when a BPCS flow control loop fails, feeding excess reactant into a reactor, high temperature initiates (or demands) an IPF shutdown of the feed.

When dangerous failures are detected in demand mode equipment, the affected equipment should be taken to a specified safe state. If a simplex subsystem is configured to fail away from the safe state, the fault alarm becomes a protective alarm and should be designed and implemented according to protective alarm practices.

An operating procedure should document the required operator response that:

- Can provide risk reduction equivalent to the disabled IPF,
- Can be executed in one-half the process safety time,
- Can be put in place immediately, and
- Maintained continuously throughout the allowable repair time.

5

ENGINEERING, INSTALLATION, COMMISSIONING AND VALIDATION

The protective management system encompasses the instrumented protective system (IPS) lifecycle. This chapter focuses on the activities necessary to complete the detailed engineering, installation, commissioning and validation (EICV) phase for the IPS. The goal is to implement the design basis, achieve the core attributes as defined by the management system, and meet the needs of the process operation. This requires meticulous attention to detail, as well expertise across a broad range of disciplines, including project management, instrumentation and electrical design, application programming, construction methods, and procedure development.

The basic objectives can be summarized as:

- To specify and procure equipment to meet the requirements of the design basis,
- To provide a detailed engineering design package for construction efforts,
- To install and commission the IPS in accordance with the hardware and software specifications and detailed design package,
- To validate that the IPS works in the operating environment,
- To provide as-built documentation to the operational & mechanical integrity phase of the lifecycle, and
- To ultimately demonstrate that the installed and operational IPS meets the design basis.

As illustrated in Figure 5.1, the design basis is issued to the engineering and construction team as the specification document. The design basis defines the project scope. Often during project execution, other non-IPS issues are discovered, including opportunities for instrumentation and control upgrades, replacements, or enhancements.

Consideration should be given to whether these are necessary changes or simply scope creep (colloquially speaking, also known as pet peeves or band wagon changes). These changes should be evaluated carefully, because simply ‘tacking on’ additional unrelated items to the IPS project may result in resources being diverted from the primary goal.

Ancillary projects and late requests can cause a loss of focus on the IPS implementation, which, if not executed properly, can result in long and expensive start-up delays and can jeopardize process safety. The additional project load can also make it difficult to execute the project at the required level of rigor, especially if the resource allocation was based on the smaller scope.

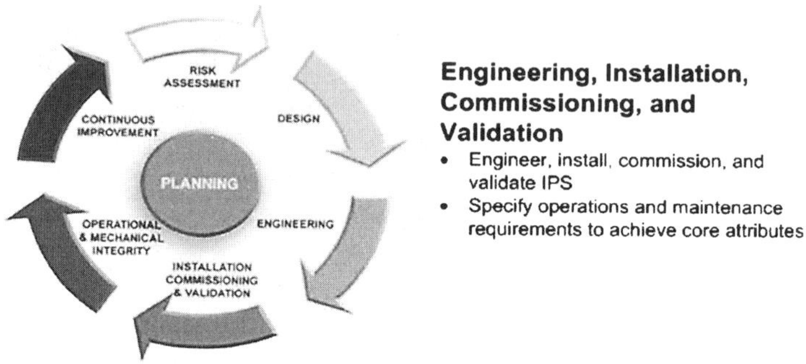

Figure 5.1. Engineering, Installation, Commissioning, and Validation Phase.

The EICV phase is the most labor intensive, schedule driven phase of the lifecycle. Many different parties rely on one another to complete simultaneous tasks in an orderly, efficient manner according to a prescribed schedule for each specific task. Until this point, time constraints have not been considered a major factor in the project execution. Depending upon the specifics of the project, this may continue to hold true, but generally during the EICV phase, schedule becomes a very important factor. New systems and most system upgrades require a process outage for installation. Since the outage is planned and controlled by the operating unit, the schedule becomes relatively fixed. Essentially, there is a limited window of opportunity to get the work completed.

Furthermore, prior to this phase, a dedicated project team has performed the majority of the work with external expertise called upon as needed. The design basis was developed in an organized and coordinated environment focused solely on the system design. As the project progresses and outside engineering and construction contractors mobilize, plant personnel have a dual responsibility: 1) to support and maintain on-stream process units and 2) to participate in the development and approval of various documents, work permits, and procedures. With more people involved in the day-to-day activities of the project, coordination between project activities and daily operations becomes crucial.

Scheduling also becomes imperative. Once the IPS equipment is specified, delivery times can be estimated. Downtime to perform necessary tie-ins or equipment upgrades can also be determined. In sold-out or seasonal markets, this information is vital to scheduling outage durations, so production losses can be minimized.

Once the IPS equipment is selected, commissioning and validation activities should be planned and scheduled. Appropriate personnel, procedures, and equipment are made available for validation and startup. Assigned personnel should be adequately trained to execute the procedures and to raise attention to any field problems, so that they can be addressed as efficiently and effectively as possible. Many of these activities will require overtime or rescheduling of work shifts to leverage efficiencies and spending efforts.

The ultimate outcome of the EICV phase is an operational IPS that meets the design basis. While the work processes presented in this chapter focus on the IPS, the same activities are generally necessary to achieve good performance from any of the instrumentation and controls installed in the process unit. The owner/operator should document the acceptable approach for other instrumented systems, as part of a comprehensive management system.

5.1 INTENDED AUDIENCE

The EICV phase requires the largest number of diverse groups of personnel yet to be utilized. Contract engineering, design and construction personnel, as well as owner/operator representatives from many disciplines now join project team members. This chapter is intended for:

- Process who define IPS operating procedure requirements and support the pre-startup safety review,
- Instrumentation and Electrical who specify the hardware and software requirements, define IPS maintenance procedure requirements, verify IPS installation, commissioning and validation, investigate and track IPS failures, and support the pre-start-up safety review,
- Operations who provide access to equipment, and resource support for commissioning and validation, identify and report failures of IPS equipment, and conduct pre-start-up safety review,
- Maintenance who support installation and commissioning of equipment, execute proof testing to validate hardware and software, identify and report failures of IPS equipment, support the pre-startup safety review, and
- Manufacturers who supply information on how to properly install, operate, and maintain their products.

5.2 INPUT INFORMATION

The design basis provides the general specifications necessary for each IPF to achieve its core attributes. It includes information related to the following issues, which should be reviewed at the beginning of the EICV phase:

- Identified hazardous events,
- Identified process operating modes,
- Functionality, including safe state and process actions, required sequencing, time delays, and process safety time,
- Relationship between process inputs and outputs, including logic, mathematical functions and any required permissives,
- Input: technology, description, redundancy, and voting architecture,
- Logic Solver: technology, description, redundancy, and voting architecture,
- Actuator: technology, description, redundancy, and voting architecture,
- Final Element: technology, description, redundancy, and voting architecture,
- Fault detection and response requirements,
- Inspection and preventive maintenance requirements,
- Proof test requirements,
- Bypass requirements,
- Reset requirements,
- Manual shutdown requirements,
- Operator interface alarms and messages,
- Spurious trip rate, and
- Required risk reduction.

The design basis establishes minimum requirements for detailed engineering, installation, and validation. The requirements are refined and finalized as equipment is specified and purchased. Documentation created during this phase should be reviewed during a functional assessment for agreement with the design basis intent.

5.3 BASIC WORK PROCESS

Many essential tasks are completed during the EICV phase, resulting in the physical realization of the design basis. Any deviation from the approved design basis should be reviewed and approved by appropriate parties prior to change implementation.

A general work process is presented to facilitate discussion of each task. Since the scope and schedule are different for each project, the work process should be adapted to accommodate the time constraints for each particular project. Steps may be combined, deleted or reordered, depending upon the unique attributes of the specific project.

At the beginning of the EICV phase, the design basis should be reviewed, including the process unit boundary and project intent (e.g., IPS upgrade, control system upgrade, or process unit modification). Any outstanding items from the design basis requiring further definition should be noted and finalized as soon as practicable in the detailed engineering work.

A schedule should be created showing when major tasks must be completed, noting when special skills or tools are required and highlighting critical path items. Any scheduled process outages required for construction should be aligned with marketing and seasonal demands for the product.

Contingency planning should be performed to take advantage of potential unscheduled outages, such as spurious trips, or scheduled periods of reduced production rates. These contingency plans should consider potential force majeure scenarios, such as labor actions or severe weather.

The EICV phase encompasses many simultaneous tasks that must be suitably scheduled and executed so as to avoid delays and cost penalties. While no opportunity for improved performance, operability, maintainability, or reliability should be ignored, changes to the design basis come at increasing cost and schedule impact as the work progresses, as Figure 5.2 illustrates.

During the EICV phase, opportunities to enhance the project often become evident, such as a means to reduce lifecycle costs, achieve a better technical solution, improve the field installation, or simplify operation and maintenance. These alternatives should be identified and brought forward to the engineering and project management personnel, who will evaluate potential benefits against the overall impact to the project scheduling and cost.

The work process, as shown in Figure 5.3, immediately splits into two distinct, yet concurrent paths. One deals primarily with the hardware specification.

This path addresses the extensive list of physical requirements and constraints, such as separation, independence, redundancy, and diversity, and includes:

- Specifying the equipment,
- Developing installation and commissioning plans, and
- Defining validation (or site acceptance test) requirements.

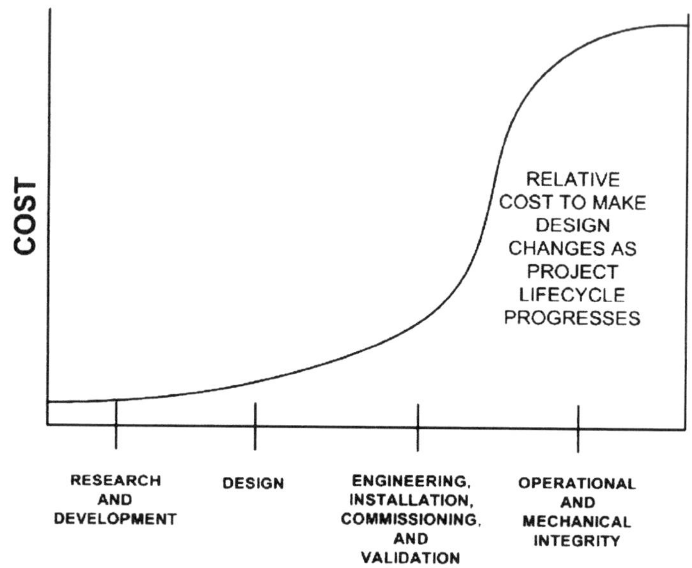

Figure 5.2. Relative Cost of Making Design Changes.

The second path addresses application software, which has specific needs and constraints that are as detailed as that of the hardware. During this phase, the software requirements are thoroughly documented and verified to ensure the software configuration achieves the operational requirements for each IPF. The software work process includes:

- Creating the software specification,
- Writing the application program,
- Defining verification and validation requirements, and
- Specifying the operator and engineering interfaces.

Detailed engineering and specification documents must achieve the intent of the design basis and the hazard and risk analysis findings. Prior to completion, the stage 2 functional assessment is conducted to ensure owner/operator requirements are being met.

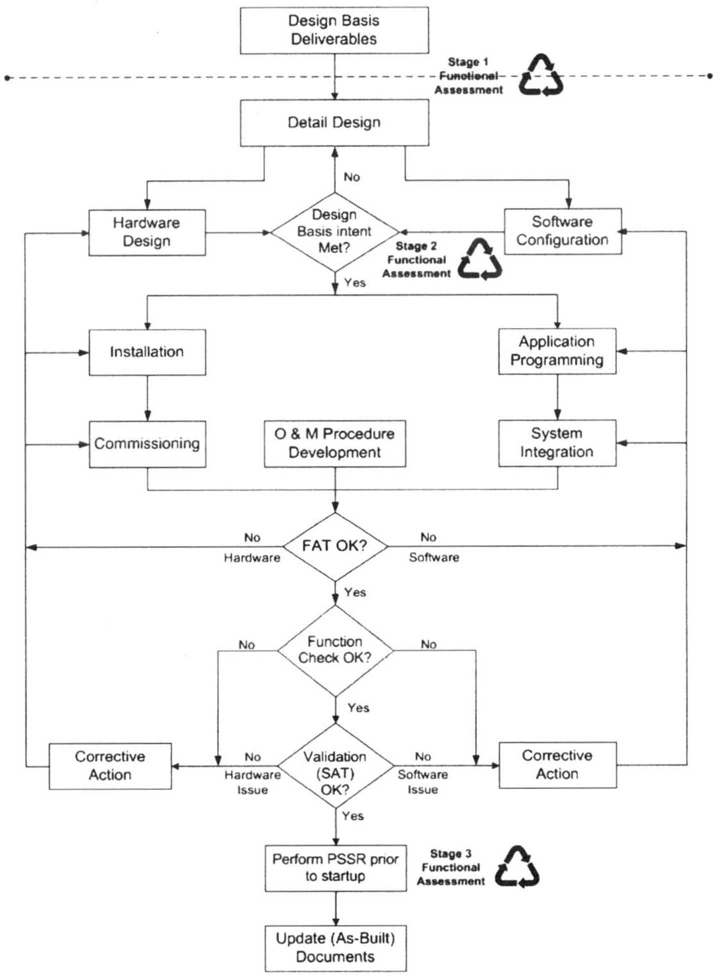

Figure 5.3. Engineering, Installation, Commissioning and Validation Work Process.

An important consideration during engineering is that the IPS will eventually become the responsibility of site personnel. As hardware and software design takes place, the resources and skills of these personnel should be taken into account. Site personnel must be capable of maintaining the IPS operational and mechanical integrity throughout its life. Consequently, thought should be given to their skills and knowledge, as well as to the content and depth of the information to be communicated to relevant personnel as the IPS progresses from project team to site responsibility.

Installation and commissioning are planned activities incorporating quality control activities to reduce the likelihood of systematic errors. Poor installation and improper commissioning are commonly cited as reasons for IPS equipment failure. Since most IPSs operate in a dormant mode, these failures can be difficult and costly to identify after start-up, so thorough inspection and verification is essential.

The application program should conform to the software requirements specification. The integration of the application program into the hardware platform should result in an operational system that communicates properly with operator interfaces and other systems.

As IPS equipment is commissioned, inspection is performed to determine that equipment is properly installed according to the installation drawings and acceptable practices. When commissioning is completed, tests are conducted to prove that the IPS operates according the design basis and equipment specification. These tests may include factory acceptance tests (FAT), loop checks, and site acceptance tests (SAT).

An FAT is often conducted to verify equipment functionality prior to field installation. A comprehensive FAT can significantly reduce the problems which must be corrected on-site, lowering implementation costs. Function (or loop) checks verify that the equipment is properly connected, configured, and operating per specification.

Each IPF is then validated using an input to output test to demonstrate and document that the installed equipment meets the requirements. Until this point, the IPS was designed and verified from a theoretical perspective. It has been shown to operate under limited and controlled test conditions. A complete input to output test of the installed equipment proves that the IPF actually works in the real world. Validation is generally performed as part of an SAT.

The SAT serves as an opportunity for operations and maintenance to see the IPS in action and to be trained on its design and operation. It demonstrates how the IPS and its equipment interact with other systems, such as the BPCS and the various operator interfaces.

Operating and maintenance procedures are developed to maintain the IPS in the "as good as new" condition. The procedures discussed in this chapter are limited to those necessary to properly commission, validate, and startup a new IPS.

Chapter 6 provides a more thorough discussion of the procedures that as the operating basis for the process unit.

A functional assessment or pre-startup safety review (PSSR) is conducted prior to process equipment start-up. PSSR approval establishes the point where the IPS design and construction is considered sufficiently complete by the owner/operator that the process equipment can be started. This functional assessment is required for new or modified instrumented safety systems (ISS) and highly recommended for IPS.

After the PSSR approval, the documentation should be formally updated to 'as-built' status. As-built documentation should incorporate changes that have occurred since the last formal document revision. This documentation provides the owner/operator with an owner's manual for the instrumentation and control system.

5.4 OUTPUT DOCUMENTATION

Once the IPS has been placed in service and the limelight surrounding the project begins to fade, the impetus to follow-through with the clean up of details often wanes. The final step in the EICV phase is critical to long-term performance of the IPS and involves the update of project paperwork and issuance of 'as-built' documentation.

Documentation should be updated to accurately reflect the installed IPS. This includes updating the design basis, as well as any documentation necessary to show that the IPF performance exceeds the required performance.

The 'as-built' documentation includes various procedures required by the operating basis to support the core attributes, such as:

- Proof test schedule and procedures,
- Operating procedures,
- Access security, and
- Management of change procedures.

While much of this information may seem readily apparent to those who have lived and breathed the project for the duration, it must also be available and understandable to those who will be expected to support the IPS in the future. Some common items that require attention are:

- As-built drawings, documentation and procedures,
- Loop sheets, specifications, and other design and installation documents,

- Additional training based on feedback from commissioning and validation efforts,
- Process safety information,
- Training plan,
- Design basis, and
- Approved Equipment List.

A good practice is to write a detailed project report, which outlines the project goals and describes any problems encountered and the methods taken to overcome these issues. This is especially useful for circumstances that require unorthodox solutions, so that the rationale for the solution is not lost.

5.5 HARDWARE

The hardware constraints in the design basis should be implemented to achieve the core attributes. These constraints include separation, independence, redundancy, diversity, diagnostics, test provisions, resets, etc. Deviation from the design basis should be evaluated relative to its impact on the IPF performance. The hardware work process involves consideration of the following:

- Application requirements,
- Field device specification,
- Logic solver specification,
- Operator interface specification, and
- Response time.

5.5.1 Application Requirements

All equipment should be selected based on known performance in the operating environment, including conditions within the process and the external environment. The extremes of all environmental conditions likely to be encountered by the IPS should be identified. The operating environment issues in Table 5.1 should be considered, as a minimum.

Fluid state	Density
Entrained solids	Contaminants
Humidity	Shock/vibration
Dust	Electrostatic discharge
Flooding	Viscosity
Corrosive atmospheres	Temperature
Radio frequency interference (RFI)	Grounding
Electromagnetic interference (EMI)	Lightning
Electrical area classification	Hazardous area classification
Cleaning media, e.g., steam	Response Time Requirement

Table 5.1. Example Operating Environment Conditions.

Many equipment failures occur due to incorrect or inaccurate specification for the operating environment. Improper specification can result in common mode failure, where redundant equipment fails to operate in the same mode.

Environmental conditioning equipment should be provided as necessary to ensure equipment performance and to minimize common mode failures. A fault alarm should be initiated when there is a detected failure of the environmental conditioning equipment.

In addition to the operating environment, physical limitations should be considered during the design, such as spacing, structural or weight limitations. Physical structures and piping layout may limit the equipment technologies that can be employed.

5.5.2 Field Device Specification

The design basis identifies the IPF field device technology. Detailed equipment specifications are generated to facilitate equipment procurement. Some equipment may be implicitly selected in the design basis, such as a preferred manufacturer for the logic solver or a field device technology that is of limited availability. However, most 'commodity' equipment is specified during the EICV phase.

Equipment should meet the design basis and should have any necessary certifications and approvals (e.g., hazardous locations, electrical codes) required for the application. Equipment should also be selected that provides good performance, can be easily proof tested to ensure continued performance, provides access security means, and supports conformance audits.

When specifying equipment that will be shared between the IPS and the control system (or other non-IPS functions), the equipment must: 1) meet the needs of the IPF and 2) meet the needs of the control system (or non-IPF). In all cases, the IPF requirements take precedence over the needs of the control system.

For example, if an IPF and a control function share field equipment, the field equipment should be specified such that it can achieve the required hazard rate.

Manufacturer constraints, including equipment manual requirements, should be understood prior to specifying field equipment. Selected equipment should be user approved for the application and should meet the core attributes defined for the IPS. The final design should take the potential failure modes of the equipment into account, specifying the necessary diagnostics and test procedures to detect and annunciate faults. ISA TR84.00.04 Annex L provides additional information concerning the selection of SIS equipment. See Appendix F of this guidelines book for more discussion on user approval.

Other operating environment needs should also be addressed, such as:

- Materials of construction for gaskets, packing,
- Hazardous event survivability,
- Hazardous area classification for equipment, conduit systems, and junction boxes,
- Reactivity issues specific to the process and the equipment material of construction,
- Corrosivity, plugging, and deposition issues,
- Separate process connections and isolation valves for field sensors,
- Winterization/insulation requirements, and
- Acceptable leakage rate for valves.

Example--Explosion proof housing required

A salt dome storage facility located on the Texas Gulf Coast had many areas that were not considered hazardous locations per the NEC. However, due to the prevailing wind direction blowing across the open brine retention ponds, the atmosphere was permeated with salt residue, resulting in rapid and extreme oxidation of metal. Explosion proof equipment housings and conduit fittings were found to maintain mechanical integrity longer in the corrosive environment than standard equipment. Consequently, housings and fittings were specified as explosion proof. Junction boxes were specified as 304 stainless steel as a minimum.

5.5.3 Logic Solver Specification

Logic solver specification requires a thorough review of the design basis requirements to ensure that the correct type and quantity of each logic solver component is ordered. Field devices should have dedicated wiring to the logic solver I/O. De-energize to trip circuits should be used where practical. When energize to trip circuits are used, additional requirements apply (see Appendix F.5.2.2).

Total digital input/output (I/O) and analog I/O should be examined to select a platform that provides the best lifecycle cost. Below 20 to 30 I/O, electrical systems based on non-programmable electronic (non-PE) equipment, such as relays and trip amplifiers, are very cost effective and have a long design life. Above 30 I/O, programmable electronic (PE) logic solvers are generally more cost effective even though they become obsolete faster than non-PE systems.

Another consideration for a PE logic solver is the distribution of I/O to minimize common cause failures. To achieve high integrity and reliability, consideration should be given to the distribution of redundant I/O across multiple modules. Electrical systems provide this distribution inherently by virtue of their design. With a PE logic solver, the distribution must be specified by detailed design. Each piece of equipment in a redundant subsystem should be allocated to a separate module.

For example, the implementation of a PE logic solver may only require 1 analog input module by total point count. However, if the analog inputs are part of a 2oo3 voting architecture, the designer should consider using 3 analog input modules to allow separation of each input to a separate module. This minimizes common cause failure and increases system reliability.

The communication equipment required for the operator interfaces, control system and other external equipment should be specified. Failure of this equipment should not result in IPS failure. Each communication type has specific protocol requirements and data transfer limitations that should be understood to minimize interconnectivity issues at commissioning. The resolution and scan time for the sequence of events or first out reporting should be specified for compatibility with IPS set points and actions to ensure that the actual sequence is recorded.

Logic solver selection should consider the ability to perform on-line maintenance. Some logic solvers do not permit on-line replacement of components, necessitating a shutdown for repair. Others provide this capability but with differing levels of complexity and risk. On-line logic solver maintenance is a common source of spurious IPS events, which is not considered in manufacturer claims as this is a systematic error.

In addition to hardware limitations, PE logic solvers have differing capabilities with regard to application software modifications. Some systems allow on-line software changes. On-line changes involve significant risk and should be carefully planned and executed. The need for on-line change should be reviewed and approved. Software should be modified off-line using a back-up of the application software and changes should be thoroughly validated prior to download. Manufacturer recommendations should be followed when software download is performed.

Other considerations include:

- Choice of voltages for discrete points,
- Sequence of event recorders and first-out indication,
- Supervised outputs in energize-to-trip service, and
- Thermocouple or multiplexer requirements.

Manufacturer specifications should also be reviewed to assess the need for commodity type connector cables, adapters, etc. While this equipment is commonly available at the local electronic supply store, the local product may not have the shielding or 'hardness' required for industrial applications. Many companies use manufacturer alliances to satisfy most of their instrumentation and control needs, but care must be taken to assure that the preferred manufacturer supplies equipment that is user approved for the application.

> Example--Improperly rated equipment
>
> A DCS was revamped with all new hardware and software. A new IPS was also installed. The project implemented new marshalling to save time and reused the homerun cabling. This resulted in the wiring being disconnected, new marshalling installed, and wiring re-terminated and checked out in less than 5 days. The terminal blocks selected by the designer were on the manufacturer's approved list and had an LED to indicate when voltage was present on the line. While the terminal blocks worked well for 120VAC signals, the LED caused a voltage drop that completely corrupted the 4-20mA transmitter signals.
>
> All 2,386 terminal blocks were ripped out and replaced at 2 am with 2 days until startup. There was nothing stated on the manufacturer's approved list about the terminal block's suitability for use with analog signals, and the designer didn't read the fine print in the catalog.

One piece of equipment which seems to consistently provoke conflict when specified and ordered is the personal computer (PC) used for engineering interface or engineering work station. There is often debate between the process control group and the information technology (IT) department, concerning who is ultimately responsible for the PC and consequently who should specify and maintain it. The needs of the IPS can result in the PC being a 'non-standard' machine in the IT world, and serious confusion can result down the road.

The IPS engineering interface has a completely different purpose from an office computer, should be specified to meet the needs of the IPS and be managed by the organization that has responsibility for the IPS logic solver. It is recommended that business software suites, such as e-mail and word processing, not be loaded on an IPS engineering interface, since these are often sources for computer viruses and system crashes. It is further recommended that only the software necessary to perform activities associated with the IPS be installed on the IPS engineering station.

> Example--IT changes PC operating system
>
> Someone from the IT department was charged with the task of upgrading and standardizing the PCs in the process unit. One day, the employee happened across a remotely located PC used as the IPS engineering station. The employee proceeded to erase the IPS interface software, utility software, and application program, because it was not on the IT approved software list. The machine was upgraded from Microsoft DOS and locked administratively so that unauthorized software could not be reloaded onto the machine.
>
> Luckily, the situation was discovered and rectified without a process upset or incident. This is an excellent example of the importance of controlling and monitoring access to IPS engineering interface, of maintaining restricted communication between the interface and IPS logic solver, of emergency retrieval procedures with software back-ups of utility software and application programs.

5.5.4 Operator Interface Specification

The IPS engineering interface should not be used as the operator interface. A separate and independent operator interface should be provided. This interface should be designed to ensure that the operator understands the IPS status and to provide a clear and unambiguous indication of when manual action is required. Consideration should also be given to specifying an IPS operator interface that is independent of the BPCS operator interface.

Protective alarms are identified and classified during various hazard and risk analysis. Emergency response plans may also identify the need for specific emergency alarms. Design and engineering typically identifies alarms to support the process and equipment operation, such as quality, diagnostic, bypass, and status.

Rationalization establishes an alarm hierarchy, so alarms are justified, documented and displayed based on priority. The interface should present the information and data in a logical manner. When PE based, the interface should allow the data to be viewed at different levels of detail, e.g., a process overview, unit overview, detail view, diagnostic view, and alarm summary.

Proposed or existing alarms should be subjected to a thorough review, which determines the alarm purpose, the event being detected, the time available for response, the expected risk reduction, and the intended operator action. In essence, a rationale should be established for each alarm, and each alarm should require action covered by an operating procedure. Workload analysis should be considered if a single operator is expected to address numerous protective alarms.

The number of alarm priorities should be minimized to reduce complexity and the potential for systematic error. A common practice is 2 to 4 alarm priorities. The more complex the alarm priority structure becomes, the more likely it is to be misunderstood and violated in the future.

Rationalization is used to determine whether a proposed alarm is truly needed. This ensures that alarms are not used as a convenient way to log the time of the event. This may include the manual start of a pump by the operator, or alarms caused by transitions between normal operating steps. Such alarms usually contribute only noise (both literally and figuratively) to the operator and may overshadow more pressing alarms.

Alarm system design should also minimize standing or chronic alarms. Standing alarms occur whenever the equipment is in a certain state and are typically caused by non-operating or out-of-service equipment. An example may be a pump, which is shutdown for operating reasons or maintenance. The change of state from running to not running may require alarming, but when acknowledged, the alarm state should clear.

Alarm settings should be reviewed to ensure that operating at minimum and maximum production rates do not cause nuisance alarms. Alarms generally should not occur when the process is operating within the normal operating range. The likelihood of nuisance alarms increases when multiple alarm levels are used for each process variable. Such practice causes the operator to spend valuable response time acknowledging alarms rather than responding to them. When an alarm flood occurs, the operator's ability to distinguish between protective and non-protective alarms is impaired.

One method to combat alarm flooding is the use of suppression to control the display of alarms. For example, global power failure leads to multiple pump shutdowns with each pump generating its own set of alarms, i.e., pump not running, low flow, loss of pressure, etc. A master "Global Power Failure" alarm can be used to annunciate the power failure and to suppress the individual pump alarms.

While alarm suppression provides benefits, it does introduce complexity into the alarm design. The suppression strategy should be thoroughly documented and reviewed to ensure that the strategy is appropriate for various events which affect the process operation. It is also requires more verification and validation to ensure that safety alarms display correctly during individual and global events.

Alarm management focuses on reducing the number of alarms. However, while alarm management seeks to reduce the number of alarms to a manageable level, the alarm system must fulfill the hazard and risk analysis requirements and support safe operation.

Alarm management should consider human factors and other situational factors that might drive the need for a specific alarm. If the analysis determines that the alarm is intended to address a system limitation or shortcoming, the alarm should not be removed until the shortcomings are address. Strong consideration should be given to the alarm intent, rather than applying strict time or consequence severity rules.

Consider repeating bypass and protective alarms (especially safety) intermittently until the condition is cleared. A common practice for bypass alarms is to repeat the alarm at least once per shift to ensure operator acknowledgement of the condition.

Alarms should be logged with a date/time stamp for use in troubleshooting and isolation of the event initiator or "first-out." The alarm log should also be reviewed to track the on-going performance of the alarm system as a whole. This review will identify problems with the alarm system basis and/or operating procedures.

For example, the alarm log may identify equipment, which is generating frequent alarms. These alarms may be associated with a process control problem and should be investigated. Alternatively, these alarms may be symptoms of other problems, such as poor alarm design, poor operating procedures or training gaps. Inspection and maintenance activities should also be considered, since the frequent occurrence and clearing of an alarm may indicate a sensor problem.

5.5.5 Response Time

The IPF must be capable of consistently detecting unacceptable process conditions and taking the process to a safe state prior to the occurrence of the hazardous event. The response time is the time interval required to fully execute the IPF from detecting the process condition to achieving the safe state of the process. The IPF response time should be less than one-half the process safety time to ensure that the IPF can perform its intended function in a timely fashion.

An IPF's response time is determined by various factors, such as detection lag, measurement error, and shutdown lag. Figure 5.4 illustrates the cumulative effect these factors can have on system response. In this example, the actual process condition exceeds the equipment design limit prior to shutdown completion due to delays in the IPF response.

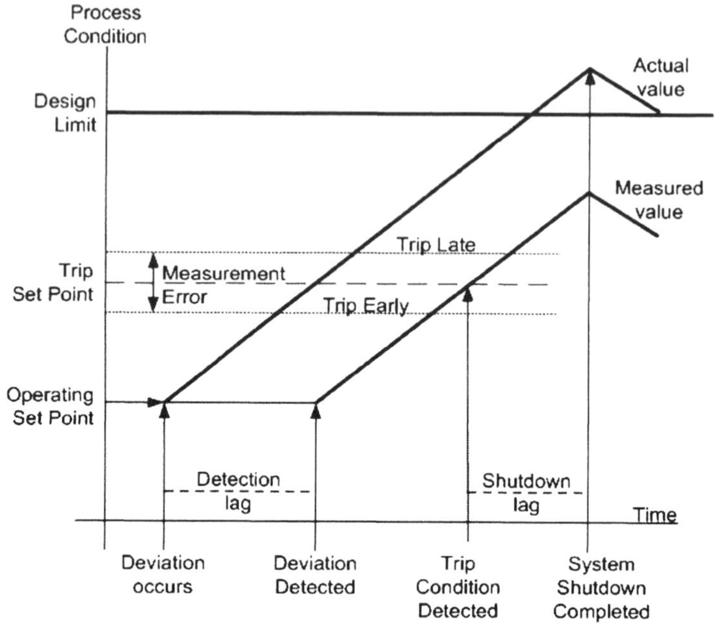

Figure 5.4. Response Time.

The time required to detect the trip condition is affected by detection lag and measurement error. Detection lag can be significantly impacted by the equipment technology, installation, and operating environment. For example, detection lag can vary from milliseconds for a switch to minutes for an analyzer loop. Measurement errors cause early or late detection of the trip condition, depending on the error direction. Some device technologies are less accurate and experience more drift than others. The measurement error can also be aggravated by data processing and conversion.

Generally, shutdown lag is the controlling factor in the overall IPF response time. Shutdown lag includes logic processing time, intentional time delays or sequencing activities, the final element response time, and the time required for the process to achieve the safe state. The final element response time can be on the order of milliseconds for shutting down a motor to minutes for closing a block valve. As discussed in Chapter 4.5.1.5, the process may respond immediately to the IPF action or may take minutes.

The successful operation of manually-initiated IPFs is heavily dependent on having sufficient process safety time. Manual systems require the operator to detect the process condition, recognize the need to take action, determine the correct action, and complete the specified action. The response time for the operator to recognize that action is required and to take the appropriate action contributes significantly to shutdown lag.

Automated systems are generally much faster and, in many cases, response time is not considered a significant factor design factor as long as the IPF is designed and implemented properly. Logic processing time is affected by the logic solver technology, accuracy, logic complexity, logic interconnectivity, the programming convention for PE logic solvers, and user implementation of software and/or hardware diagnostics.

If safety configuration by the user is required, the scan rate achieved by the fully integrated system may be substantially slower than that reported by the manufacturer. The scan time for safety PE logic solvers is less than 500 ms with many achieving less than 200 ms. Hardwired systems, including non-programmable relay, trips amp, and solid state systems, can achieve response times less than 50 ms.

5.6 SOFTWARE

Programmable equipment typically relies on three layers of software to execute the IPF logic: embedded, utility, and application program. The embedded and utility software is generally supplied by third parties and their compatibility with the system hardware should be evaluated during the user approval process. This evaluation is important regardless of whether it is supplied by the hardware manufacturer or other third party. The application program is created by the user to execute specific IPF logic. The application program development may rely on function libraries provided with the utility software. The use of pre-built functions should also be approved.

Embedded software is the operating system supplied by the manufacturer and is not accessible for modification by the application programmer. Embedded software is also referred to as firmware or system software.

Utility software consists of the tools used for the creation, modification, verification, and documentation of application program. The utility software includes the compiler and other tools necessary to create a functional system. To improve the ease of application program development, the utility software often includes pre-built logic that executes specific functions.

This section focuses on the application software development. A software specification is used to translate the design basis into a program structure and sequence. A detailed software specification is likely to reduce configuration errors and provides a validation tool. Three types of software languages are considered: fixed program, limited variability, and full variability. The software work process consists of the following:

- Software specification,
- Utility software selection,

- Software language selection,
- Application program development,
- Application program verification, and
- Management of change.

5.6.1 Software Specification

The design basis outlines the requirements for each IPF operation. Various documents are developed to support the design basis and detailed design effort, including cause and effect matrices and written narratives. This information is used to develop a software specification.

As with the hardware specification, the software specification is a collection of information, such as input and output data types, a register and flag list, and logic diagrams or flow charts defining the sequencing of the logic and the required separation of IPF and non-protective functions. The software specification should define the desired program modularity and comment expectations. This improves program traceability and will make later verification, validation, and management of change easier.

The software specification should be sufficiently detailed that the application programmer does not have to decipher or interpret the logic needs. Incorrect interpretation can result in improper sequencing execution, improper shutdown actions, and conflicting logic.

Further, the software specification should include how fault detection should be executed for field instrumentation, such as out of bounds (e.g., failed high or failed low), out-of-range, drifting or stuck measurements, wrong position, and not in commanded state. It should define where the fault alarms are displayed and under what conditions.

The software specification should outline the expectations for communication to the BPCS and to the operator interface, so that the required communication points can be provided and properly configured. The specification should consider each logic path, assessing the potential impact of bad data from the various input points. If bad data can cause erroneous program action, means should be implemented within the application program to verify the data prior to taking action with the data.

IPF resets and start-up permissives should also be described including any required time delays or conditions. Proper planning of resets and permissives eliminates the need to manually bypass IPFs for start-up.

Example--Manual bypass of low flow IPF

> A project implemented low discharge flow shutdowns on 15 pumps. When it came time to start-up the process, none of the pumps would start because they were "tripped" due to low flow. There was insufficient time or cabling to install a time delay. Instead, a temporary bypass switch was hardwired into the motor controls. The operator had to go into the motor control circuit (MCC) building, bypass the pump, go back to the unit to start the pump, wait for the pump to establish flow, and then remember to remove the bypass in the MCC building. This was not a good solution, because it left the pump protection vulnerable to operator error.

The software specification should also define any on-line test logic, including field device or logic solver testing, to ensure tests are performed correctly, success or failure is properly reported, and the IPS is returned to the working condition after test.

Prior to issuance for programming, the software specification should be reviewed by someone who is familiar with the intended hardware platform and IPF application. This becomes more important when complex or high coupled functions are specified. Many programmers (or users) only understand the function operation at a high level and may not fully appreciate the complex real time problems that can occur in the software code. A comprehensive review by an experienced user of the hardware and software should reduce the likelihood of problems (i.e., software bugs) being identified during software validation.

It is good practice to conduct a review of the proposed program structure with an owner/operator representative, who will be expected to take responsibility for the system when completed. This representative should also be present during software tests, such as the FAT. This familiarizes the owner/operator with the program and its interactions with other instrumentation and control systems. It also ensures that the program is achieving the desired functionality and operability.

5.6.2 Utility Software Selection

Typically, the logic solver manufacturer provides the utility software, but third party software can be used when proven to work with the hardware. The user must assess whether the selected software is consistent with the hardware platform and with other aspects such as:

- Anticipated size and complexity of the application program,
- Software specification,
- Modularity needs, and
- Programmer skills.

It is important to treat the utility software more carefully than 'something' that comes with the logic solver. It is far more significant than a computer program on an office personal computer. Unfortunately, software manufacturers, often in an attempt to get products to the market faster, release software with known problems and many unknown and unpredictable ones. Regardless of how good the hardware is, it will not operate correctly if the utility software quality is poor.

The utility software supports program development and verification through configuration, library, compiler, and emulation tools. Utility software errors can result in significant problems with the hardware operation, including IPF failure. Some errors can be detected through data verification. Others are more complex and can evade detection becoming part of the real time system. For example, library functions are used repeatedly throughout an application program. A single function error potentially causes the failure of multiple IPFs due to incorrect logic or data.

Confidence in the software can only be gained using software tools that have undergone extensive testing. The workbench should be fully tested by the manufacturer and, preferably, by the owner/operator to ensure a high degree of confidence in the software integrity. Customized functions developed by the programmer using the manufacturer's workbench should also be tested by the developer and by an independent person.

5.6.3 Software Language Selection

There are three software languages used for the programmable equipment in IPS applications: fixed program, limited variability, and full variability. Each language has different requirements and present to the user different levels of complexity. In all cases, the software should implement the software requirements specification and should be subject to rigorous verification and validation.

5.6.3.1 Fixed Program Language

Fixed program languages (FPL) are used in proprietary dedicated function equipment, such as analyzers, transmitters, positioners, and turbo-machinery monitors. FPL equipment does not provide an interface for modifying the equipment software. Rather, the manufacturer provides an interface, which allows the user to adjust a limited number of parameters associated with its operation, such as the transmitter range. Thus, the language at the user level can be considered "fixed."

The manufacturer provides data sheets, user manuals, or technical guides to describe how the equipment operates and what parameters can be adjusted to configure the equipment for the specific application. FPL equipment typically undergoes extensive testing at alpha and beta level to ensure the equipment operates as intended.

5.6.3.2 Limited Variability Language

Configurability is significantly more flexible with limited variability than with fixed program equipment. Limited variability language uses pre-developed logic together with a range of simple mathematical functions. These functions can be configured in different combinations as necessary to fulfill the software requirements specification.

The flexibility of limited variability is also one of its greatest disadvantages. Some functions are simple and deterministic, providing a high degree of confidence in IPF functionality and program quality. Others are complex and even non-deterministic, resulting in a program that is only as good as the verification and testing allows. For complex programs, the full range of program behavior may never be understood or fully tested even when the good engineering practices recommended in this chapter are applied.

Limited variability languages include ladder logic, function block diagrams, and sequential function charts. These languages allow the user to combine functions to configure the system to execute the application logic. The software manufacturer should provide a user manual, which defines the basic functional components contained in the software library. The user manual should stipulate if any of the functions should not be used for safety or critical applications. The manual should provide information on how the engineering tool is used to configure the functions into application software.

5.6.3.3 Full Variability Language

Full variability languages are general purpose programming languages, such as C++ and Pascal. This type of language uses an operating system, which provides system resource allocation and a real-time multi-programming environment. These languages can be used to create the application software with full flexibility regarding the logic construction. These languages are highly flexible, allowing the individual programmer to create unique programs for the specific application, such as simulating process operations. For programmable electronic equipment, full variability languages are typically found in the embedded software that operates the equipment.

ISA 84.01/IEC 61511 provides requirements for the SIS software. The standard recommends the use of limited variability languages for application programming. While the standard does not prohibit the use of full variability language, it does require that application software relying on full variability be developed according to IEC 61508 Part 3 work processes.

5.6.4 Application Program Development

The application program is developed to achieve the software requirements specification. In general, the application program consists of the logic sequences, permissives, limits and expressions that control the appropriate input, output, calculations, and decisions necessary to achieve the desired IPF functionality. The application program should be developed in a modular fashion and safety and non-safety logic should be separated.

Program development requires careful attention to detail when translating the software specification into an application program. Some specific areas that require special attention are program organization, documentation, verification, and storage.

5.6.4.1 Program Organization and Documentation

The software specification defines how the code should be organized. The program should be developed in a modular manner that allows isolation of specific IPF logic. In general, this organization reduces the potential to introduce errors in the program, facilitates troubleshooting during the FAT, commissioning and the SAT, and assists in management of change during long-term operation and maintenance.

Safety and non-safety programs should be separated from each other and labeled as safety and non-safety. Such separation, along with comments explaining programming intent and references to the software specification, assists future programmers in locating specific logic for modification or verification. The application program should be produced in a structured way to achieve:

- Modularity,
- Testability,
- Capacity for safe modification,
- Traceability, and
- Application program documentation.

Often, the initial programming is performed by someone very familiar with all aspects of the software, while future maintenance may be done by personnel with a less intensive knowledge of the software capabilities. Programmer notation within the program is very important to explain the reasoning behind a certain programming approach. The software requirements should contain the following information:

- Application program description,
- Library functions used,
- Logic sequencing,
- Identification of IPFs and non-IPFs,
- Identification of any logic functions that should not be used,
- Revision control expectations,

- Expected traceability to the design basis and software specification, and
- Inputs and outputs (including internal or memory variables).

Hard copies of the application program are not necessary, but the information listed above should be kept readily accessible to anyone who has the need to modify, verify, or validate the program.

5.6.4.2 Application Program Storage

Software does not wear out, because it is not susceptible to environmental conditions, process conditions, or external stresses. It can however degrade with time due to inadequate software revision management and testing. The system software can become obsolete, necessitating upgrades in the utility and application software. Upgrades may cause incompatibility between the software layers or between the hardware and software.

Software tools and data storage media have a limited technological life. As computer technology evolves, data storage and retrieval media must also evolve. Internal practices and procedures should address how stored information will not become lost. This is especially true for the application program, which must remain accessible as technology changes. The owner/operator should ensure that there is a migration path and suitable equipment for application program storage and retrieval. Electronic copies of the as-installed program and the previous version should be kept in a safe and secure location, in accordance with the facility disaster recovery procedures. Periodically these copies should be accessed to ensure that they can be retrieved and restored.

5.6.4.3 Verification

Emulation tools are available to facilitate early software verification. Many manufacturers provide emulation tools as part of the utility software. These tools can be a valuable aid in early software verification. Emulators usually run on a PC and allow the programmer or tester to force input coils or registers to off-normal states or values. The emulation tool then follows the logic through its functionality to the output coils or registers.

Emulation is a valuable debugging tool, because it allows extensive testing to be performed prior to download into the hardware platform. However, it can be difficult to fully emulate complex programs, especially those utilizing timers, as the nature of the emulator and the clock speed of the actual emulation computer may impact their operation.

Emulation testing should not be considered a substitute for software validation in the specified equipment.

5.6.5 Managing Changes

Utility software should assist in the control of unauthorized changes to the application program. The software selected should have version management capability and should preferably be capable of tracking modifications to the function level. The program should be structured such that a revision block is available as a part of the commented logic. Changes should be noted with the pertinent information, such as date, change, reason for change and affected subroutines, sections, etc.

5.7 FACTORY ACCEPTANCE TEST

A factory acceptance test (FAT) is conducted to determine and document the equipment hardware and software operates according to its specification, covering functional, fault management, communications, support systems, and interface requirements. The FAT demonstrates that the hardware and/or software work according to specification. It is particularly important for highly integrated systems due to the complexity of the potential problems that could occur. For programmable systems, the FAT demonstrates that the application program satisfies the requirements of the software specification and works within the specified hardware.

Schedule, cost, personnel resources, and equipment readiness should be considered when determining when an FAT is appropriate for a particular project. The FAT may be conducted at the manufacturer's facility, owner/operator workshop, or a third-party location. When possible, the FAT should involve integrated hardware, software, and communication tests.

Extensive hardware and/or software validation can take significant time, depending upon the size of the IPS. Cost is therefore an issue to consider, both monetary and required resource investment. An FAT often ties up the manufacturer's resources and testing facilities, as well as dedicated owner/operator representatives. To some, this additional cost may seem unnecessary, since each IPF still requires validation by an SAT. To others, early testing is considered an excellent investment of both time and money.

For projects on a tight time schedule, the physical installation of the hardware may be a critical path item. In these cases, the decision may be made to combine the FAT and site validation. This can pose significant risk to the project schedule and budget. Identifying equipment problems during site validation shifts focus away from the field installation and commissioning to troubleshooting the equipment. This can cause significant commissioning and validation delays with numerous personnel waiting around and doing very little until the equipment is ready.

A comprehensive FAT is the most cost effective means to ensure that the equipment operates correctly and is of high quality and integrity. The SAT is almost invariably a chaotic and tense time, so it is not the optimum time to be searching for IPF equipment 'bugs,' whether hardware or software-related. The best approach from a project management standpoint is for the hardware and software to come to the field fully tested, so field performance troubleshooting can focus on installation issues.

The FAT should be witnessed by someone familiar with the design basis intent, but who was not responsible for assembling the hardware or programming the software under review. In general, independent review is provided by an owner/operator representative, who will take responsibility for the system when completed, and by a technical specialist who is familiar with the hardware and software expectations.

5.7.1 FAT Procedure

An FAT procedure should outline the equipment and functionality to be tested. The FAT should evaluate the hardware and software organization, understandability, and annotation, because this affects the ability of the owner/operator to safely modify the equipment in the future. It should also include peripheral equipment, such as sequence of events recorders, communication modules, and remote I/O communication.

The hardware manufacturer generally conducts a functional test of the equipment in accordance with their quality control plan. The equipment components are connected and powered up and the basic functionality, fault management, and communication ability is validated. This quality control test is not a substitute for validation.

PE logic solver manufacturers sometimes refer to the hardware test as a "rack and stack" test. It is typically used to verify the operation of the PE logic solver hardware and diagnostics, including the processors, power supplies, communications modules, and I/O modules. The application program and its compatibility with the hardware is validated separately.

A test plan and procedure should be developed to ensure comprehensiveness of the test execution. The procedure should, at a minimum, describe the following:

- Define test method(s),
- Functionality to be tested,
- Pass/Fail criteria, and
- Test Documentation

The FAT should fully demonstrate that the hardware and software achieves its specification and the design basis. As with all testing, the intent is to demonstrate and document that the equipment operates as intended. The test outcome should give the test team confidence that the equipment behaves as expected under the full range of foreseeable conditions, including misuse and error. The FAT procedure should cover:

- Measures to prevent unauthorized access to or alteration of equipment,
- Compatibility of any application software with the embedded software,
- Required equipment response time (e.g., scan rate for PLC),
- Fault management by simulating errors (e.g., disconnecting power cables or communication cables),
- Data parameters, such as scaling factors, numerical definition (e.g., integer, floating point or fixed), read/write capability, engineering units, and filter factors,
- Data correctness, including initial data values, set points, timers, counters, and I/O loading,
- Functional or logic requirements using hardware and/or software specification,
- When applicable, measures taken to minimize impact of bad data (e.g., invalid, undefined, or erroneous values) from being communicated or written to the IPS,
- Communications equipment and associated software, including fault management, data protection and data validation features, and
- Non-IPS equipment connected to the IPS does not interfere with the IPF operation.

Hardware and software testing is generally accomplished using simulators or test panels to drive inputs and emulate outputs. If operator input is required on any IPF, the interaction of the operator interfaces and the IPF software should be tested.

Operating or maintenance procedures addressing system diagnostic or fault alarms should be checked for ease of understanding and implementation. Potential faults should be simulated by giving the equipment erroneous input and output conditions. One company's approach is to challenge the operators and maintenance technicians to try to lock up, bypass, or otherwise "get around" the system. The lessons learned from this experience can be used to improve the procedures and to add cautions about certain actions.

5.7.2 Deficiency Tracking

A test log is maintained during verification to ensure that the equipment is thoroughly tested. The test log should document deviations found, deviation impact, the modification required to correct identified deviations, and the current deviation status. Consideration should be given to the overall organization of the hardware and software components, as well as the traceability of those components to the specification.

Identified deficiencies and corrections should be traceable from the test log to the hardware and software component and its specification. When a particular verification activity is completed, the test log should be reviewed to determine the seriousness of the identified deficiencies and open items. Serious items may result in the need to conduct verification again after the problems are corrected. Less serious open items may be passed forward to the next verification or validation activity. Some owner/operators monitor the program deficiency rate as the hardware and software transitions from one verification activity to the next to confirm the overall system quality.

A word of caution: while the FAT provides an excellent opportunity to correct any errors prior to the field validation or SAT, debugging should be done with the mindset of the overall expected outcome. It is extremely easy to become focused on finding and exterminating a single 'bug', but the fix may result in new problems in the system as a whole.

Each modification should be examined to determine the impact of the change on the system. Changing a register or flag may result in some IPFs not operating properly or may cause operation in an incorrect sequence. Some fixes may cause program instability. Based on the extent of the identified problems, the FAT may need to be repeated to check for new problems.

5.7.3 Test Documentation

Test documentation should be completed prior to release of the system for process plant implementation. The test documentation should include the following:

- Test parameters,
- Deviation found,
- Deviation impact, and
- Deviation resolution.

Tracking deviations to completion ensures a quality program. When the application program contains faults, the IPS may fail impacting process equipment safety and reliability. Hardware failure is dependent on the random failure of components. IPS designers use known failure rates, redundancy, testing, and diagnostics to design high integrity hardware architectures. In contrast, the application program failure is highly dependent on the engineering tool and programmer, whose failure rates are uncertain. The only way to identify these failures is thorough testing and documentation.

5.8 INSTALLATION PLANS

The IPS should be installed according to installation drawings and detailed specifications. Many infant mortality failures are induced by poor installation and commissioning. A poorly installed system may have hidden problems that are not revealed until a particular set of circumstances occurs. Consequently, safe operation depends of quality installation.

A task list and schedule is critical for efficient installation. The task list should cover the installation scope and activities. It should identify who is responsible for executing and verifying each work task and who should be notified when problems are encountered. The installation planning should include the communication of the schedule and schedule changes to affected parties.

The construction effort must be carefully planned to optimize any potential synergies within the overall project scope, especially if the installation is part of a major retrofit or a grassroots facility. Such an effort will help to ensure the lowest installed cost by coordinating activities to avoid interference with other crafts and minimize rework. A typical example is the installation of in-line equipment that must later be removed, so piping can be flushed. Good planning and communication efforts can minimize this unnecessary rework.

Applicable safety, installation, and maintenance manuals should be reviewed and understood by the installer. Problems with the physical installation should be brought to the attention of the person designated in the plan. This would include issues such as adequate clearance per the National Electrical Code (NEC) for junction boxes and marshalling panels, any requirements for rear access to equipment intended to be enclosed or wall mounted, and special wiring or grounding requirements detailed in the manuals. Any problems discovered by the construction personnel should also be brought to the attention of the project team for resolution to minimize rework and future problems.

> <u>Example--Local panel exposed to intermittent steam release</u>
>
> A local control panel for a piece of equipment was located and set in the field. The panel was appropriately designed for the class/division requirements of the area and the equipment was commissioned and operable. Just a few feet away and a bit higher in elevation, there was an unrelated steam blow-off valve, which operated intermittently. At one point during the checkout, a technician was present at the local panel when the blow-off valve operated.

While there was no immediate safety concern, the valve operation was startling, to say the least, to personnel at the panel. The drift from the valve was directed toward the panel, resulting in possible long-term corrosion issues. In this instance, it was considered prudent to relocate the blow-off valve.

Another significant installation issue is the replacement or installation of equipment in the control room. Work inside an existing, operating control room requires a good knowledge of what the operators must have to continue process unit operation, while new equipment is being installed around them. This is often done in stages, especially if the installation consists of replacing obsolete equipment with new and reusing the existing field or inter-connection cables. Provisions must occasionally be made to allow the temporary relocation of some process loops during the migration period, so it is helpful to review the needs of the operators prior to beginning the physical installation to avoid unforeseen surprises.

When installation is considered complete, a field walk down with visual inspection should be performed and punch lists developed to track deficiencies. Punch list items should be tracked to completion. A few of the more common inspection items are:

- Packing or transportation materials removed,
- Physical equipment damage,
- Tagging and labeling, including equipment, wiring, cables, and terminal strips,
- Conduit seals poured,
- Doors and covers replaced,
- Solenoid screens installed, and
- Tubing and fittings properly supported to prevent crimping or bending.

The installation integrity should be verified with a complete I/O check. This is typically done in stages, i.e., by junction box to field device, junction box to house, or any cross-connected jumper wiring in the house, etc.

5.9 COMMISSIONING PLANS

Equipment commissioning is a vital step in the EICV phase. During commissioning, the IPS is powered up, and utilities and field instrumentation are brought on-line. Careful planning is required to facilitate the equipment commissioning effort to ensure that resources and procedures are available when needed and efficiently deployed.

Commissioning activities are project-specific and can be quite extensive for large projects. A dedicated team is recommended for the commissioning effort. This increases commissioning efficiency by ensuring a consistent approach and a cohesive understanding of the current status.

The commissioning plan should be broken down by equipment system, such as electrical, instrument air, logic solver, etc. Each equipment system should be identified along with the items required to commission it. This allows manpower needs, including discipline specialties, to be defined. The required resources should be listed in a master schedule.

Once the schedule is completed, the commissioning plan should be communicated to relevant personnel, such as regular operations and shift maintenance personnel. Procedures should be developed which outline the necessary steps for IPF commissioning. These procedures should incorporate verification steps, such as check-offs, site review, and sign-off. Discrepancies identified during commissioning should be tracked to completion.

- Procedures should be considered for the following:
- Hydro testing,
- Dry out requirements,
- Equipment run in,
- Blinds and/or car seals to be removed/installed,
- Electrical tie-ins and power up sequences,
- Operator and maintenance procedures (developed to date) are correct, and
- As-built checking for agreement with P&IDs.

Commissioning demonstrates that the equipment is operating according to its specification and is ready for validation. Commissioning includes the following activities:

- Instruments are properly calibrated, including verification of sensor ranges and engineering units used in the IPS logic and operator interfaces,
- Loop checks from the field devices to the logic solver,
- Energy sources are operational,
- Equipment and wiring are correctly installed,
- Operator and engineering interfaces are operational,
- Field devices are operational, and
- Logic solver is operational.

Participation in the commissioning effort is an excellent opportunity for hands-on training with the new or modified equipment systems. This is especially important for those personnel who will be responsible for the day-to-day operation of the equipment. Therefore, some consideration should be given to provide a 'rotating' position on the team to allow personnel to have the opportunity to work with the commissioning team.

5.9.1 Commissioning Sequence

Figure 5.5 illustrates a typical sequence of commissioning activities for an IPS project. In the case of retrofit projects, existing utilities should be analyzed to determine that the demands from the modified IPS are within the system design tolerances.

Any new or modified utilities should be commissioned first, including power distribution equipment or substations, UPS and battery back-up systems and any grounding system(s) or tie-ins. It is important that the electrical equipment systems be commissioned prior to power-up of the IPS equipment.

Once the power systems are operating satisfactorily, other utilities such as instrument air, nitrogen, and steam should be brought on-line. Any heat tracing and winterization precautions should be checked to ensure proper operation.

The utility pressure should be set appropriately for each final element using a regulator. This is especially important if partial stroke testing is being implemented. Ideally, each of these should be commissioned at normal operating conditions and left in service. Again, there can be a major difference between using instrument air from bottle racks, which is clean and dry, versus site instrument air supply, which may operate at a wetter condition with substantially more particulates.

When the utilities are commissioned and operating satisfactorily, the logic solver brought on-line. This should be done following all manufacturers recommendations. It should be verified that the equipment is functioning and communicating with the IPS operator interface. The logic solver should then be allowed to run unimpeded to allow self-checking, system, and external diagnostics to determine if components are operating properly. Many manufacturers will include power-up / start-up assistance as a part of their standard package or as an optional service. In certain cases, the manufacturer participation may be required to preserve the equipment warranty.

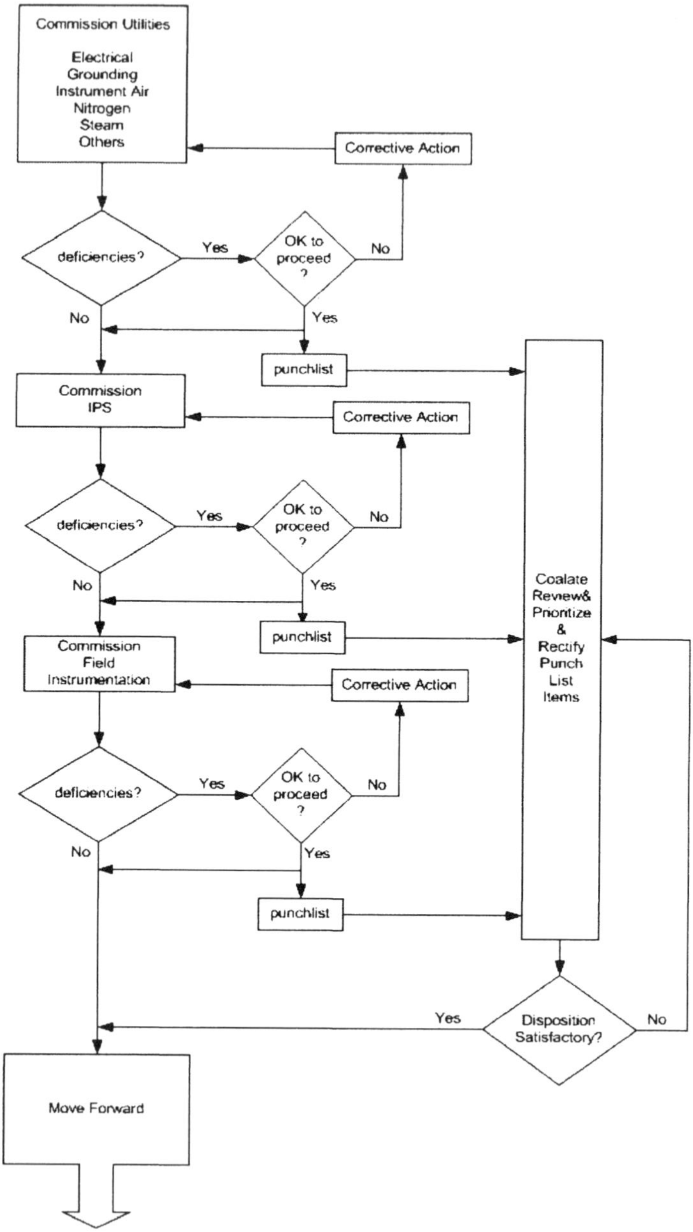

Figure 5.5. Commissioning Activities.

Field instrumentation is generally commissioned last. The installation should be visually inspected for completeness, including verification that all process connections and root valves are blocked-in. The schedule for bringing the equipment on-line, such as opening of the root valves or process connections, should be defined as part of the commissioning plan.

Prior to applying loop power to field I/O, it is important to ensure that all I/O are properly connected. Each loop should be powered one-at-a-time, checking circuit integrity. The following illustrates why this is important:

> Example--Rushing the job
>
> During a major control system upgrade, it was decided to power up a 3000 point BPCS to meet a milestone schedule when many field devices were not terminated (i.e., wire coiled up in conduit fittings). This resulted in US$2000 worth of blown fuses and a significant project delay.

When possible, it is advisable to bring instrumentation on-line and allow it to operate, even without process present. This provides some burn-in time for the instrumentation and controls and can occasionally be a useful as a troubleshooting tool.

> Example--Completing check-out during start-up
>
> A parallel reaction train was being added and was supposed to be left blocked-in and off-line, until the unit was restarted and operating smoothly. The instrumentation for the new reactor train, including a number of thermocouple indications in the catalyst reactor beds, was commissioned prior to restart of the facility, and the thermocouple indicators read ambient temperatures with the train blocked in.
>
> However, as the facility startup began, it was noticed that these temperature indications began to creep up, some, especially in the lower sections of the bed, beginning to approach operating temperatures, yet the new train was still blocked in and off-line. It was determined that the outlet block valve was leaking process gas back into the catalyst bed. The forward process flow was diverted for a time during the startup, so that the valve could be pulled and repaired without production loss.

5.9.2 Commissioning Activity Follow-up

Detail punch lists should be compiled of any items that are found to be incomplete, incorrect or unsatisfactory. These punch lists should be reviewed by project management and compiled into a master list, with resource requirements, due dates, and realistic completion targets. The status and disposition of these items should be discussed and agreed to by all participants prior to the start of the SAT to prevent potential delays in the IPS validation and start-up. Any modifications that are deemed essential to a safe and successful SAT and start-up should receive high priority. Those items considered as 'clean-up' type issues may lag in the schedule, if consensus is reached.

While much effort is put into the planning and execution of the installation and construction, no one is infallible and the commissioning schedule should allow time for correction of a reasonable number of minor concerns. If it is not feasible to correct a major deficiency prior to the SAT, this should be communicated to all involved parties. Contingency plans should then be evaluated to determine if a viable, safe option exists.

In any case where deviations from the original commissioning plan occur, the revised plan and its impact on any subsequent activities should be reviewed and approved by the project team. Once a system has been commissioned, it should be considered 'live,' and this information communicated to all disciplines working in the area.

Even with the best intentions and efforts, last minute changes to items can occur. Changes should be clearly communicated and documented by all involved parties. When major change is needed to make the IPF work properly, the design basis should be reviewed to ensure that the original intent is met and the deviation should be reviewed and approved by the project team. The requirements for formal Management of Change (MOC) documentation are defined as part of the protective management system.

5.10 VERIFY OPERATOR AND EXTERNAL INTERFACES

Communication from the IPS to other systems is critical to safe and reliable operation. Operator interfaces receive status information from the IPS, consisting of indications, alarms, and notifications. External equipment is used to record time stamped information for sequence of events and first out reporting.

The failure of interface and external equipment should not impact the capability of the IPS to execute its safe state action, or the ability of the operator to view the state of the process. Otherwise, this equipment is part of the IPS and must be designed and managed to achieve the required risk reduction. The IPS response to loss of communication should be thoroughly tested to ensure appropriate action is taken on the loss of any critical variable.

All interfaces should be testable without adversely affecting the operation process. The test plan should identify the required resources, the extent to which the system will be checked (boundaries), and the criteria for a successful test. The following list outlines some of the more common items tested:

- System action on loss of communications,
- System action on 'freeze' of information,
- Data highway integrity checks,

- IPS system diagnostics appropriately alarmed,
- Hardwired alarms--properly wired and alarmed,
- Soft linked alarms--appropriately configured and alarmed,
- Time stamps information correctly, and
- Hierarchy, navigation, symbols and color standards for operator interface displays correctly and adheres to owner/operator practices.

5.11 VALIDATION

The installed and commissioned IPS should be validated through inspection and proof test to meet the design basis requirements. Validation activities are used to demonstrate different aspects of the IPF operation, including loop checks, calibration procedures, and SAT.

5.11.1 Loop checks

A loop check or function check is a method used to verify the IPF operation on a loop-by-loop basis. Loop check is initiated after the utilities and IPS equipment are commissioned. It is often conducted in a piece-meal fashion, as loops are installed and commissioned one at a time. Loop checks verify the equipment is properly connected, configured, and operating per specification. These checks include verifying set points, ranges, diagnostics, failure directions, and operator interface information and alarms. For PE logic solvers, the loop check also assesses the status of any punch list items remaining from earlier software verification. For non-PE logic solvers, the loop check verifies the logic.

On the field side, the focus is on the physical aspects of the installation, such as wiring, status of any punch list items remaining from the commissioning effort, and adherence to construction specifications. Other aspects include field inspections to determine:

- Tags and labeling are correct,
- Painting is complete,
- Conduit seals are poured,
- Covers are replaced,
- Barriers/warning sign are in place,
- Process connections are properly installed and root valves closed,
- Tubing is appropriately supported,

- Bug screens, heating, and other environmental controls are installed, and
- Drain plugs are installed.

5.11.2 Site Acceptance Test

The Site Acceptance Test (SAT) demonstrates the correct operation of each IPF for independent owner/operator representatives. In general, it is witnessed by an IPF technical specialist and a production (or manufacturing) representative. The SAT validates that the IPS meets the design basis intent, is installed in accordance with construction, installation and software requirements, and is ready for process equipment startup. This validation involves a complete input to output test of each IPF. It should be executed to a high level of rigor to ensure the IPF operates correctly in each required operating mode as installed and commissioned.

The SAT provides the opportunity for personnel, who will interface with the system on a daily basis, to become familiar with it via hands on experience. The SAT records provide the baseline for subsequent proof tests. As such, the testing and sign-off procedures should be strictly adhered to and any deviations should be managed according to an SAT plan.

5.11.2.1 SAT Plan

The SAT validates IPS functionality and ensures the IPS is fully operational prior to startup of the process unit. Again, as with all validation efforts, an execution plan should be developed, outlining the required resources, the expected level of involvement of each participant, the protocol to be followed during the test, the order in which the IPS or sub-systems are to be tested, and the extent of each test. The plan should also define when and to whom failure should be reported.

IPS projects tend to generate IPF information in various forms, each type intended to support various engineering, procurement, construction and commissioning activities. The IPS documentation may include the following:

- Instrument specification sheets for procurement,
- Logic flow diagrams and/or Boolean drawings for application program testing,
- Cause & effect matrices and loop drawings for maintenance troubleshooting, and
- IPF set point lists for operating procedures.

This information should be both consistent and accurate, and one set of documentation should be considered as master for the SAT, while the others are used as reference materials.

Field clean-up of deficiencies found during the commissioning or loop check phases should be reviewed prior to start of the SAT. This reduces the potential for unforeseen delays during the SAT execution. All known deficiencies should be reviewed with those performing the test and the witnessing, so that there is no question as to the SAT thoroughness or completeness.

5.11.2.2 SAT Execution

During the SAT, emphasis should be on testing each IPF, not correcting deficiencies. It is expected that most, if not all, problems have been identified during earlier verification activities, such as the application program testing, hardware integration testing, commissioning and function checks. If earlier verification activities were thoroughly performed, the SAT progresses smoothly.

When problems are encountered, a minimal amount of time (if any) should be spent troubleshooting the problem. The focus should remain on the testing aspects. The tendency to try to 'fix' the problem in the field or change 'just this one parameter' in the software should be discouraged. Problems should be noted in a log and correcting the deficiency should be assigned to an individual, group or team. By proceeding in this manner, a logical, wider looking view can be taken and any potential impacts of the changes to other systems considered and evaluated.

Various aspects of the IPS should be validated as a part of the SAT, including but not limited to the following:

- Setpoints and ranges,
- Status of sensors and final elements,
- Operator interface,
- Diagnostic indications, such as out of bounds, deviation, or not in commanded state,
- Indication of that any automated logic changes, such as voting degradation and/or fault handling,
- Indication of where the process is in its sequence, if applicable,
- Indication that an IPF has taken action,
- Indication of IPF bypass,
- Operation of manual shutdown facilities,
- Operation of resets,
- Indication of IPS support system loss,
- Failure of environmental conditioning equipment, which supports the IPS,

- Response time, and
- Criticality requirements, such as valve shutoff tightness and closure speed.

The outcome of a successful SAT provides an auditable documentation trail, which proves that the designed and constructed IPS operates according to the design basis and specification. Discrepancies identified during the SAT should be tracked to completion. SAT documentation should incorporate signoff sheets identifying the personnel who conducted tests or served as verifiers for various work activities.

Site personnel should be aware that IPS implementation is nearing completion. Until this point, revision control and management of change has been considered a project task. When the IPS is approved for service, site safety, permitting, and facility management of change procedures for in-service systems will apply. SAT approval indicates that necessary parties agree that the IPS operates as required in the operating environment and is ready for the process unit startup. SAT documentation should include a formal notice of turnover to the site management.

5.11.3 Start-up Procedures

The need for planning and development of procedures to carryout various one-time tasks, such as commissioning, factory and site acceptance testing and deficiency tracking, have been discussed throughout this chapter. Other procedures are required to maintain the seven core attributes throughout the IPS lifecycle. These procedures are discussed in detail in Chapter 6.

This section focuses only on those procedures that should be in place prior to process equipment startup. These procedures are developed to support the consistent performance of plant personnel in the identification and response to IPS equipment faults and failures, as well as their response to alarms and trips generated by the IPS. These procedures will be refined, as experience is gathered, but they are needed for pre-startup operator training and maintenance activities.

Operating personnel should be knowledgeable in the equipment start-up, alarm, shutdown, and emergency response procedures. Maintenance personnel should understand how to troubleshoot, document, repair, and test IPS equipment. These procedures should be developed as early as possible in the EICV phase, and each step toward the completion of the EICV phase (commissioning, FAT, and SAT) should be used to validate their clarity and to further refine them.

Maintenance procedures should be written with sufficient detail that personnel who do not have intimate knowledge of the process equipment can troubleshoot minor faults expeditiously. Remember that while the long-term maintenance and care of the IPS does require specialized knowledge and training, card faults do not respect holidays, vacations or off-hours. Having a pool of qualified personnel capable of dealing with small problems quickly and effectively can stop small issues from becoming major incidents, resulting in equipment failure or downtime.

Operating procedures required prior to start-up include the description of the operator's response to IPF alarms and trips. Each operator should be trained on how to respond to an alarm or trip, what the expected outcomes of each action should be, and what to do in cases where the expected outcome is not realized. Field validation testing, such as the SAT, can be used as a testing ground for these procedures, allowing them to be further developed and clarified prior to process equipment start-up. As operating experience is obtained, these procedures should be further revised to incorporate lessons learned.

5.11.4 Hot Cutover

Facilities are maximizing run times between scheduled turnarounds, which often results in the need to transition from an existing IPS to a new IPS, while the process is on-line and operational. This practice is commonly referred to as "hot cutover."

The nature of the verification and validation shifts when hot cutover is performed. A thorough FAT of the equipment is necessary as validation will be performed on-line. The typical approach is to install and commission each new IPF one at a time so that it runs in parallel with existing IPF. The equipment operates in parallel, until the new equipment is proven to operate correctly. The transfer of the new equipment to 'live' status must be performed with caution to avoid spurious operation. This often requires the use of bypasses and compensating measures for the transition period.

During the cutover period, the operator will potentially have responsibility for the live data and operation of two systems. This may result in logistical issues, requiring additional operators on shift during the transition. Communication and knowledge sharing become critical during the cutover phase, as operators must know where to look for accurate information regarding the IPS status.

5.11.5 Pre-Startup Safety Review

The pre-start-up safety review (PSSR) is an assessment used to confirm prior to start-up of the process that:

- Equipment is installed and demonstrated to operate per design intent,
- Adequate safety, operating, maintenance, and emergency procedures are in place,
- Appropriate hazard and risk analysis or management of change reviews have been conducted and their recommendations addressed, and
- Training of relevant personnel has been completed.

A PSSR is required for new or modified IPS equipment. The assessment involves an inspection of the design basis, drawings, operating, maintenance, and test procedures, installed equipment and action plans. The PSSR satisfies the Stage 3 functional safety assessment requirements in ISA 84.01/IEC 61511. For SIS, the assessment team should include at least one person who is independent of the project team and has sufficient knowledge and experience in the particular SIS application to judge its adequacy.

5.12 MANAGEMENT OF CHANGE

Typically, PSSR approval is considered the formal acceptance by the facility of a new or retrofitted IPS. Changes prior to the PSSR are typically reviewed and approved through project management of change and revision control processes. These changes are resolved during the EICV phase, with punch list items being generated and closed throughout the various verification and validation steps.

After the PSSR, the facility management of change process is used to review modification requests and to track open action items to completion. Identified issues should be tracked and documented to provide a record of approved changes for incorporation into facility procedures and documentation and for periodic hazard and risk analysis revalidation.

Any change that deviates from the design basis requires management of change review, including assessment of any software, hardware, or procedure changes. This is done to ensure that the change does not conflict with the intended risk reduction strategy. Also, as management of change is one of the core attributes of the IPS layer, the requirements for the level of rigor associated with the review and approval should be commensurate with the change. Changes must also be analyzed to identify potential hazardous events introduced by new equipment, interaction of equipment within the system, and interaction between systems.

6

OPERATIONAL AND MECHANICAL INTEGRITY

Once an instrumented protective system (IPS) has been engineered, installed, commissioned and validated against the design basis requirements, and a successful startup has been accomplished, the intense scrutiny on achieving major milestones in the schedule begins to wane and the need to focus on daily operation becomes paramount. The operational and mechanical integrity phase involves the day-to-day execution of work practices and procedures to ensure consistent long-term IPS performance.

The operating basis is essential to the sustainable operation of the IPS throughout its lifecycle and its evolution is an ongoing process that draws upon the facility's unique safety and operating culture. It also builds upon and enhances the procedures, training and documentation developed in previous lifecycle stages. The ultimate goal is to provide a means to communicate the information necessary for the safe and reliable operation of the facility, given the capabilities of personnel and the layout and design of the equipment that they will be interacting with.

A comprehensive operational and mechanical integrity program is only useful if personnel have the fundamental knowledge to understand the intent of the program and have the means and capability to execute its procedures as written. Procedure documentation is more than just the act of putting words on paper, it involves the systematic review of the steps required to execute a job task, including the examination of human factors and ergonomics.

The operating basis provides the operating premise for the process equipment and the bounds for safe operation. For SIS, the operating basis addresses ISA 84.01/IEC 61511 Clauses 16 and 17. The operating basis defines the procedures and training requirements for a variety of activities including:

- Managing failure conditions during plant operation, inspection, preventive maintenance, and proof testing,
- Controlling and monitoring the use of bypasses,
- Investigation of process demands, spurious trips, and dangerous failures,
- Minimum required inspection and preventive maintenance practices to maintain the mechanical integrity of the equipment,

- Minimum required proof testing to ensure correct operation of equipment,
- Minimum requirements for proof testing following modification and repair, and
- Change management, including specific provisions for access security, configuration management, planned modification, temporary modification, and decommissioning.

When the IPS specification and engineering work is completed, plant personnel are expected to take a more significant role in the implementation and support the IPS. Operations and maintenance personnel must understand what it means for the equipment to be in the "as good as new" condition. They should also work to improve IPS performance when practical. It is vital that personnel be trained on any special equipment requirements prior to closing out the project and turning the IPS over to the site operations.

The facility safety and operating culture should be considered when establishing the management system. The core attributes, namely independence, integrity, functionality, reliability, auditability, access security, and management of change, are maintained throughout the IPS lifecycle, as shown in Figure 6.1. Once an IPS is designed and implemented, independence, integrity, functionality and reliability become inherent attributes of the installation, which require constant attention to detail. This chapter and others have referred to this attention to detail as, the level of rigor. Auditability, access security, and management of change are the attributes that demonstrate management system rigor. The facility safety and operating culture must be capable of supporting the specific needs of the IPS design basis.

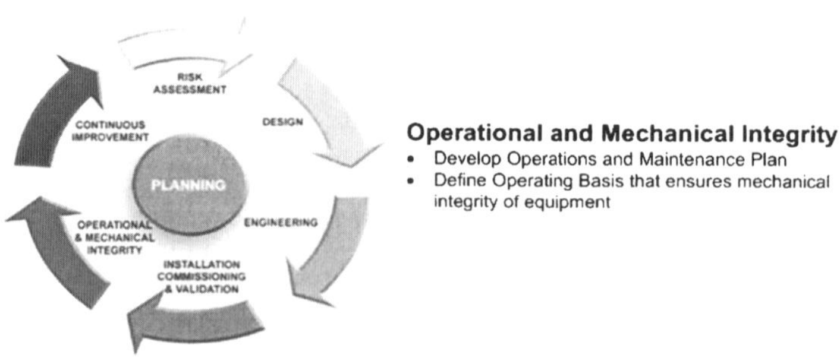

Figure 6.1. Operational and Mechanical Integrity Phase.

6.1 INTENDED AUDIENCE

This chapter addresses the following areas:

- Instrumentation and electrical who maintain the Approved Equipment List, and investigate and track IPS equipment failures,
- Operations who maintain up-to-date operating procedures, ensure operators are trained on procedures, control access to IPSs to only trained and authorized personnel, provide access to equipment and resource support for proof tests, identify and report IPS equipment failure, and investigate and report IPS trips,
- Maintenance who ensure maintenance procedures are kept up-to-date, ensure maintenance personnel are trained for assigned tasks, manage the IPS configuration, and identify and investigate IPS equipment failure, and
- Manufacturers who provide support for the troubleshooting, repair, maintenance, and testing of their products.

6.2 INPUT INFORMATION

The operating basis consists of procedures that govern IPS management, operating, maintenance, and engineering activities, and training programs. The following input information is generally needed to define the operating basis:

- IPS classification,
- Design basis, including the process, instrumentation and electrical requirements,
- Approved equipment list and associated documentation, including equipment manual (or safety manual), installation and maintenance manual, and other relevant information, as necessary,
- Training recommended by specialty manufacturers,
- As-built engineering and installation drawings, and
- Operating basis internal practices.

6.3 BASIC WORK PROCESS

As the project progresses toward completion, specific procedures and training programs are developed to address IPS management and work practices. The work process and procedures should consider how the disciplines work together to maintain safe and reliable operation. Figure 6.2 provides an example of a simplified work process, illustrating the interrelationship between operational and maintenance activities.

The content and depth of the information communicated to various personnel should be based on the intended role of the individual in addressing process risk. It is especially important that these documents be in place as the responsibility for the IPS transitions from the project team to operations and maintenance control. The work practices and procedures developed to support the operating basis should be written with the intended audience in mind. Consideration should be given to the level of technical knowledge expected of the reader.

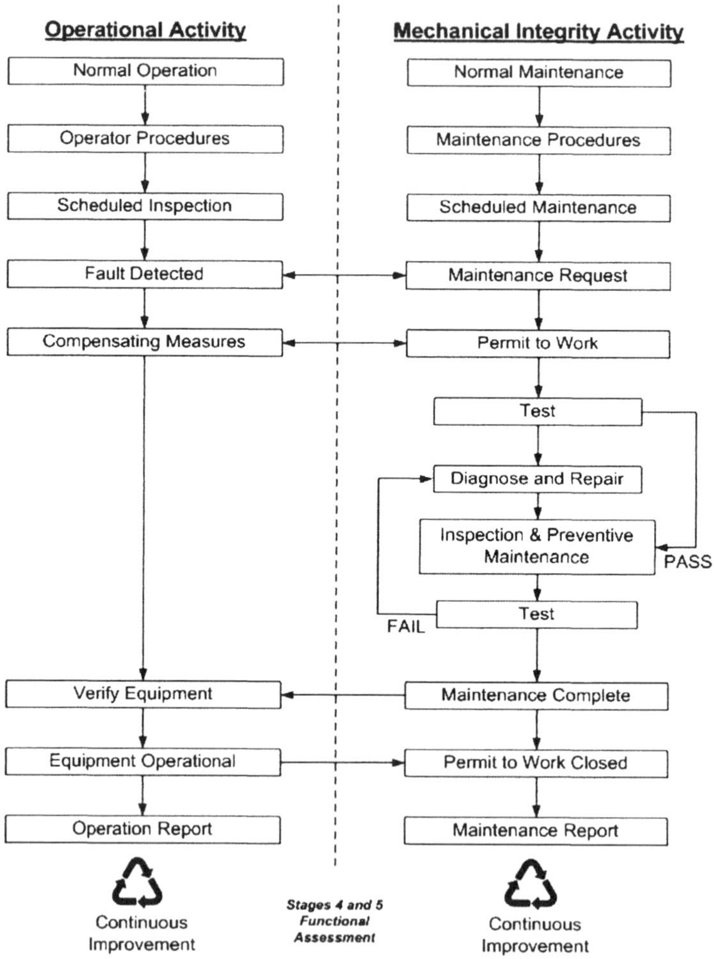

Figure 6.2. Operational and Mechanical Integrity Work Process. (adapted from IEC 61511)

Process engineering and operations are primarily responsible for defining the content of IPS operating procedures, with assistance from other groups, depending upon the technologies involved. These procedures should cover IPS specific information and should explain to the operator the correct use of bypasses and resets, the required response to IPS alarms and trips, when to execute a manual shutdown, and provisions for operation with detected faults. These procedures, along with analogous ones developed by maintenance/reliability engineering for maintenance activities, make up the backbone of the operating basis.

Proof tests require both operation and maintenance procedures. Proof tests include checking not only the IPS functionality, but also any IPS alarm and indication. Similar tests should be periodically performed on the overall IPS health, including main processors, communications links, power, and grounding systems. Each test serves as an opportunity for personnel to see the IPS in action and to validate the procedures associated with its operation.

Many procedures are denoted by department, as in "operating procedures," "maintenance procedures," or "documentation procedures." However, the creation, maintenance and implementation of the operating basis should be viewed as the responsibility of all parties and should be understood by everyone involved with the IPS. Consideration should be given to establishing a single responsible party for IPS practices, ensuring a consistent and comprehensive IPS approach, regardless of the resource allocation. Engineering, operation, and maintenance practices should be developed with the facility's risk goals in mind. It is the responsibility of the owner/operator to manage this risk throughout the IPS lifetime.

6.3.1 Developing IPS Procedures

Safe and reliable operation demands personnel who are committed to performing work procedures consistently and accurately. Each work procedure should be evaluated to identify areas for improvement. The performance related to high likelihood errors can be improved by incorporating work verification, installing IPS, and performing audits. The potential for human error should be assessed based on three questions (Reason 1997):

1. Hands on: How likely is the activity to involve direct human contact with the system and offer the opportunity for human error to cause a failure?
2. Criticality: What activity, if performed less than adequately, poses the greatest risk to the system?
3. Frequency: How often is the activity performed on a day-to-day basis?

All procedures should be written with their intended audience in mind and with an appreciation for the specific technical knowledge of the reader. The procedures should be clear and concise, with minimum complexity. Procedures should provide information in different formats, such as text, graphics, and flowcharts, to accommodate different learning styles. Where multiple languages are spoken, consideration should be given to developing procedures and training materials in each language to ensure critical information is not lost in translation.

If the intended audience does not understand them or feels that they are too complex, the procedures will not be followed. In an operating and maintenance environment, people often tend to follow the path of least resistance and, if they perceive a difficult path, they may find an easier, though not necessarily correct, or safe, one.

Table 6.1 provides a listing of people, situations, and system related errors. Slips, such as omissions and lapses, are common, yet critical errors. Many have assembled a bicycle, piece of furniture or other kit according to instructions, only to be bewildered later, when a pile of parts and screws remain after the project is complete. Incorrect equipment line-up is a common operator error, while incorrect assembly, installation, and repair are common maintenance errors.

Each procedure should define the individuals, departments, and/or job functions responsible for the development, approval, upkeep, distribution, and revision management of the procedures themselves. Work procedures are most successful when they are broken down into steps or tasks intended to achieve specific results.

Task lists, checklists, hierarchical outlines, or task analysis can be used to create procedures, which are easy to understand and use. Task analysis offers a more rigorous organization than other methods. It often uses a three or four column format delineating major steps, providing detailed work tasks, caution notes and comments.

People-Oriented Errors
Slips (lapses, omissions, execution errors) **Capture error** **Identification error** **Impossible tasks** **Input or misperception errors** **Lack of knowledge** **Over-motivation or under under-motivation** **Reasoning error** **Task mismatches**
Situation-Oriented Errors
Environmental **Stress** **Timing**
System-Oriented Errors
Errors by others **Procedural** **Violations**
Human Errors in System Design
Mistakes **Specification errors** **Communication breakdown** **Lack of competency** **Functional errors** **Common errors in instrument design**

Table 6.1. Categories of Human Error. (Mostia 2003)

The choice of technique is highly related to the complexity of the procedure. Task lists are generally restricted to very simple work instructions, requiring few steps and decisions. Longer instructions should be written in checklists or in hierarchical (i.e., outline) format to break the work process into smaller logical steps that are generally executed in series to obtain the specified result. For example, a series of maintenance steps for a transmitter would include activities such as checking the transmitter range, verifying the deviation alarms, and validating the trip set point. Each step has specified pass-fail criteria, which is evaluated and recorded.

When many decisions are required, graphical techniques for presenting the steps of the procedure, such as flow charts, should be considered. Flow charts break down the procedures into small logical steps and provide an effective means to illustrate decision blocks where the answer choice, e.g., a "yes or no," affects the action to be taken.

Regardless of the format chosen, the goal is to ensure that safe and reliable operation is achieved through the detection and correction of failures. The IPS procedures should be written with sufficient detail to achieve the performance specified in the design basis. Just as the core attributes affect the design basis, they are also significant to effective procedure development.

Procedures should focus on the ways in which the core attributes, namely independence, integrity, functionality, reliability, auditability, access security, and management of change, are maintained to the suitable level of rigor. Well written procedures provide instructions, improve communication, reduce training time, and improve work consistency.

Personnel should contribute their experience and knowledge to the continuous improvement of procedures and practices. Cooperation of multiple parties is necessary to ensure that the IPS requirements match the capability of personnel. Procedures used in combination with training and regular performance feedback achieve predictable work results.

6.3.2 Auditing IPS Procedures

Periodic auditing of the operating, maintenance, and engineering IPS procedures should be performed. Audits should follow a protocol that ensures procedures are up-to-date, personnel are familiar with the procedures, and the instructions are being followed. The auditing cycle is generally in the 3-5 year range. More frequent auditing may be required if there are numerous or repeated findings.

The audit should review records, information, and documentation to determine whether procedures are being adhered to. Audit findings should be addressed in a timely manner and tracked to completion. Shortcomings identified in the audit should be addressed with an action plan that establishes a schedule and assigns responsibility for correcting deficiencies to specific personnel or departments.

6.3.3 Revising IPS Procedures

IPS procedures should be under revision control. Procedures should be periodically reviewed to ensure that the procedures are up-to-date and reflect current work tasks and expectations. Changes to IPS procedures, whether technical or editorial, should be reviewed and approved.

6.4 OUTPUT DOCUMENTATION

The operational and mechanical integrity phase is about execution. In this phase, operations and maintenance develop and maintain procedures that support safe and reliable operation. Operating procedures are needed to ensure effective response to IPF indications and alarms. Operating and maintenance procedures are needed to support the inspection, preventive maintenance, and proof test activities necessary to maintain the IPS equipment in the "as good as new" condition. To ensure performance, IPS equipment failures and trip must be investigated, tracked, and periodically assessed to identify areas for continuous improvement.

The design basis establishes minimum requirements for the operating procedures. When changes to the design basis are approved, the operating basis should be updated accordingly. Operating procedures are reviewed on an annual basis to ensure that they are up-to-date and reflect how work is performed. Operations planning and work processes should include:

- Restricting access to IPS equipment,
- Review and approval of compensating measures,
- Review and approval for exceeding the allowable repair time,
- Approval and notification for IPS equipment bypass,
- Approval and notification for past due proof tests for IPS equipment,
- Recording and reporting of fault alarms,
- Recording and reporting protective alarms,
- Recording and reporting process trips and trip failures,
- Maintaining procedures in an up-to-date manner, and
- Maintaining sufficient resources to support compensating measures during IPS bypass.

Maintenance procedures are developed and maintained to support the I&E requirements specification. Maintenance procedures are reviewed periodically to ensure that procedures reflect the current specification, installation and configuration. Maintenance planning and work processes should consider the following:

- Restricting access to IPS equipment,
- Scheduling inspection and preventive maintenance for IPS equipment,
- Scheduling IPS equipment tests to meet proof test interval,
- Recording and reporting past due proof tests,
- Recording and reporting equipment failures,
- Managing changes to the installation and equipment,
- Maintaining procedures in an up-to-date manner, and
- Maintaining sufficient resources, tools, and spare equipment to meet allowable repair time requirements.

6.5 OPERATING PROCEDURES

Operating procedures, also known as standard operating procedures (SOP) or standard operating instructions (SOI), provide instructions and guidance for safe and reliable operation during normal operating modes, including:

- Steady-state,
- Start-up,
- Shutdown,
- Maintenance,
- Instrument Testing,
- Preparation for Maintenance, and
- Process equipment bypassed or out of service.

The procedures should define how operating personnel interact with the IPS during each operating mode. Controls, alarms, graphics, and indications should be described to ensure the operator understands the information being displayed and what is expected in response to abnormal and emergency conditions.

Operating procedures should include safe operating procedures, which provide the following information:

- Hazardous event being prevented by the IPS,
- IPS description,
- Set points and IPS actions,
- Correct use of bypasses and resets, including the process conditions necessary for safe use, and
- Response to IPS alarms and trips,
- When and how to execute a manual shutdown, and
- Provisions for operation with detected faults, including compensating measures necessary to maintain safe operation.

Safe operating procedures may be integrated within the facility's general operating procedures or maintained as a separate set of procedures. When the procedures are integrated into the general procedures, a special symbol or format should be used to annotate those related to safe operation. Operating procedures should be in place prior to process equipment start-up, updated before any change is placed in service, and kept current throughout the IPS life.

6.5.1 Operator Response to Hazardous Events

The process industry often relies on operators to make critical decisions with regard to plant operation. Most owner/operators establish minimum education, skills and/or experience requirements for employment as an operator. This background is then supplemented through site-specific training on operating practices and procedures.

Caution should be applied when expecting operators to take responsibility for executing an IPF. Rigorous procedures and training are necessary to consistently achieve risk reduction (Appendix B). Based on information and/or observation, the operator makes the decision to take a specific action among a set of actions that are possible, such as bringing the process to safer operating conditions, initiating manual shutdown of selected parts of the process, performing manual equipment isolation, or reducing production rates. An operator is generally expected to be capable of executing any of the following actions:

- Detect and respond to changes in the operation of the control system by monitoring trends in production, quality, and input/output signals,
- Perform production or quality critical actions based on alarms or indications,
- Record information and data according to procedure,
- Respond to detected faults of IPS equipment by implementing compensating measures and initiating a maintenance order, if necessary,
- Respond to protective alarms to prevent hazardous event,
- Respond to abnormal and emergency operation and conditions,
- Record and report IPS trips, and
- Initiate alarms that require personnel to shelter-in-place or move to a safe area (see Appendix B.6).

Human factors should be considered, assessing how effective an operator can be during the hazardous event propagation. Work load, human reliability, or human factor analysis can be used to assess task expectations related to the work environment. Instrumentation and controls used as part of the supervisory function (see Appendix B) should be designed and managed to achieve the desired risk reduction.

The amount of time available for operator response is a critical design parameter. The operator should be able to recognize the need to take action, accomplish the action, and have the process achieve the safe state in less than one-half the available process safety time. If there is insufficient process safety time, the probability of success is low. If action in the field is required, this also potentially exposes the operator to the hazardous event.

Procedures should describe the alarm functionality, the specific actions required when the alarm activates, the alarm priority, means of annunciation and the expected response time for operator action. The alarm functionality description should include various set points required for each operating mode, e.g., start-up, different rates, different products, etc. Operator actions required to change the IPS mode of operation should be included, if appropriate.

The required operator response to IPS alarms and indications should be specified. Safe operating procedures should define the specific steps implemented by the operator to achieve or maintain a safe state of the process to reduce the risk of an identified hazardous event. Procedures should also define the required response to detected equipment faults and process trips. The procedures should define the "never exceed never deviate" conditions requiring a manual shutdown.

Manual shutdown switches should be used when possible as part of normal shutdown procedures so the operators gain experience in their use and the switches are tested periodically. Manual switches should be proof tested as necessary to ensure equipment performance.

6.5.2 Operator Response to Failure

The procedure should also outline any IPS fault alarms and the required operator response when received. The description should include compensating measures to be put in place until the IPF has been returned to service. Procedures should define how long the process can continue to operate with a known failure. Limitations related to the process operation should also be provided to ensure that maintenance is not performed under unsafe operating conditions.

6.5.3 Compensating Measures and MTTR

When a fault is detected, a decision must be made to either continue process operation or to take the process to a safe state. Continuing operation relies on compensating measures, which address the additional risk posed by operating the process with a degraded or disabled IPF. Compensating measures must be continuously present during the at-risk period and fully capable of operating within one-half the process safety time. Such measures should be in addition to the protective functions used to reduce the risk below the owner/operator risk criteria.

Compensating measures may include, but are not limited to:

- Operating at reduced rates,
- Implementation of tighter operating limits,
- Implementation of additional safeguards,
- Provision for an independent alarm with operator response,

- Continuous process monitoring with alternate means of shutdown, and
- Limiting operating modes.

Procedures should be established that describe the compensating measures applied when IPF faults are detected, as well as the approved duration of use. This duration is known as the allowable repair time. For cases, where the actual repair time is anticipated to exceed the allowable repair time, management of change (MOC) review and approval should be conducted. This review typically examines the compensating measures put in place and assesses the risk for continuing on-line repair. If the risk is judged to be unacceptable, process shutdown should be initiated, unless the risk can be reduced by other measures.

6.6 BYPASS MANAGEMENT PROCEDURE

Bypasses are sometimes necessary to allow on-line maintenance and proof test, reducing equipment downtime and improving process reliability. However, bypasses increase the likelihood of systematic errors (e.g., left in bypass, inappropriate use of bypass). This increased systematic error must be off-set with procedures, administrative controls, and access security provisions.

Bypasses may be implemented to prevent the operation of a single IPF device, an IPF subsystem, or an IPF. Acceptable means of bypass should be defined by internal practices. Bypass facilities should be included as part of the IPS design basis and validated prior to start-up. The failure of the bypass equipment should not cause a dangerous IPF failure. If bypass failure causes a dangerous failure, this should be considered during the design basis and risk reduction verification.

Depending on how the bypass is executed, the IPF may be impaired, but still capable of acting due to the IPF subsystem being designed to be fault tolerant against dangerous failure, or disabled and incapable of responding as specified. Hardwired bypasses (e.g., jumpers) should not be used due to the high potential for systematic error associated with their use.

While bypasses may be necessary, their use must be controlled. Bypasses should not be used to allow the process to operator outside the normal operating limits. Configuration management and access security play significant roles in the long-term performance of the IPS.

Regardless of the bypass means, there is generally an increased risk. An assessment should be performed to identify compensating measures, which safeguard the process during system impairment. Compensating measures are documented in operating procedures and are implemented for the bypass duration. The ability of operator to maintain the compensating measures may limit the allowable repair time.

The operator should be informed by alarm and/or by procedure, prior to placing any IPS in bypass. When circumstances necessitate continuation of a bypass across shifts, a formal method should be used to ensure that the presence of the bypass and its purpose is communicated to the next shift. When a bypass alarm is implemented, "ring back" functionality should be considered. When a bypass extends across shifts, the bypass should "ring back" or repeat so that each shift operator formally acknowledges the bypass.

Administrative procedures should be used to control access to bypasses and to ensure that compensating measures are in place to maintain safe operation during bypass. The initiation of a bypass should be approved, logged, monitored, and audited. A site bypass procedure should describe the requirements to initiate a bypass, including permit, approval, documentation, and notification requirements.

6.7 MAINTENANCE PROCEDURES

Maintenance procedures, commonly called standard maintenance procedures (SMP) or standard maintenance instructions (SMI), are a primary tool for assuring maintenance activities are carried out appropriately. Procedures should provide maintenance instructions, practices, and guidelines used for IPS equipment inspection, preventive maintenance, and testing to ensure the IPS mechanical integrity. Maintenance procedures should be in place before process equipment is placed in service, updated before any change is implemented, and kept current throughout the IPS life.

An internal maintenance practice should provide overall requirements for maintenance procedure scope and content. Each IPS should have a set of maintenance procedures covering the mechanical integrity requirements unique to that specific IPS and its IPFs. Separate work processes are often used for on-line versus off-line maintenance.

Work processes and resource planning should define how maintenance activities will be managed across shifts, off shift, or on weekends and holidays. Access and bypass permits, as well as approval and notification requirements, should also be specified for maintenance activities.

IPS equipment is covered under a management system that is often known as "mechanical integrity." For IPS, a rigorous mechanical integrity program is necessary to maintain the IPS equipment in the "as good as new" condition. Mechanical integrity procedures define the inspection, preventive maintenance and proof test activities necessary to assure the equipment integrity and to determine when equipment requires replacement or upgrade. Inspection and preventive maintenance should be sufficient to ensure equipment is not run to failure. Equipment performance should be monitored to improve the mechanical integrity program and to prevent dangerous failure.

Maintenance procedures should be analyzed using a suitable, standardized method to determine the coverage comprehensiveness of the test procedure, ensure adequate test coverage for all dangerous failure modes, and ensure the potential for systematic (human) errors have been considered in the procedure. These methods may vary depending upon the complexity of the task and may include failure mode and effects analysis (FMEA), job step analysis, task analysis, or equivalent. While a test should be comprehensive, if it is too difficult or complex, there is a greater likelihood the test will not be completed properly.

Maintenance procedures should be available and should describe the work tasks in a step-by-step manner with clear pass/fail criteria. As with other procedures, responsible personnel or departments, the required permits and notifications, the required test equipment and tools, and any appropriate hazard or safety warnings should be identified. Procedures should provide the work process steps necessary to successfully complete equipment commissioning and validation. Validation should be performed whether repair is done on-site or by the manufacturer.

Maintenance procedures should describe any related functions, such as IPS alarms, bypass switches, manual shutdown buttons, and resets. Procedures may be modularized as desired with procedures written for individual pieces of IPF equipment, each IPF, a set of IPFs, or the entire IPS. Procedures should be comprehensive and clearly convey the work expectations and requirements. Maintenance records should be signed and dated by the person(s) conducting the work.

Those assigned responsibility for conducting work according to a test procedure should be sufficiently competent to understand and implement the procedure as written. The procedures should include an inspection of the physical installation to provide visual confirmation that equipment is in satisfactory condition. Preventive maintenance activities should also be described.

IPS equipment should be periodically proof tested to demonstrate and document that the equipment is operating according to the design basis and specification. Proof tests can be performed on-line or off-line. On-line test procedures should be carefully planned, documented, and validated, because minor mistakes during on-line testing can potentially lead to process upsets or spurious trips. Off-line testing is inherently safer, but given the current trend of increasing run time between process facility turnarounds, it is becoming increasingly difficult to determine the "as good as new" equipment status without some on-line testing.

When automated diagnostics detect a fault, the IPF is configured to initiate 1) an automatic shutdown, 2) a protective alarm, or 3) a fault alarm. The required configuration is defined in the design basis and is determined by the equipment choice, subsystem fault tolerance against dangerous failure, the nature of the failure (e.g., dangerous versus degraded), and the availability of compensating measures. Continued operation requires compensating measures to ensure safe operation during the allowable repair time. When applicable, operating procedures should provide restrictions on the maintenance activities, e.g., prohibited during certain operating modes.

Maintenance procedures should cover in detail how maintenance is performed safely while the process equipment is operating. A key parameter for on-line repair is the allowable repair time established in the design and operating basis. The allowable repair time provides the maximum time that the equipment can be out of service prior to initiating management of change activity. The management of change review is performed to determine whether the compensating measures are sufficient for the extended period, additional measures are required, or manual shutdown executed. The review should also address the priority status for the repair activity.

Procedures should be in place to assure that all test equipment used on the IPS is properly maintained, calibrated (certified, if necessary), and fully operational. Calibration cycles of test equipment should follow manufacturer's recommendations and methods to assure the accuracy of the equipment. Test equipment that is found to be out of calibration, past established calibration dates, poorly maintained, or in poor physical condition should not be used on IPS systems.

The test procedure should include return to service provisions to assure proper transfer of IPS equipment responsibility from maintenance to operations. The operator should confirm by process condition and/or equipment observation that the IPS equipment is on line. Operations should approve work completion closing the work permit. Additional supervisory sign off may be appropriate in some cases.

The proof test interval is determined by good engineering practice, manufacturer's recommendations, operating history, insurance requirements, industry standards, operational constraints and the risk reduction requirements. The test interval is specified in the design basis. Since operational issues can affect the test window, meeting the exact test interval may be difficult at times. Maintenance procedures should discuss the allowable test interval variation, including management approvals for delayed testing.

Management should be made aware of the risks involved with delay of IPS testing and approve deferments on a case-by-case basis. Variances from the allowable test interval should be documented providing the corrective action plan with compensating measures and an approved variance request.

6.8 TRAINING

A thorough understanding of the IPS operation is critical to ensuring that the IPS is operated and maintained to meet the required performance. Consideration should be given to potential language barriers to effective learning. If multiple languages are spoken, safety and emergency information should be communicated in other languages as necessary to ensure personnel understand work process requirements and expectations. If personnel do not understand how the IPS equipment is expected to operate, a procedure change, variance, or deviation may seem acceptable, yet yield an undesirable outcome.

Personnel should be trained on facility procedures, such as safe work practices, evacuation and response procedures, access permit requirements, and management of change. Personnel should receive specific training related to their assigned responsibility. Personnel training should be verified as complete during the pre-start-up safety review (see Chapter 5.11.5) for any new or modified IPS. New personnel should complete training on the IPS operation prior to taking responsibility for the process equipment.

Once an IPS is operational, knowledge and skills should be maintained through an on-going training program. For best results, facility training should emphasize the fundamental criticality of IPS operation. Means for evaluating the training program effectiveness should be implemented. Training should be revised to resolve deficiencies. Knowledge and skills based testing can be used to validate training effectiveness, as necessary. When knowledge and skills do not match expectations, consideration should be given to improving training content, depth, or frequency to obtain the desired level of competence. Training records should be maintained.

6.8.1 Operations

Periodic training is essential to ensure that operations personnel are aware of the hazardous events the IPF protects against, how the IPS operates, and what operator actions, if any, are required. Operators should understand what process condition is being measured and what action is taken by the IPF or operator when the process condition is detected.

Consideration should be given to using process simulators to aid operator training on complex processes. The simulator displays abnormal conditions of various types and provides an interface to allow the operator to respond to the simulated event. By practicing responses, the operator gains abnormal situation awareness under controlled conditions. Simulators provide the opportunity to assess the quality of operator procedures. Lessons learned during simulation can be incorporated into procedures, reducing the potential for hazardous events.

Training should include a periodic review of IPS operating procedures. Training should ensure that operations personnel understand the following:

- Actions of the manual shutdown facilities and the proper response to IPS alarms,

- Operation of bypasses (their function, how to activate, when to use, their allowable duration, compensating measures, and bypass alarms),
- Approval requirements for equipment access and IPS bypass,
- IPS reset requirements,
- On-line testing functions,
- Operator action required after a trip, and
- Failure documentation and reporting requirements.

6.8.2 Maintenance

Training should familiarize maintenance personnel with the hazardous events the IPS protects against and the expected IPS operation. Personnel assigned responsibility to perform maintenance and testing on the IPS equipment require the knowledge and experience necessary to perform the procedures correctly. Training should ensure that maintenance personnel understand what permits and notifications are required to work on or to bypass IPS equipment. Training should cover task expectations, such as documentation, reporting, and failure investigation.

6,8.3 Engineering

Engineering personnel who design and engineer new IPS installations or who are assigned a support role for an existing IPS should undergo training on relevant facility policies and procedures. Training should discuss the core attributes and how each IPS is designed and managed to achieve the attributes.

Since engineering often serves as the originator of IPS change, it is important that these individuals possess a fundamental understanding of what constitutes a change and how changes are evaluated. Training should include a periodic review of engineering IPS specifications, procedures, and practices and any other general facility procedures related to IPS.

Other specialized training may be required based on job responsibility. For example, those engineers who have responsibility for the verification calculations should, at a minimum, understand the calculation fundamentals, such as the basic assumptions and limitations. Examples of other specialized training include consequence modeling, quantitative risk assessment, application programming, and equipment configuration.

6.8.4 Auditing

Auditing is required to assure that personnel are receiving training as required, and that the training is up-to-date with latest practices, comprehensive and technically appropriate for the task. Furthermore, it is important to verify the training is occurring at the designated time intervals, and records of training are being maintained. The audit cycle is typically in the 3-5 year range. Shortcomings identified in the audit should be prioritized and addressed in a timely manner.

6.9 MANAGING CHANGES

Change is inevitable and equipment occasionally needs to be replaced, repaired, and upgraded. The process facility may be expanded, leading to additional hazardous events requiring new IPFs or placing new requirements on existing IPFs. Process and operational changes should be reviewed through management of change to determine how these changes affect the IPS design and operating basis.

For IPS, the purpose of a change management procedure is to control changes to the IPS equipment and to ensure:

- Appropriate analysis is conducted prior to change implementation,
- Approval is obtained from affected parties,
- Change is consistent with current practices,
- Documentation is completed and consistent with field application, and
- Risk is not adversely affected.

Effective change management requires the use of administrative and physical means to prevent unauthorized or inadvertent changes. Since the design basis development involved input from many disciplines, changes should be assessed and approved by similar disciplines. Such evaluation is needed for any change other than replacement in kind, such as:

- Adding new IPS equipment,
- Removing IPS equipment,
- Changing the IPS equipment specification,
- Changing the brand or model of IPS equipment,
- Modifying the IPS equipment installation details,
- Changing the IPS equipment calibration or trip set-point,
- Changing IPS equipment software, and
- Modifying IPS inspection, preventive maintenance, and proof test procedures.

The following sections discuss various aspects of change management, including access security, management of change, and configuration management.

6.9.1 Access Security

Provisions for access security are an essential part of the design basis and are provided to support IPS change management and configuration control. IPS performance is limited by the level of rigor used to protect the IPS from unauthorized bypass, modification and change. IPS access security procedures are established to assure that only authorized personnel access the IPS hardware and software and that only approved changes are made to the IPS. These procedures generally depend on a combination of administrative and engineering controls.

Administrative controls identify those people and/or job functions that have authorization to access the IPS hardware and software. Personnel training and supervision ensures that only qualified personnel are allowed access to IPS equipment. Engineered controls restrict access to equipment using physical means, such as keys for rooms/cabinets or passwords for software changes. Biometric means using retinal scans, fingerprints, etc. can be used to restrict access to specific personnel.

The success or failure of an administrative or engineering control is entirely dependent on how well the responsible parties want them to work. The key or password is often readily available to anyone authorized to use it, so a key or password alone is insufficient to provide access security. A combination of administrative and engineered controls are generally required to achieve sufficient IPS access security.

Administrative processes, such as access permits, should be required to ensure that only approved activities are conducted. The effectiveness of administrative and engineered controls is limited by the safety culture and attitudes of plant personnel. Personnel who respect the process, its hazards, and the role of the IPS in achieving safe operation typically will not attempt to access the IPS in an unauthorized manner.

Training related to access security requirements is essential to effective change management. Personnel should understand the need for the security and be expected to adhere to the access security provisions. Personnel should also understand that a failure to follow procedures will be treated seriously. Procedures and training should specifically address the investigation of improper access or modification and the requirements for follow-up activities.

6.9.2 Management of Change

At some point in the IPS lifecycle, change will be required in the hardware or software. Changes are initiated for a variety of reasons and plant management of change (MOC) procedures should be used to identify those changes potentially affecting the IPS integrity, operability, functionality, maintainability, or reliability. Facility MOC procedures should be implemented as soon as the equipment has been approved for start-up. MOC procedures should include the evaluation of the potential impact of a proposed change on the core attributes of the IPF.

> <u>Example--Process change results in incident</u>
>
> A chemical company produces a resin using a caustic-based reaction. When high pressure is detected in the reactor, the operator initiates a "reactor kill system" which injects an acidic solution into the reactor, thereby neutralizing the reaction.
>
> The chemical company decides to begin producing an acid-based resin in the same reactor. During one of the acid runs, the high pressure alarm activates. The operator initiates the kill system injecting the acidic "kill" solution into the reactor. The acid injection causes a runaway reaction that ruptures the vessel, destroys the surrounding building walls and most of the roof, kills one worker, and injures multiple others.
>
> The root cause investigation determined that when the hazards associated with the acid-based resin were reviewed, personnel listed the high pressure SIF as a safeguard. The change from a caustic to an acidic reaction was approved without a detailed analysis of the actions of the high pressure SIF. The hazard and risk analysis was not sufficiently detailed to determine whether the existing risk reduction strategy was sufficient.

Other operational and mechanical integrity issues also drive the need to change an IPF design. For example, facility management may set an operational goal to extend turnaround intervals. If the design basis assumes proof testing is done off-line every 2 years and the turnaround interval is extended to 4 years, the impact of the longer test interval on the IPF performance must be determined prior to extending the turnaround beyond 2 years. Such a change often drives the requirement for additional equipment redundancy, provisions for on-line testing, or diagnostics to compensate for the longer test interval.

The following activities are required before making changes to the process or its equipment:

- A hazard and risk analysis to determine whether the change impacts the IPS requirements, especially the core attributes,
- Review and update of the design basis to reflect the state of the modified IPS,
- Update operation and maintenance procedures and train personnel on the change, and
- Verification of the IPF change to ensure it meets the design basis intent and has no unforeseen impacts to other IPF actions.

6.9.3. Hardware Configuration Management

During the creation of the design basis, user approved equipment is integrated into an overall IPS that achieves the desired functionality and performance. Architecture, redundancy, diagnostics, and proof test intervals are defined based on the expected capability of the chosen equipment. IPS documentation should specify the approved equipment, along with any calibration, configuration, or installation requirements.

Throughout the IPS lifecycle, it is reasonable to expect that equipment will need to be replaced or upgraded due to failure or obsolescence. The upgrade may be as simple as getting the latest transmitter version from the manufacturer or as complex as upgrading the logic solver operating system. Configuration management is intended to ensure that when equipment is repaired or replaced that the "changed" equipment still meets the intent of the design basis.

A configuration management process is required for IPS equipment. Replacement in kind requires a like for like change in hardware and software. Any equipment substitution must undergo an approval process. Configuration management procedures should specify when additional analysis by the instrumentation and electrical department is necessary prior to equipment substitution.

Equipment configuration details should be under revision control and clearly documented in the maintenance procedures. The configuration details cover the various repairable/replaceable components and how they are configured to operate in accordance with the design basis. After repair/replacement activities are completed, the equipment configuration should be verified and the equipment proof tested to demonstrate that the IPF operates as intended.

6.9.4 Embedded Software Management

The IPS logic solver often relies on programmable electronic systems (PES). Software modifications can occur at three levels: firmware, operating system, and application. Any firmware (i.e., operating software) change, whether a manufacturer system upgrade or an entirely new operating system, requires careful planning and assessment of potential risks. Compatibility tests should be completed which ensure uncompromised functionality of all IPFs prior to process equipment start up with modified firmware or operating system.

6.9.5 Application Program Management

After acceptance and start-up, the application program should rarely require modification and, when it does, it should be done under management of change. In fact, changing the application program should be treated as seriously as modifying the process. While changing the program may be fast and inexpensive, the change can have a profound impact on plant safety. Application program changes represent configuration changes for the overall IPS operation or to individual IPFs, so the software should be under revision control.

Due to the relative ease of making program changes, access to the application program should be controlled using administrative controls, passwords and other security measures. Access should be monitored and audited. The operator should never have the authority or the capability to modify the application program. Override or bypass of the application program should be approved by management of change and supported by a written procedure.

Any proposed software changes should be assessed to determine the potential impact to other IPFs being executed by the IPS. When changes are necessary, the modification should be evaluated in light of the software specification. An impact analysis should be performed to ensure that the change does not impact any other IPF.

Changes should be recorded in a written log to allow auditing of the program version against the log of approved changes. Unless the change can be sufficiently isolated to a specific program module, the entire program should be tested. If isolation is possible, only the affected program module(s) are tested.

6.9.6 Decommissioning

Decommissioning an IPS, or even a single IPF, involves more than just the removal of equipment. Decommissioning represents a substantial change to process operation, consequently, management of change (MOC) approval is required prior to decommissioning. The MOC review should consider the impact of the decommissioning on the process risk as a whole. The IPF should remain in service until a hazard and risk analysis determines that the IPF is not required for risk reduction or for other operational purposes.

For example, a process change may result in an inherently safer process eliminating the hazardous event that the IPF was designed to address. However, due to the current IPF design, activation may also address other hazardous or non-hazardous events. Over time, operations may have become dependent on the IPF to perform these additional housekeeping or process quality functions. Therefore, each protective and non-protective action should be assessed to ensure required functionality is retained.

In cases where an entire IPS is being decommissioned, the IPFs should be assessed collectively and individually to determine any collateral impact to process operations. When such IPFs are discovered and their continued operation deemed necessary, they should be maintained within the IPS or another appropriate location (e.g., BPCS, relay system, etc.).

For those instances where an older or obsolete IPS is being replaced with a newer system and there is no intent to modify the functionality, a procedure should be provided to assure the existing IPS is removed from service to minimize confusion. Documents and training programs should be appropriately updated.

6.1 0 MONITORING PERFORMANCE

Throughout the process equipment life, numerous assumptions are made about the IPS equipment used to achieve or maintain a safe state of the process when unacceptable process conditions occur. The hazard and risk analysis made assumptions about the initiating cause frequency, hazardous event rate, and IPF risk reduction. These expectations led to a design basis where IPF operational and mechanical integrity requirements were specified.

Periodically the actual performance should be compared to the expected performance to determine whether the IPS equipment is suitable for continued use as is or whether improvement should be initiated. There are five facets of IPF performance to monitor:

- Process demands,
- Detected faults,
- Dangerous failures,
- Spurious operation, and
- Personnel conformance to work practices.

When the above are identified, root cause analysis should be conducted to (1) describe what caused the identified failure, (2) determine the failure impact (3) identify the underlying reasons the failure, and (4) implement corrective actions. Consideration should be given to changing the design, installation, operation, and maintenance practices to reduce the likelihood of failure re-occurrence.

Failure reports can be collected from across a facility or a company and used to identify patterns of failure, indicating systematic or common cause problems. One means of monitoring failures is provided by the CCPS/AIChE Process Equipment Reliability Database (PERD) initiative. This program develops and distributes failure classification taxonomies. Refer to Appendix E for more discussion.

6.10.1 Process Demands

An initiating cause may lead to a process demand on IPFs and perhaps other IPLs. A process demand is undesirable, because it signals the occurrence of abnormal operation where the process equipment is not achieving desired performance and continued operation threatens safe operation. A process demand causes the IPF to take action on the process when the specified conditions are reached.

Following a process shutdown, information should be collected to support the trip investigation. This includes the operating conditions preceding the trip, any unusual maintenance or operating activities, the as-found state of the IPS equipment after the trip, and statements from the operator and other involved personnel. It is important to look at the incident sequence to determine whether it matches what is described in the hazard and risk analysis.

The successful operation of an IPF in response to a process demand should be a celebrated rather than reviled event. The hazard and risk analysis identifies the potential hazardous event that is prevented by the IPF. While a process shutdown is hardly desirable, a shutdown is highly preferable to equipment damage, loss of containment, or catastrophic incident. The IPF should be appropriately credited for preventing the process demand from propagating. Process demands should be monitored and tracked, so that the frequency of occurrence can be reduced if possible. By reducing the demands, process shutdowns are similarly reduced.

6.10.2 Detected Faults

Detected faults are found through automatic diagnostics or through the operator's normal observation of the process and its equipment. The response to detected faults is defined in the design basis and operating procedures to ensure appropriate action is taken in response to the fault. Operation with a known fault requires compensating measures, notification of appropriate personnel, and initiation of repair activities. On-line repair activities should be completed within the allowable repair time, unless approved by management of change.

When IPS equipment is out of service for repair, the equipment availability is reduced. For example, if the allowable repair period is 72 hours and on-line repair is required once per year, the IPF availability is reduced to a maximum of 99.18% = (8760 – 72)/8760, assuming 100% uptime of the process.

The number of IPS equipment detected faults between proof tests should be monitored, tracked, and reported. IPS equipment should demonstrate robustness in the operating environment. Equipment experiencing frequent faults, whether degraded or critical, should be targeted for performance improvement. Faults should be investigated to determine whether the root causes can be addressed by changes in the equipment specification, installation, operation, etc. If there are no practical means to reduce the number of faults, the design basis should be changed to address the loss of IPF availability.

6.10.3 Dangerous Failures

A dangerous failure is a failure that causes the process to be put in a hazardous state (continuous mode IPF) or puts the IPF in a condition where it may fail-to-operate when required (demand mode IPF). Dangerous failures can be identified when process excursions result in demands on failed IPS equipment that does not operate correctly to stop the event propagation. Dangerous failures identified by process demand are typically investigated as part of the near miss or incident investigation.

Other types of demands are placed on IPS equipment as well, that can identify dangerous failures. These include proof tests that demand the equipment to operate according to its specification, and manual shutdowns that place demands on final elements. Failures occurring during these demands should also be investigated and reported.

Any repeated equipment failure should be investigated to determine whether there is a systematic error in the equipment specification, installation, or operation. Failures early in the equipment life may indicate infant mortality issues. Increased failures late in the equipment useful life may indicate a transition into the wear out period, triggering equipment replacement (see Appendix D). If nothing can be done to reduce the likelihood of equipment failure, the design and/or operating basis should be modified to address the actual equipment performance versus the expected performance.

6.10.4 Spurious Operation

IPS equipment sometimes fails in such a way that the IPF partially or completely takes action on the process without need. The impact of an IPS's unintentional operation should be identified during risk assessment and addressed during design. Frequent spurious equipment operation may increase the likelihood of a hazardous event due to increased process demands on other equipment.

Spurious trips have other significant implications from an operational perspective. Operations personnel cannot be expected to maximize uptime and production, if they are constantly struggling with the IPF in order to keep the process unit operational. Unnecessary shutdowns can cause process problems, such as off-spec products, environmental impacts, large flaring events, waste material generation, and loss of plant production. Business impact can be quite substantial. For example, a trip of an ethylene recirculation compressor can cost $500,000 or more.

Spurious operation also leads to a process start-up. Industry data has shown that process equipment start-up is a high risk process operating mode. There are many reasons why start-up is a high risk mode. Start-up after a spurious shutdown is particularly a problem, since the trip was not caused by a process problem and everyone feels pressure to get the process equipment back to operation.

The poorer the IPS reliability, the less operations personnel will trust the information it generates. This may reduce the operator's response to fault alarms, process alarms, and/or protective alarms, even to the extent of disabling them. Consequently, it is important to monitor and track IPS spurious operation.

Spurious events should be investigated to determine how the frequency can be reduced.

Choices made during the design basis influence the frequency of spurious operation. High reliability IPFs rely on user approved equipment implemented in fault tolerant against spurious failure architectures. When safe failures are identified and spurious operation was avoided by design, the IPF should be appropriately credited for preventing the spurious operation. The cost of the avoided spurious event should be estimated and reported. Successful prevention of spurious trips demonstrates the value of investments in fault tolerant against spurious failure architectures.

6.10.5 Personnel conformance to work practices

Periodically, the performance of personnel, equipment and procedures should be reviewed against expectations. Personnel should be knowledgeable in the chemicals they work with, including the potential hazards that can occur when chemicals are released. Operations and maintenance plans (Section 6.4) are established to support the IPS operational and mechanical integrity. These plans include the review, approval, and notification requirements for various management system activities.

The stage 4 functional assessment examines the trends in the various records and reports to identify performance gaps and to make recommendations as necessary to continuously improve performance. Hazard analysis reports and management of change records demonstrate that the in-place risk reduction strategy is sound, changes are being adequately evaluated prior to implementation, and recommendation action plans are addressed in a timely manner. The mechanical integrity documents demonstrate that equipment is being inspected and maintained in accordance with the design basis. The number of days behind schedule maintenance should be tracked and reported.

Equipment is installed and demonstrated to operate according to the design basis by examining mechanical integrity records and investigation reports. Mechanical integrity records should demonstrate that equipment operates in an as "good as new" condition. Personnel interviews, records and reports should indicate that training and procedures are adequate, considering personnel competency, personnel staffing, and management oversight.

7

CONTINUOUS IMPROVEMENT

The historian John Lewis Gaddis defines strategy as "the process by which ends are related to means, intentions to capabilities, and objectives to resources" (Alden 1996). The goal of the process industry is no incidents, no harm to people, and no damage to the environment. This goal can only be reached by going beyond compliance and establishing safety as a strategic business value. Success requires alignment of personnel, procedures and equipment, leading to more efficient operating performance.

Operating performance is influenced by the culture of the operating facility. According to Jim Collins in *Good to Great: Why Some Companies Make the Leap...and Others Don't* (2001), "Culture is leadership, the discipline of people, thought, and action." An organization's safety culture is represented by shared underlying values and assumptions that drive individual behavior and management attitude and expectations concerning individual behavior. In 1993, Great Britain's Health and Safety Commission (HSE 1997) stated the following:

> "The safety culture of an organisation is the product of individual and group values, attitudes, competencies, and patterns of behavior that determine the commitment to, and the style and proficiency of, an organisation's health and safety programs. Organisations with a positive safety culture are characterised by communications founded on mutual trust, by shared perceptions of the importance of safety, and by confidence in the efficacy of preventive measures."

Achieving operating excellence requires that the outcome of the chosen risk reduction strategy meet or exceed expectations in a cost effective manner. Cost effectiveness is often interpreted by front-line personnel as minimum cost, time, and resources with "minimum" being defined by today's budget. Doing more with less generally does not lead to safe or reliable operation and it does not support continuous improvement.

An organization's culture is ultimately driven by what management indicates is important; what is measured; and what is rewarded. Safe operation can only be sustained where it is recognized that the direct costs of an incident represent the tip of an iceberg, as shown in Figure 7.1. Hidden from view are the indirect costs and long-term business damages resulting from unsafe operations.

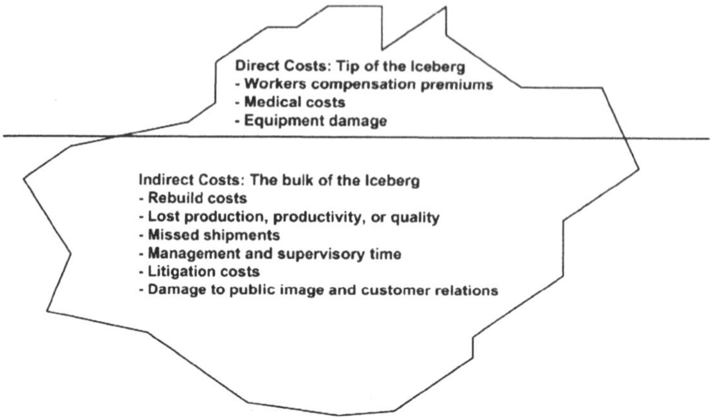

Figure 7.1. Iceberg Illustrating the Direct and Indirect Costs of Injuries

When the true cost of an incident is understood, it becomes very clear that being cost effective is much more than simply today's budget. Success requires that personnel believe that investment in reducing risk further is encouraged and rewarded. Market leaders recognize that this investment provides benefits that far outweigh its costs. Operating excellence seeks to prevent incidents, because it is good for business and it is the right thing to do.

Management must establish clear performance expectations and provide the resources necessary to meet expectations. Resource deployment is not about quantity, but about quality, discipline, and rigor. Organizational discipline expects compliance with work processes as the foundation for a safe and reliable culture. The protective triangle (Figure 7.2) has been used throughout this book to emphasize the interconnection of people, equipment and work practices required to support a strategy where "ends are related to means, intentions to capabilities, and objectives to resources" (Alden 1996).

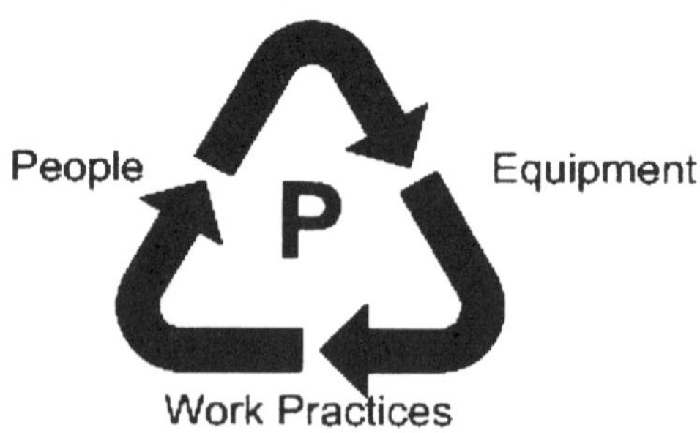

Figure 7.2. Protective Triangle.

Continuous improvement, as shown in Figure 7.3, ensures that everything ties back together and continues the lifecycle. Performance is monitored and gaps between expected and actual performance are closed in a timely manner. Investments are made to reduce risk further when practical. Continuous improvement evolves the design and management practices to ensure relevance to current needs, consistency with good engineering practices, and the installed equipment is designed, maintained, inspected, tested, and operating in a safe manner.

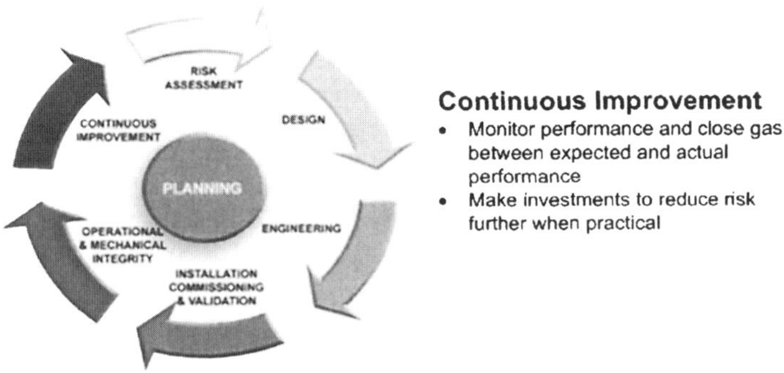

Figure 7.3. Continuous Improvement Phase.

7.1 INTENDED AUDIENCE

Personnel assigned responsibility for the IPS should understand the continuous improvement expectations. No one wants to change for the sake of change. Continuous improvement should leverage areas where instrumentation and control investments are proven to reduce residual risk, mechanical integrity requirements, and spurious events.

Understanding expectations ensures that day-to-day personnel decisions achieve long-term safe and reliable operation. In the absence of understanding, decisions can be made today that could have a negative impact on future operation. Establishing, documenting, and regularly communicating owner/operator goals encourages everyone to work together to achieve the desired performance.

This phase of the protective management system requires the support of all personnel assigned responsibility for a lifecycle activity.

This chapter is intended for:

- Management who ensure that IPS installations are maintained, inspected, tested, and operated in accordance with good engineering practices, provide personnel with resources, tools, and training necessary to support safe and reliable operation, and ensure that the behavior and actions of its personnel are consistent with a culture that encourages continuous improvement in IPSs.
- Process safety who monitor hazard and risk analysis, near miss, incident, failure, and trip reports for recommendations related to IPSs.
- Instrumentation and electrical who monitor manufacturer technical advisory bulletins, failure reports and recommend replacement and upgrades as required.
- Operations who provide feedback on the IPS performance through trip tracking, near miss, and incident investigation.
- Maintenance who track and investigate equipment failures, report findings to other affected personnel, and recommend modification to IPS as necessary to maintain mechanical integrity.
- Manufacturers who track changes related to their products, communicate issues related to their product to designated personnel, provide information concerning product enhancements or obsolescence, and provide assistance in the migration of existing hardware and software to new hardware and software.

7.2 INPUT INFORMATION

Information is collected and documentation is created throughout the life of the IPS. Over time, this information should build a stronger case for the claimed IPS performance. Figure 7.4 provides an overview of the recommended information that should be maintained throughout the IPS life.

For safety systems, this information constitutes the process safety information, operating procedures and mechanical integrity program documentation. The collected information should be under revision and management of change control. The documents should be available to support any hazard and risk analysis, assessment, verification, audit, or investigation.

An organization's culture drives its management decisions and organizational work processes. The workplace culture establishes conditions where deviations cause safe and reliable operation to become vulnerable to errors and violations.

Latent failures occur due to organizational decisions that promote errors and violations, such as high work load, unreasonable schedules, poor training, and inadequate tools. Routine violations can become part of the culture when an owner/operator rarely sanctions deviations or rewards compliance (Reason 1997).

Continuous improvement requires the following:

- Management leadership and goals,
- Internal practices and work processes,
- Personnel responsibility and accountability,
- Employee involvement,
- Worksite analysis,
- Clear and documented risk reduction strategy,
- User approved equipment,
- Up-to-date IPS documentation, procedures, and training,
- Performance monitoring and failure investigation,
- Plans for improvement and gap closure, and
- Knowledge sharing environment.

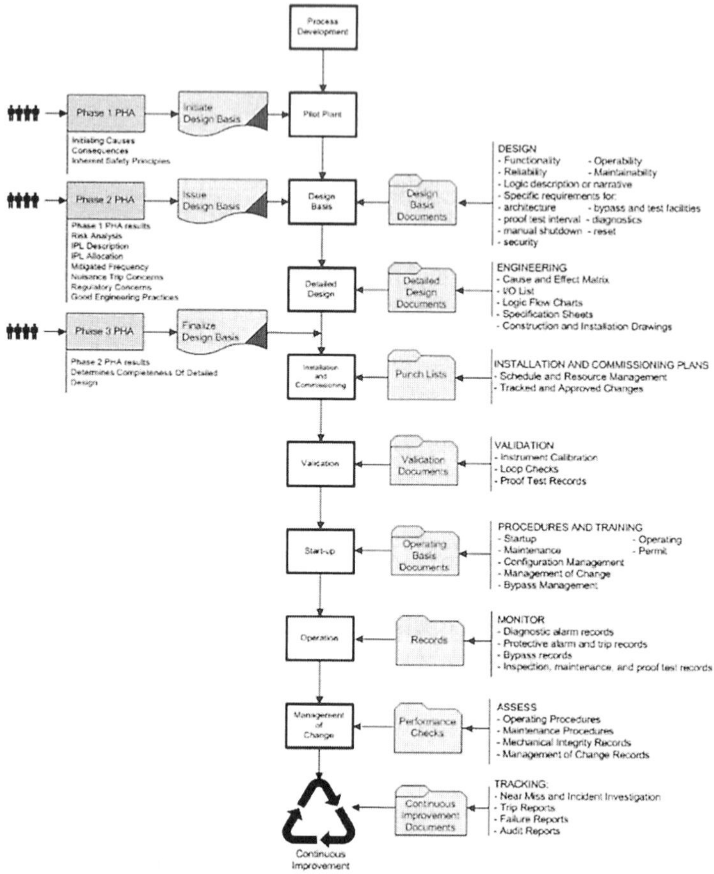

Figure 7.4. Lifecycle Illustrating Information Collected at Each Phase.

7.3 BASIC WORK PROCESS

Balancing safety and production goals can be a tenuous, delicate and complex act. It is undeniable that safety and production are compatible. It is indisputable that investments in safety yield long-term benefits. However, these benefits are not as obvious nor do they produce the rapid results associated with production investments, which generally have a high certainty of providing a measurable, positive effect within a short time frame.

For protection and safety, many of the benefits are less tangible. When successful, the IPS is blamed for a process outage; when it fails, it is blamed for the incident. The hazard and risk analysis describes the hazardous event prevented by the operation of each instrumented protective function (IPF). When an IPF operates as required, the IPF should be given credit for the event avoided by its successful operation, including potential fatalities, injuries, environmental releases, equipment damage, and financial losses. Also, the IPF should be credited when fault tolerant design allows equipment failure to be detected without causing disruption in the process operation.

Figure 7.5 provides a summary of the decision making process, illustrating how available resources must be allocated across safety and production goals. Decision makers often have defensive filters which affect the receipt and interpretation of information (Reason 1990). Today's business climate puts pressure on personnel in a variety of forms, such as production forecasts, budget cuts, resource reductions, or colleague retirement.

In the absence of a strong safety culture, production and budget pressure can result in a culture of denial where the decision maker's defensive filter refuses to acknowledge any evidence that does not support production or budget plans. Risk assessment can become skewed with credible safety recommendations and concerns being dismissed without appropriate consideration. Erroneous assumptions concerning equipment and procedure robustness lead to complacency and an acceptance of increased risk. Often, this is done in the absence of dependable documentation, information, and data, or a rigorous mechanical integrity program.

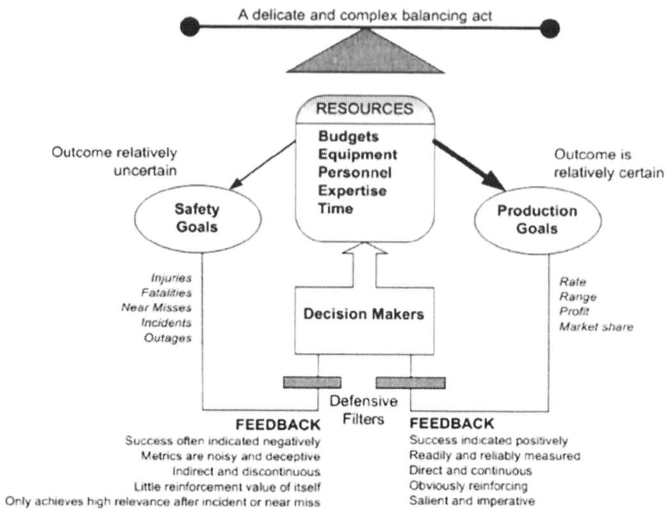

Figure 7.5. Overview Illustrating the Complexity of the Decision Making Process. (adapted from Reason 1990)

Good engineering practices should be applied in preventing process safety incidents. Internal practices should be benchmarked against those of market sector peers or other process industry companies. Periodic gap analysis should be conducted to determine if existing equipment is designed, maintained, inspected, tested, and operated in a safe manner. Based on observed performance and benchmarking information, action plans for improvement should be developed and implemented.

7.3.1 Understanding History

A series of chemical incidents occurred during the 1970s and 1980s. These incidents are so legendary that they are often referenced by city only: Flixborough (1974), Seveso (1976), Mexico City (1984), and Bhopal (1984). These catastrophic incidents awakened the world to chemical industry risk and are summarized in *Lees' Loss Prevention in the Process Industry (Mannan 2005).*

The first process safety regulations were issued in Europe in the 1970s in direct response to the impact on the communities of Flixborough and Seveso. Nearly a decade later, the tragedy caused by the Mexico City explosion and the Bhopal chemical release resulted in process safety regulations being issued in the United States and many other countries. Industrial societies responded worldwide by publishing numerous codes, standards, and practices on a variety of process safety topics.

Despite these incidents occurring more than three decades ago, similar errors and root causes still exist today. Trevor Kletz discussed this problem in his book,

Lessons from Disaster: How organizations have no memory and accidents recur (Kletz 1993). He presents numerous cases where an incident occurs and is repeated just a few years later. Kletz finds that organizations have poor memory due to many factors, such as insufficient failure investigation, inadequate communication and distribution of investigation findings, lack of information retention and little training concerning previous events.

A safety culture does not rely on balance sheet improvements to justify IPSs. It understands that the potential for incidents is an inherent part of the process design and that, without focused effort, incidents invariably occur.

7.3.2 Benchmarking Current Status

The owner/operator should understand how internal practices compare with recognized and generally accepted good engineering practices. This so-called benchmarking establishes the owner/operator's position with regard to industrial and market peers. This is often the most painful part of the continuous improvement process, as it tends to shed light on the shortfalls and inadequacies of the protective management system as a whole.

Some owner/operators will likely find that their design and management philosophy is out of alignment with aspects of various good engineering practices. When unacceptable risk is identified, an action plan should be established with short-term and long-term measures sufficient to reduce the risk below the owner/operator risk criteria. Investments to improve safety and reliability of the process operation yield long-term economic returns.

7.3.3 Defining Gaps

An increasing number of owner/operators are finding themselves operating within a regulatory framework which often does not provide prescriptive requirements. Instead, the requirements are a moving target based on the somewhat fuzzy concept of "good engineering practice." Keeping up and complying with the basic requirements concerning process safety and IPSs can seem taxing enough. How does an owner/operator move forward with continuous improvement when it seems that even the immediate goals are a moving target? To some, it may seem challenging enough just to maintain the status quo; let alone to embrace more changes.

Continuous improvement operates on the principle that finding failures and errors is the beginning of a learning process. Minimizing their reoccurrence requires an understanding of how these failures or errors developed. Continuous improvement, as well as the application of good engineering practice, should be viewed as an on-going process rather than an endpoint. Work processes, information, resources, and skills should be analyzed periodically to identify weaknesses that limit performance and to recommend improvements, as necessary. Information systems, whether computerized or manual, should provide personnel with up-to-date information in a format that is easy to understand. Information should be revision controlled, yet accessible.

Many factors affect IPF performance expectations and the designs used to attain them, such as the economy, market trends, and technology, along with legal and political issues. A strong safety culture expects ownership and accountability for safe and reliable performance of the process equipment over its life. Management should support periodic evaluation of the existing equipment to determine that it is designed, maintained, inspected, tested, and operating in a safe manner.

Changes in operability, functionality, reliability, or maintainability expectations may require the implementation of different design/or management practices. Proof test, failure investigation, alarm, trip, audit reports, etc. provide valuable insight into personnel and management system performance. Operating excellence requires that the root causes of unacceptable process reliability and equipment performance be identified and resolved. Improving equipment mechanical integrity requires a culture that values maintenance.

A need for improvement may be identified through various management system activities. As shown in Figure 7.4, various reports and records are generated throughout the lifecycle. Chapter 6 discussed the investigation of individual events and responses to reduce their re-occurrence in specific applications or services. Continuous improvement processes should include periodic examination of overall available information to identify, trend, and correct systematic problems.

A gap analysis (stage 5 functional assessment) is performed to compare the observed IPS performance to the expected performance. The gap analysis should determine that:

- Equipment is operating according to design intent,
- Safety, operating, maintenance, and emergency procedures are appropriate for competency and risk reduction expectations,
- Hazard and risk analysis or management of change recommendations are addressed in a timely manner, and
- Training of relevant personnel is adequate for current expectations.

Significant issues may be identified during the analysis. Management system failures are often reflected in multiple performance metrics. Systematic problems may be identified, such as poor adherence to policies, procedures, and practices or insufficient inspection and preventive maintenance. If IPS equipment is not maintained, it is likely that other equipment is suffering from the same inadequate maintenance. The cumulative maintenance deviations, whether intentional or unintentional, may cause a breakdown of multiple protection layers.

Team effort is often needed to evaluate IPS requirements and performance. Some organizations establish a formal structure where identified personnel participate as site representatives on a core team. The core team evaluates changes in the good engineering practices and makes recommendations for modifying internal practices.

Whenever work processes are modified, a shift in emphasis often leads to changes in the way team members perceive the process, its associated risks, and various protection layers and IPSs. This shift may result in recommendations for additional risk reduction or IPS. These recommendations and other continuous improvement efforts complete the lifecycle moving the process toward safer and more reliable operation.

7.4 OUTPUT DOCUMENTATION

An effective protective management system yields regulatory and good engineering practice compliance. It protects the owner/operator's reputation and shareholder value. It engages personnel, encourages their participation, and assures that practices and procedures support safety and reliability.

The continuous improvement phase is the most difficult phase of the protective management system for many technical and management personnel to grasp. Its goal is to reduce risk further, but there are practicality limits. Continuous improvement is often initiated by lessons learned and recommendations and result in the continuation of the lifecycle.

Gaps in safe and reliable operation are generally symptomatic of gaps in areas that move the process unit away from efficient operation. Understanding and reducing risk can provide substantial benefits, such as fewer injuries and illnesses, reduced loss of containment events, fewer transportation incidents, and reduced liability. This phase results in:

- Recommended changes to equipment, practices, or training,
- Incorporation of lessons learned as appropriate into personnel training, equipment selection, installation practices, and mechanical integrity practices,
- Improved process safety and reliability performance,
- Fewer incidents and near misses,
- Regulatory and good engineering practice compliance,
- Better owner/operator reputation and shareholder value, and
- Better personnel participation.

7.5 DETERMINING PATH FORWARD

The key aspect of continuous improvement is charting the course to achieve it. Over time, various options will be presented to upgrade hardware, software, or human interface systems. Proposed changes should be reviewed using a management of change process to identify how the change affects other functions or systems. Areas for improvement should be addressed with an action plan, which typically prioritizes recommendations based on consequence severity and risk gap.

Action plans should define objectives, milestones, and timelines. Action plans should be periodically assessed to determine whether there is a need to accelerate the schedule or broaden its objectives. For example, a planned IPS upgrade may be accelerated when the manufacturer withdraws support for critical equipment. To be successful, action plans should be communicated to affected personnel so they understand it and commit to it.

Implementing upgrades aimed at improving long-term operational effectiveness takes time to complete, depending on the complexity and degree of change involved. As the IPS is changed, operating plans and targets should consider any additional risk borne by the process during the transition. Once the design basis changes are underway, the operating and mechanical integrity basis should be reviewed and needed revisions implemented.

There are many barriers to improvement, including:

- Poor data integrity and quality,

- Poor information availability and consistency,
- Lack of broad understanding of facts and procedures,
- Poor or missing internal practices and procedures,
- Poorly understood compliance expectations,
- Inadequate revision control or notification of changes, and
- Lack of comprehensive training on data, information, procedures, and practices.

To overcome these barriers, personnel must be provided with more than just another initiative or mandate for change. Continuous improvement must be part of an organization's culture, beginning at the highest management level and continuing to the front-line operator. Personnel should feel that safe and reliable operation is an institutional value and that they won't lose their jobs or be held back for speaking out. Front-line personnel must believe that continuous improvement is supported by all levels of management. Everyone should understand that employment is conditional on safe work performance.

To succeed, personnel must be aware of the potential risk and be committed to do what is necessary to maintain and continuously improve operational and mechanical integrity. The path forward encompasses many detailed tasks, but generally includes the following:

- Assign responsibility and hold personnel accountable,
- Audit to ensure practices and procedures are followed,
- Question norms and reduce risk further when practical,
- Integrate business and process safety goals,
- Track performance, address bad actors, and celebrate success, and
- Learn and remember.

A

DEFINITIONS

Abnormal Operation--Operation outside the normal operating limits necessitating corrective action by either a protective system or by trained personnel to achieve or maintain a safe state of the process.

Access Security--Core attribute of a protection layer, involving the use of administrative controls and physical means to reduce the probability for unintentional or unauthorized changes.

Administrative Controls--Procedural mechanisms for controlling, monitoring, and auditing human performance, such as lock out/tag out procedures, bypass approval processes, permit systems.

Alarm Management--The set of processes and practices for determining, documenting, designing, monitoring, and maintaining alarm messages.

ALARP (As Low As Reasonably Practicable)-- The concept that efforts to reduce risk should be continued until the incremental sacrifice (in terms of cost, time, effort, or other expenditure of resources) is grossly disproportionate to the incremental risk reduction achieved.

Allowable Time To Repair--Length of time that has been determined by hazard and risk analysis to be acceptable for continued process operation with degraded or disabled equipment. Time is often constrained by operations ability to maintain the necessary compensating measure.

Allocation--The process of distributing the risk reduction requirements to protective systems.

Alpha-Test--First functional test of new equipment, usually limited to laboratory tests that verify functionality.

Application Program--A generic term for a computer program written for a particular application, possibly unique to an installation.

Application Software--See "Software Types."

Approved--The equipment, activity, or practice is deemed suitable for use in the application and operating environment by the owner/operator. This determination generally involves an evaluation of analysis and testing information and demonstration of performance through operating history.

Approved Equipment List--An equipment list developed by a user approval process, which determines that the equipment is suitable for use in the operating environment and in control, protective, or safety applications.

Approved Substitution--Replacement of equipment with an item that meets one of the following requirements:

- Is specifically permitted as a substitute or duplicate item by internal practice (i.e., the approved manufacturers list has more than one brand and/or model number which can be used interchangeably for a replacement item).
- Is approved as an equivalent substitute by the appropriate plant or owner/operator personnel, meets user approval, and is covered by existing training and procedures.

Architecture--The physical organization, interconnection, or integration of the equipment of a system that operates according to the design basis.

As good as new--Equipment is maintained in a manner that sustains its useful life. "As good as new" often refers to the initial condition after proof test, where the probability of failure at time 0 is zero and the failure rate expected during the useful life is unchanged.

Assessment--A review of evidence to determine and document that the design, maintenance, inspection, testing, and operating strategy is consistent with current practices and expectations.

Asset Protection Level (APL)--Integrity level assigned to an asset protection system. The asset protection level may be alternatively defined by a required risk reduction or availability target.

Asset Protective Systems--Instrumented protective system that reduces the risk associated with property damage and business interruption. The management system may define "IPS classes," e.g., Class II IPSs are asset protective systems.

Attribute--A characteristic that can be assessed to determine the degree to which an objective is achieved.

Audit--An inspection of evidence to determine and document that management system work processes and procedures are in place, up-to-date, and being followed.

Auditability--Core attribute of a protection layer, relating to the ability to inspect information, documents and procedures, which demonstrate the adequacy of and adherence to the design, inspection, maintenance, testing, and operation practices used to achieve the other core attributes. Management system rigor is often established through the audit trail and approval requirements.

Audit Trail--Systematic documentation of activities that allows an auditor to confirm compliance with required or desired behavior.

Availability (Instantaneous)--The probability that equipment is capable of operating under given conditions at a given instant of time, assuming the required external resources or support systems are provided.

Availability (Mean)--The fraction of time that equipment is capable of operating in the desired manner. It is the mean of the instantaneous availability over a given time interval.

Barrier--Protection layer consisting of physical structures, such as dikes, berms, bunds, and explosion barriers, which minimize the consequence severity due to physical design.

Basic Process Control System (BPCS)--System that is separate and independent of the safety instrumented system and whose failure to operate does not affect the core attributes of any safety instrumented function. The BPCS responds to input signals from the process and its associated equipment, other programmable systems, and/or from an operator and generates output signals causing the process and its associated equipment to operate in the desired manner.

Bathtub Curve--Typically a plot of equipment failures as a function of time and is used to characterize the equipment lifecycle.

Benchmark--A point of reference from which measurements may be made or from which other things can be measured. A program that is used to compare the operation of two or more systems is known as a benchmark program.

Benign Failure--Failure which does not affect the ability of equipment to perform its design function. Benign failures are often further classified as degraded or incipient failure.

Beta Test--Equipment test involving installation in the operating environment so that its performance can be monitored and compared against expectations.

Burner Management System--An instrumented protective system dedicated to combustion safety that monitors fuel conditions, verifies the presence of pilot and main flame, and ensures proper light-off (e.g., purge cycle and ignition). It does not control the air-to-fuel ratio, firing rate, boiler feed water, or other related functions. This system has the objective of preventing erroneous operation and damage to fuel preparation and fired equipment.

Burn-In--The equipment is subjected to higher than typical stress to identify early failures. Often, referred to as infant mortality.

Bypass--An action taken to override, defeat, disable, or inhibit a protective system. These actions prevent operation of the protective system.

Capability--Ability of equipment to operate as specified.

Car Seal--A metal or plastic cable used to fix a valve in the open position (car seal open) or closed position (car seal closed). Proper authorization, controlled via administrative procedures, must be obtained before operating the valve. The physical seal should have suitable mechanical strength to prevent unauthorized valve operation.

Catastrophic--A loss with major consequences and unacceptable lasting effects, usually involving significant harm to humans, substantial damage to the environment, and/or loss of community trust with possible loss of franchise to operate.

Catastrophic Incident--An incident involving significant harm to humans, substantial damage to the environment, and/or loss of community trust with possible loss of franchise to operate.

Catastrophic Release--Major uncontrolled emission, fire, or explosion, which presents serious danger to personnel, imminent and substantial endangerment to public health or the environment.

Cause--An event, situation, condition that results, or could result, directly or indirectly in an incident.

Checklist--A written procedure in which each item is marked off (or acknowledged on a computer screen) as status is verified.

Chemical--Any element, chemical compound, or mixture of elements and/or compounds.

Claim Limit--The maximum integrity level in which equipment can be used without additional fault tolerance against dangerous failure. The limit occurs due to random and systematic failures.

Classification--Process of categorizing equipment based on event consequence.

Common Cause Failure--Failure of more than one device, function, or system due to the same cause.

Common Mode Failure--Failure of more than one device, function, or system in the same manner, causing the same erroneous result.

Communication--Data exchange between various systems and devices. This includes communication between the instrumented protective system (IPS) and basic process control system (BPCS), the IPS and various interfaces, such as the operator, IPS, and maintenance/engineering interfaces.

- **External Communication**--Data exchange between an IPS and a variety of other systems or devices. These include shared operator interfaces, maintenance /engineering interfaces, data acquisition systems, host computers, etc.

- **Internal Communication**--Data exchange between the various devices within a given IPS. These include bus backplane connections, the local or remote I/O bus, etc.

Competence--Personnel assigned responsibility for a work process activity should possess the skills, training, and ability necessary to execute their assigned tasks in conformance with approved practices and procedures in a manner acceptable to the owner/operator.

Compliance--The determination that information, documents, activities, tasks, products, etc. meet the relevant specifications, standards, or good engineering practices.

Component--One of the parts of a system, subsystem, or device performing a specific function.

Compensating Measures--Planned and documented means for managing risk during periods of process operation with known failures or problems, resulting in increased risk.

Configuration--The functional and/or physical characteristics of the hardware and/or software required for the equipment to operate according to the design basis.

Configuration Management--The systematic application of management policies, procedures, and practices to assess and control changes to the hardware and/or software of a system and to maintain traceability of the configuration to the design basis throughout the system life. Configuration management is a specialized form of management of change.

Consequence--The undesirable result of an incident, usually measured in health and safety effects, environmental impacts, loss of property and business interruption costs.

Consequence Analysis--The analysis of the potential impact of an incident independent of its likelihood.

Consequence Severity--Measure of the expected impact of an incident.

Continuous Mode--A dangerous IPF failure causes a hazardous event without further failure.

Control--A mechanism used to regulate or guide the operation of a machine, apparatus, process, or system.

Control Function--Function, which causes the process to operate within the normal operating limits.

Controlled Document--Documents covered under a revision control process to ensure that up-to-date documents are available and out-of-date documents are removed from circulation.

Control Layer--Protection layer, which implements standard operating procedures, basic process control systems, and process alarm systems to maintain the process within the normal operating limits.

Control System--System, which implements one or more control functions and includes sensors, controllers, final control elements, operator interfaces, and necessary support systems.

Core Attribute--Fundamental property of a protection layer. The core attributes are independence, functionality, integrity, reliability, auditability, management of change, and access security.

Corrective Maintenance--Maintenance performed to repair a detected fault.

Cost-Benefit Analysis--A decision aid that involves quantifying the costs and benefit of each alternative in dollar terms based on the preferences of individuals affected by the decision, and choosing the one that maximizes the net benefit (i.e., total benefit minus total cost utilizing the time value of those costs and benefits).

Coverage--See "Diagnostic Coverage."

Covered Process--Any process subjected to governmental and/or owner/operator requirements.

Critical Failure—Failure resulting in the loss of the equipment's ability to operate as specified. These failures may be further classified as safe or dangerous.

Cross Reference List--A list showing all the locations in an application program at which a selected list of data objects is operated upon.

Dangerous Failure--Failure affecting equipment within a system, which causes the process to be put in a hazardous state or puts the system in a condition where it may fail-to-operate when required.

Data--A representation of facts, concepts, or instructions in a formalized manner suitable for communication, interpretation, or processing by human or by automatic means.

De-Commissioning--The permanent removal of one or more protective functions from service.

De-Energize To Trip--Circuits where the final elements are energized under normal operation and the removal of power source (e.g., electricity, instrument air) causes the instrumented protective system to take its defined action.

Degraded Failure--Equipment failure that is gradual or partial, resulting in performance that is less than "as good as new." The equipment is still able to operate as required, but the degraded condition may develop into a critical failure if repair action is not taken.

Demand--See "Process Demand."

Demand Mode--Dormant or standby operation where the IPF takes action only when a process demand occurs.

Demand Rate--The number of demands divided by the total elapsed operating time during which the demands occur.

Dependent Failure--Failure of multiple devices due to a common cause.

Design Basis--Collection of documents and information that clearly defines how the protective system is designed to achieve the risk reduction strategy and the core attributes of the protection layer.

Design Life--The expected equipment life due to its design and durability, as well as its capability to continue meeting the design specification. The design life may end with obsolescence.

Detected Failure--Failure found through diagnostics or through the operator's normal observation of the process and its equipment. Synonyms include announced, revealed and overt.

Detected Fault-- See "Detected Failure."

Diagnostic Coverage--Fractional decrease in the probability of dangerous failure resulting from automated diagnostics that report faults to the operator and take a specified action on fault detection.

Diagnostic Interval--Time period between the operation of diagnostics used to detect equipment failures.

Diagnostics--Hardware and software installed to automatically identify and report specific failure modes.

Distributed Control System--A system which divides process control functions into specific areas interconnected by communications (normally data highways) to form a single entity.

Diverse--Use of independent and different means to perform the same function. Diversity may include the use of different physical methods, technology, manufacturers, installation, maintenance personnel and/or environment.

Early Failure--Failure identified early in the installed equipment life. These failures are typically due to manufacturing defects, assembly errors, installation or implementation errors. Many of these failures are discovered and corrected during commissioning. The remainder are typically discovered during and shortly after startup. Failures in this time period are generally dominated by systematic rather than random failures.

Emergency Shutdown System (ESD)--Term has been used historically to describe a variety of different types of instrumented systems used for protection within the process industry: 1) a safety instrumented system, 2) an emergency isolation and/or de-inventory system, 3) an instrumented protective system, and 4) the equipment used by the operator to respond to safety alarms. In this book, the term references those systems implemented in the limitation layer, such as systems for emergency isolation and/or de-inventory.

Enabling Condition--Operating conditions necessary for an initiating cause to propagate into a hazardous event. Enabling conditions do not independently cause the incident, but must be present or active for it to proceed.

Energize To Trip--Circuits where the final elements require the power to take or maintain the safe state.

Environmental Integrity Level (EIL)--Integrity level assigned to an environmental instrumented system. The environmental integrity level may be alternatively defined by a required risk reduction or availability target.

Environmental Protective System--Instrumented protective system that reduces the risk associated with substantial environmental damage.

Equipment Boundary--Demarcation of the equipment, defining devices, components and interfaces needed for the equipment to function according to its specification.

Equipment Manual--Compilation of installation, configuration, and mechanical integrity requirements for equipment approved for Instrumented Protective System (IPS) applications.

Error--1) Discrepancy between a computed, observed, or measured value (or condition) and the true, specified, or theoretically correct value (or condition) or 2) Failure of planned actions to achieve their desired goal.

Event--An occurrence involving the process caused by equipment performance, human action, or by an occurrence external to the process.

Event Tree Analysis--Method used for modeling the propagation of an initiating event through the sequence of possible incident outcomes. The event is represented graphically by a tree with branches from the initiating cause through the success or failure of independent protection layers.

Expected Life--The length of time a device is expected to function according to its specification.

Factory Acceptance Test--Conducted to determine and document equipment hardware and software operates according to its specification, covering functional, fault management, communication, support systems, and interface requirements.

Failure--Termination of the ability of equipment to operate as specified.

Failure Cause--The root cause of equipment failure.

Failure Frequency (Instantaneous)--The expected number of failures between time t and time $t+\Delta t$. The number counted may include first, second, or more failures of the same equipment. The failure frequency can be integrated between 0 and t to determine the expected number of failures within time t.

Failure Mode--A symptom, condition, or effect by which a failure is observed. Failure modes are often divided into three categories: critical (spurious and dangerous), degraded, and incipient.

Failure Mode And Effects Analysis (FMEA)--A qualitative analysis method that identifies equipment failure modes and determines their impact on the equipment operation.

Failure Mode, Effects and Criticality Analysis (FMECA)--An extension of FMEA that includes a risk estimate (criticality) based on the failure probability and consequence severity.

Failure Rate--Limit when Δt goes to 0 of the expected rate at which equipment failures occur in the time interval t to $t+\Delta t$ given that no failures have occurred until time t.

Failure-To-Operate--Failure that inhibits, or sufficiently delays, the actions required to achieve or maintain a safe state of the process when a process demand occurs. A failure to operate on demand has a direct and detrimental effect on safety.

Fault--Abnormal condition resulting in degraded operation or critical failure.

Fault Tolerant--Voting architecture that allows an equipment subsystem to continue to operate in the presence of one or more hardware or software faults.

Fault Tree Analysis (FTA)--Method used to analyze graphically the failure logic of a given event, to identify various failure scenarios (so called cut-sets), and to support the probabilistic estimation of the event.

Field Experience--The collection of data and information about the performance of equipment based on actual use in a similar application.

Field Sensors--See "Sensor."

Final Element--Device that takes action on the process or process equipment. For an instrumented protective function (IPF), the final element takes action on the process to achieve or maintain the safe state. The final element boundary includes the signal connection to the logic solver and the devices required to take action on the process.

First Out--Refers to an annunciator feature that indicates the process condition that initiated the instrumented protective function response.

Fitness for Service--Management system used to assess the current condition of equipment to determine whether it is capable of continuing operation within equipment specification.

Flame Arrestor--A device fitted to the opening of an enclosure or to the connected piping of a system of enclosures and whose intended function is to allow flow but prevent the transmission of flame from either a deflagration or detonation.

Fractional Dead Time--The proportion of the total relevant time that equipment is in the failed state.

Frequency--Number of occurrences of an event per unit time (e.g., 1 event in 1000 years = 1×10^{-3} events/yr).

Frequency Modifiers--Factors, such as probability of personnel being exposed to a hazard, probability of ignition, and probability of injury or fatality, used to estimate the frequency of harm caused by an incident.

Functional Assessment--An independent review of information, such as documentation, procedures, equipment, reports, and records, to determine whether it reflects current practices and achieves the expected performance.

Functional Independence--Equipment associated with other functions can be removed and the function under evaluation can still receive necessary inputs, make decisions based on those inputs, and take action as necessary. Shared equipment is not functionally independent and its failure rate must support the overall claimed hazard rate or risk reduction.

Functionality--Core attribute of a protection layer, defining the required operation of the protection layer within the overall risk reduction strategy.

Functional Requirements--Detailed description of the relationship between the inputs and outputs of a system.

Functional Safety--Part of the overall safety, relating to the process and the BPCS, which depends on the correct functioning of the safety instrumented systems (SIS) and other protection layers.

Functional Safety Assessment--Review of evidence to judge the functional safety achieved by one or more protection layers.

Function Test--See "Proof Test."

Grandfather--Review of documentation and information associated with an existing protective system (i.e., designed and constructed in accordance with codes, standards, or practices prior to the issuance of a new good engineering practice) to determine whether the equipment is designed, maintained, inspected, tested and operated in a safe manner. If the result of the review is satisfactory, the owner/operator may chose to maintain the existing system as is.

Grounding--The process of connecting one or more conductive objects to ground so that each is at the same potential as the earth. By convention, the earth has zero potential. In practice, grounding is the process of providing a sufficiently small resistance to ground so that a static hazard cannot be created at the maximum credible charging current to a system.

Hardware--Physical equipment directly involved in performing functions, as opposed to software, procedures, rules, and associated documentation.

Hardware Configuration--See "Architecture" or "Configuration."

Hardware Fault Tolerance--See "Fault Tolerance."

Hardwired--That portion of the logic which is executed by electrical circuits comprised exclusively of hardware.

Harm--The physical injury suffered by the people exposed directly or indirectly to the hazardous event.

Hazard--Chemical or physical condition that has the potential for causing harm, environmental damage, or economic loss.

Hazard Analysis--Used to determine how causes (or failures) lead to hazardous events of undesirable consequences.

Hazard and Operability Study (HAZOP)--A systematic, qualitative technique used to identify hazardous events and operational problems using a series of guidewords (e.g., high, low, reverse) to examine deviations from normal process conditions.

Hazard and Risk Analysis (H&RA)--The identification of hazardous events, the analysis of the mechanisms by which these undesired events occur, and the estimation of event frequency and consequence severity.

Hazard and Risk Analysis Team (H&RA Team)--Personnel assigned responsibility for executing a hazard and risk analysis in accordance with an appropriate procedure.

Hazard Classification--Identified hazards are categorized based on the event consequence.

Hazard Evaluation--Analysis of the significance of hazardous events associated with a process or activity, using techniques to identify weaknesses in design and operation.

Hazard Identification--The identification of causes that lead to hazardous events and an estimation of the event consequence.

Hazardous Event--An event that leads to an undesirable consequence.

Hazardous State--A state of the process where a hazard can occur.

Hazard Rate--See "Failure Rate."

Hazard Scenario--An unplanned event or sequence of events, which results in undesirable consequences.

High Integrity Protective Systems (HIPS)--Special application of safety instrumented systems, whose purpose is to prevent over-pressure of process equipment, such as pressure vessels and pipelines. HIPS are installed when it is not practical to mitigate over-pressure using a pressure relief device.

Historical Data--Data recorded from actual past experience.

Human Error--Mistakes by people, such as designers, engineers, operators, maintenance personnel, or managers, which may contribute to or result in hazardous events and incidents.

Human Factors--Designing machines, operations, and work environments to match human capabilities, limitations, and needs.

Human/Machine Interface (HMI)--See "Operator Interface."

Human Reliability Analysis (HRA)--A method used to evaluate whether required human actions, tasks, or jobs will be completed successfully within a required time period. Also, used to determine the probability that no extraneous human actions detrimental to the system will be performed.

Impact--The ultimate result of a hazardous event, expressed, depending on the particular application, in terms of injuries, fatalities, environmental or property damage, business interruption, etc.

Impact Analysis--Activity of determining the effect that a change to a function, or device, will have to other functions, or devices, in that system, as well as other systems.

Incident--An event, or series of events, may result in one or more undesirable consequences, such as harm to people, damage to the environment, or asset/business losses.

Incident Investigation--The management process by which underlying causes or root causes of undesirable events are uncovered and steps are taken to prevent similar occurrences.

Incipient Failure--The equipment operates within specification but its current condition could result in a degraded or critical failure if corrective action is not taken.

Independence--Core attribute of a protection layer, in which the performance of a protection layer is not affected by the initiating cause of the hazardous event or by the failure of other protection layers.

Independent Protection Layer (IPL)--A device, system, or human action, which meets the core attributes to the necessary level of rigor and is capable of preventing an initiating cause from propagating to a hazardous event.

Independent Review--Performed by qualified personnel who are not directly associated with the work activity being verified, assessed, or audited.

Inherent Risk--The risk associated with the process design.

Inherently Safer Design--A protection layer where the process design is executed to reduce or eliminate hazardous events through effective use of process technology, design methods, and/or operational techniques, and this reduction or elimination is permanent and inseparable.

Inhibit--See "Bypass."

Initiating Cause--A failure, error, situation, or condition that results, or may result, in the propagation of a hazardous event.

Initiating Event--The minimum combination of failures (or errors) necessary to start the propagation of a hazardous event. It can be comprised of a single initiating cause, multiple causes, or initiating cause(s) in the presence of enabling conditions.

Input Modules--A part of an Electrical/Electronic /Programmable Electronic System (E/E/PES) that acts as the interface to external devices and converts input signals into signals that the E/E/PES can utilize.

Input/Output Modules--See "Input Modules," "Communication: Internal Communication," and "Output Modules.

Installation Requirements--Mechanical and stress considerations when installing field devices, e.g., piping stress exerted on directly coupled devices or the use of thick-walled piping to provide protection against process connection damage.

Instrumentation and Electrical Requirements Specification-- Documents how functions are physically allocated to systems and how the functions are designed, maintained, inspected, tested, and operated to achieve the process requirements specification.

Instrumented Protective Function (IPF)--A protective function allocated to an instrumented protective system that provides the risk reduction necessary to reduce the risk of an identified hazardous event below the owner/operator risk criteria.

Instrumented Protective System (IPS)--Composed of a separate and independent combination of sensors, logic solvers, final elements, and support systems that are designed and managed to achieve a specified risk reduction. An IPS may implement one or more instrumented protective functions (IPFs).

Instrumented Safety Function (ISF)--A safety function allocated to an instrumented safety system that provides the risk reduction necessary to reduce the risk of an identified hazardous event below the owner/operator risk criteria. ISF may be further classified based on the event consequence and risk reduction.

Instrumented Safety System (ISS)--Composed of a separate and independent combination of sensors, logic solvers, final elements, and support systems that are designed and managed to achieve a specified risk reduction for an identified hazardous event. An ISS may implement one or more instrumented safety functions.

Integrity--Core attribute of a protection layer related to the risk reduction reasonably achievable given its design and management. Integrity is limited by the rigor of the management system used to identify and correct equipment failures and systematic errors.

Integrity Level (IL)--Represents one of four discrete ranges used to benchmark the integrity of each PIF and the PIS, where IL 4 is the highest and IL 1 is the lowest.

Interlock--A system that initiates or prevents a predefined action in response to a specified condition. An interlock may be classified based on consequence severity derived from the specified condition.

Investigation--The management process by which underlying causes of events are uncovered and steps are taken to prevent similar occurrences.

Job Safety Analysis--A procedure that systematically identifies: (1) job steps, (2) specific hazards associated with each job step, and (3) safe job procedures associated with each step to minimize incident potential.

Layer of Protection Analysis (LOPA)--Method for evaluating the effectiveness of protection layers in reducing the frequency and/or consequence severity of hazardous events.

Lessons Learned--The concept that knowledge gained from historical experience should be made available to relevant parties and incorporated into internal practices.

Life Safety Systems--Term used in the semi-conductor industry to address systems installed for toxic gas protection, as well as the detection and response to pyrophoric material releases.

Likelihood--A measure of the expected probability or frequency of occurrence of an event.

Limitation--Protection layer, which implements fire & gas systems, emergency dump systems, fire protection systems, water/steam curtains, deluge systems, and emergency shutdown systems to reduce the consequence severity of the hazardous event.

Logic Solver--That portion of an instrumented system performing one or more logic functions.

Low Demand--See "Modes of Operation: Demand Mode."

Maintainability--Ease by which equipment can be modified to correct faults, to improve performance, to maintain the core attributes, or to adapt to a changed operating environment.

Maintenance--Activities conducted to ensure equipment performance is kept within a set of prescribed limits.

Maintenance/Engineering Interface--The hardware and software provided to allow proper maintenance or modification. It can include instructions and diagnostics which may be found in software, programming terminals with appropriate communication protocols, diagnostic tools, indicators, bypass devices, test devices, and calibration devices.

Management of Change--Core attribute of a protection layer, where a formal process is used to review, document and approve modifications to equipment, procedures, raw materials, process conditions, etc., other than "replacement in kind," prior to implementation.

Management System--A program or activity involving the application of management principles and analytical techniques to ensure the core attributes of each protection layer are met.

Man/Machine Interface (MMI)--See "Human/Machine Interface."

Mean Time Between Failure (MTBF)--The average time between critical failures of a repairable device.

Mean Time Between Failure Dangerous ($MTBF^D$)--The average time between dangerous failures of a repairable device.

Mean Time Between Failure Safe ($MTBF^S$)--The average time between safe failures of a repairable device.

Mean Time to Failure (MTTF)--The average time before a device's first failure.

Mean Time to Failure Dangerous ($MTTF^D$)--The average time before a device's first dangerous failure.

Mean Time to Failure Safe ($MTTF^S$)--The average time before a device's first safe failure.

Mean Time to Repair--The average time to identify a device failure, to repair it, and return it to normal operation.

Mechanical Integrity--Management system assuring equipment is inspected, maintained, tested and operated in a safe manner.

Mitigated Risk--See "Residual Risk."

Mitigation--Action causing a consequence to be less severe.

Mitigative Layer--Protection layer, which includes mechanical equipment, such as pressure relief devices, blowout panels, and vacuum breakers, and instrumented systems, such as life safety systems, high integrity protective systems, and reactor kill systems--this layer is designed to reduce the frequency and/or consequence severity of the hazardous event.

Near Miss--An event or series of events that could have resulted in one or more identified undesirable consequences, but did not.

Normal Operation--Process operation within the design envelope. Normal operation includes any planned operating mode, such as steady-state, reduced rates, maintenance, testing, start-up, and shutdown.

Nuisance Trip--See "Spurious Trip."

On-line--Process equipment is operational (i.e., running, producing product).

Off-line--Process equipment is not operational (i.e., shutdown).

Off-line Testing--Device testing while the associated process equipment is not in operation. For instance, a compressor is designed to take gas from a low-pressure state to a higher pressure state. Off-line testing would be performed during the time the compressor is not running.

On-line Testing--Device testing while the associated process equipment is in operation. For instance, a compressor is designed to take gas from a low-pressure state to a higher pressure state. If the compressor is operating (compressing gas) while tests are performed on a transmitter, this is an on-line test of the transmitter.

Operability--The degree to which equipment enables operations to perform various tasks and activities necessary to operate the process in accordance with the operating procedures.

Operational Bypass--See "Bypass."

Operating Basis--Collection of documents and information that clearly defines how the instrumented protective system (IPS) is to be operated and maintained to achieve the risk reduction strategy and the core attributes of the protection layer.

Operating Environment--Where a device is intended to be used, such as external environmental conditions, process operating conditions, communication robustness, process and system interconnections, and support system quality.

Operating Mode--Any planned state of process operation, including modes such as start-up after turnaround or after emergency shutdown, normal operation and shutdown, temporary operations, and emergency operation and shutdown.

Operating Procedures--Written, step by step instructions and associated information (e.g., cautions, notes, warnings) for safely performing a task within operating limits.

Operator--An individual responsible for monitoring, controlling, and performing tasks as necessary to accomplish the productive activities of a system.

Operator Interface--Means by which information is communicated between the operator and a system, e.g., computer displays, indicating lights, pushbuttons, horns, or alarms. The operator interface is sometimes referred to as the human machine interface (HMI) or as man-machine-interface (MMI).

Output Modules--Part of an E/E/PES that acts as the interface to external devices and converts output signals into signals that can actuate field devices.

Override--See "Bypass."

Partial Testing--Method of proof testing that checks a portion of the failures of a device, e.g., partial stroke testing of valves and simulation of input or output signals.

Pass/Fail Criteria--Pre-established criteria that define the acceptability of equipment operation relative to the design basis and equipment specification.

Permissive--Logic that requires a specified condition be satisfied prior to allowing further process action. Permissives are often used to reduce the potential for human error to cause hazardous event. For example: 1) to add chemical to a process, a specified temperature must be achieved, 2) a lubrication system must be in operation before a pump can be started, or 3) specific valves must be closed before others can be opened. Permissives may be classified based on consequence severity derived from the specified condition.

Personnel--Humans who are being compensated for performing services for the owner/operator either directly or through approved subcontract.

Predicted Data--Data generated using correlations, factored estimation procedures, and analogies to predict failure rates.

Pre-Startup Safety Review--An assessment used to confirm prior to start-up of the process that equipment is installed and demonstrated to operate according to the design basis; adequate safety, operating, maintenance, and emergency procedures are in place; appropriate hazard analysis or management of change reviews have been conducted and their recommendations addressed; and training of relevant personnel has been completed.

Prevention--The act of causing an event not to happen.

Preventive Layer--Protection layer implementing protective instrumented systems that are intended to achieve or maintain a safe state of the process to reduce the frequency of the hazardous event.

Preventive Maintenance--Maintenance practice where equipment is inspected and maintained to achieve a constant failure rate. Preventive maintenance is typically executed on the basis of a fixed schedule, dictated by manufacturer's recommendation or by accumulated data from operating experience.

Prior Use--Evidence demonstrating that equipment is suitable for use in a specific application (e.g., safety instrumented system) and can achieve the required risk reduction, based on evidence of its performance in a similar operating environment.

Probability--Expression for the likelihood of occurrence of an event or an event sequence during an interval of time or the likelihood of the success or failure of an event on test or on demand. Probability is expressed as a dimensionless number ranging from 0 to 1.

Probability of Failure on Demand Average (PFD_{AVG})--The average probability of a device failing to respond to a demand within a specified proof test interval. It is also the average unavailability.

Process--Any activity involving chemicals including any use, storage, manufacturing, handling, or the on-site movement of the chemicals, or combination of these activities.

Process Demand--A process condition (or event) that requires a protective system to take action to achieve or maintain a safe state of the process.

Process Hazards Analysis (PHA)--An organized effort to identify and evaluate hazardous events associated with processes to enable their control. The PHA normally uses qualitative techniques to analyze the risk.

Process Hazards Analysis Team (PHA Team)--Group of personnel representing a variety of disciplines, who are knowledgeable in the process and operation under review.

Process Industry--Used loosely to include facilities which manufacture, handle, and use chemicals.

Process Interlock--Instrumented system that typically performs protective functions using devices that normally operate as part of the process operation.

Process Requirements Specification--Documents operability, functionality, maintainability, and reliability requirements for the process and finalizes the risk reduction strategy.

Process Risk--Risk arising from an identified hazardous event in the absence of protection layers. Sometimes called unmitigated risk.

Process Safety--A discipline that focuses on the prevention and mitigation of fires, explosions, and accidental chemical releases at process facilities.

Process Safety Information--A compilation of information and documentation pertaining to the process hazards, technology, and equipment

Process Safety Time--Time period between a failure occurring in the process or its control system and the occurrence of the hazardous event.

Process Safety Management--A program or activity involving the application of management principles and analytical techniques to ensure the safety of process equipment.

Process Shutdown System--Term used by the offshore industry to describe a system containing protective instrumented systems and safety instrumented systems.

Programmable Electronic (PE)--Based on computer technology, consisting of hardware and software components. This term covers micro-electronic devices based on one or more central processing units (CPU) together with associated memories. Examples of PE devices used in the process industry include the following:

- Smart sensors and final elements,
- Programmable electronic logic solvers,
- Programmable controllers,
- Programmable logic controllers, and
- Loop controllers.

Programmable Electronic System (PES)--System used to implement control, protective or monitoring functions based on one or more programmable electronic devices. The PES includes all elements of the system such as power supplies, input devices, data highways and other communication paths, and output devices.

Programmable Logic Controller (PLC)--Computer used to implement specific functions, such as logic, sequencing, timing, counting, and arithmetic, to control through digital or analog inputs and outputs, various types of processes or equipment.

Programming--Process of designing, writing, and testing a set of instructions for solving a problem or processing data.

Proof Test--A physical inspection and witnessed test, or series of tests, executed to demonstrate that the equipment operates according to the design basis and is maintained in the "as good as new" condition.

Proof Test Coverage--Expressed as the percentage of failures that can be detected by the proof test. A complete proof test should provide 100% coverage of the failures.

Proof Test Interval--See "Test Interval."

Protection Layer--A physical entity supported by a management system, which is capable of preventing a hazardous event from propagating into an undesired consequence.

Protective Alarm--Protective function which initiates an alarm causing the operator to take a documented, trained and tested response and provides the required risk reduction for an identified hazardous event.

Protective Control--Protective function which normally operates during the identified operating mode where the hazardous event could occur and provides the required hazard rate or risk reduction.

Protective Equipment--Equipment installed to address identified hazardous events. Some owner/operators call this equipment "protective critical equipment."

Protective Function--Function implemented to achieve or maintain a safe state of the process when unacceptable process conditions are detected to reduce the risk of an identified hazardous event. The protective function may be further classified in terms of the consequence severity and risk reduction requirement.

Protective Instrumented Function (PIF)--A protective function allocated to a protective instrumented system with an integrity level (IL) necessary to achieve the required risk reduction for an identified hazardous event.

Protective Instrumented System (PIS)--Composed of a separate and independent combination of sensors, logic solvers, final elements, and support systems that are designed and managed to achieve a specified integrity level. A PIS may implement one or more protective instrumented functions (PIFs).

Protective Layer--See "Protection Layer."

Protective System--A system that implements one or more protective functions.

Proven in Use--Demonstration based on an analysis of operational experience that a specific device configuration has a sufficiently low enough likelihood of dangerous failure that the device achieves the required integrity.

Qualitative Analysis--Method of evaluation based on the knowledge and experience of personnel who are familiar with the process and its operation.

Quantitative Analysis--Method of evaluation based on numerical data and mathematical analysis.

Quantitative Risk Analysis (QRA)--The systematic development of numerical estimates of expected frequency and/or consequence of potential incidents associated with a facility or operation based on engineering evaluation and mathematical techniques.

Random Failure--Failure whose occurrence is unpredictable, which results from various degradation mechanisms in the hardware.

Recommendation--A suggested course of action intended to prevent the occurrence (or recurrence) of an incident or to mitigate its consequences.

Redundancy--Use of two or more devices, systems, or layers to perform the same function.

Reliability--The probability that equipment operates according to its specification for a specified period of time under all relevant conditions. It is one of the core attributes of a protection layer.

Relief Device--A device that is designed to open during emergency or abnormal conditions to prevent rise of internal pressure in excess of a specified value.

Repair Time--The time necessary to identify and repair a fault. The average detection and repair time is called the mean time to repair.

Replacement in kind--Substitution of a device or system with an identical item or with an "approved substitute" that has similar attributes, functionality and failure modes. Refer to "Approved Substitution."

Residual Risk--Risk that exists in the presence of the protection layers.

Response Layer--Protection layer, which implements emergency response systems to notify personnel and/or the community to shelter-in-place or to evacuate to safe zones and initiates emergency response activities.

Response time--Time required to fully execute the protective function from detecting the process condition to achieving the safe state of the process.

Review--See "Assessment."

Risk--A measure of human injury, environmental damage, or economic loss in terms of the event frequency of occurrence and the severity of the injury, damage or loss.

Risk Analysis--The estimation of the risk of each hazardous event considering its frequency of occurrence and consequence severity and the identification of safeguards to reduce this risk below the owner/operator risk criteria.

Risk Assessment--The process by which the results of an analysis are used to make decisions, either through relative ranking of various risk reduction strategy options or through comparison with risk criteria.

Risk Based Inspection--Management system for identifying failure mechanisms and determining the inspection strategy for equipment based on the expected equipment degradation rate and the consequence severity if mechanical integrity is lost or equipment does not perform as expected.

Risk Criteria—Qualitative or quantitative measures used to determine whether a risk posed by identified hazardous event is tolerable (or acceptable) versus intolerable (or unacceptable).

Risk Management--The systematic application of management policies, procedures and practices to the task of preventing hazardous event.

Risk Matrix--Qualitative or semi-quantitative representation of the risk criteria. The owner/operator creates a matrix using broad categories to define the tolerable likelihood (or frequency) and consequence severity of identified hazardous events.

Risk Reduction Factor--A measure of how much a protective function reduces the frequency of the hazardous event. For demand mode systems, it is often related to the average probability of failure on demand (PFD_{AVG}). $RRF=1/PFD_{AVG}$.

Robustness--The mechanical ruggedness, over-range tolerance, and ease of recovery after fault. For example, equipment is robust if it can sustain physical abuse in the field or overpressure during process upsets.

Root Cause--The most basic underlying system-related reason that initiated the chain of events that led to an incident; the root cause indicates why an incident occurred.

Root Cause Analysis--A method used to (1) describe what happened prior to and during the hazardous event, (2) determine how the event propagated, (3) identify the underlying reasons the event occurred, and (4) implement corrective actions.

Safe Failure--Failure affecting equipment within a system, which causes, or places the equipment in condition where it can potentially cause, the process to achieve or maintain a safe state.

Safe Failure Fraction--Fraction of the hardware failures resulting in a safe failure or in a dangerous detected failure.

Safeguard--Any engineered system or administrative control that would likely interrupt the chain of events following an initiating cause or mitigate its consequence.

Safe Operating Procedure--Specific steps that are implemented by operators to achieve or maintain a safe state of the process to reduce the risk of an identified hazardous event. Some owner/operators call safe operating actions "critical operating procedures."

Safe State--A state of the process where a hazardous event cannot occur.

Safety--The expectation that a system does not, under defined conditions, lead to harm to people, either directly or indirectly.

Safety Alarm--Safety function which initiates an alarm causing the operator to take a documented, trained and tested response and provides the required risk reduction for an identified hazardous event involving a catastrophic release.

Safety Control--Safety function which normally operates during the identified operating mode where the hazardous event could occur and provides the required hazard rate or risk reduction involving a catastrophic release.

Safety Configured Logic Solver--General purpose industrial grade programmable electronic logic solver which is specifically configured to detect failures in the inputs, outputs, main processors, and communications and to take specified action on dangerous detected failure.

Safety Critical Devices--See "Safety Equipment."

Safety Critical Systems--See "Safety Systems."

Safety Equipment--Equipment installed to address identified hazardous events involving catastrophic releases. Some owner/operators call this equipment "safety critical equipment."

Safety Function--Function implemented to achieve or maintain a safe state of the process when unacceptable process conditions are detected to reduce the risk of an identified hazardous event involving a catastrophic release.

Safety Instrumented Function (SIF)--A safety function allocated to a safety instrumented system with a safety integrity level (SIL) necessary to achieve the required risk reduction for an identified hazardous event involving a catastrophic release.

Safety Instrumented System (SIS)--Composed of a separate and independent combination of sensors, logic solvers, final elements, and support systems that are designed and managed to achieve a specified safety integrity level. An SIS may implement one or more safety instrumented functions (SIFs).

Safety Integrity Level (SIL)--Represents one of four discrete ranges used to benchmark the integrity of each SIF and the SIS, where SIL 4 is the highest and SIL 1 is the lowest.

Safety Interlock System--See "Safety Instrumented System."

Safety Manual--Compilation of installation, configuration, and mechanical integrity requirements for equipment approved for SIS applications.

Safety Margin--The amount provided beyond the minimum necessary.

Safety-Related System--See "Safety Instrumented System."

Safety Requirements Specification (SRS)--Compilation of information and documentation that constitutes a design basis for the safety instrumented system. It may include logic narratives, Input/Output (I/O) list, cause & effect matrix, logic flow charts, system overview drawings, integrity level calculations, etc.

Safety Shutdown System--See "Safety Instrumented System."

Safety System--Systems, which implement one or more safety functions. May also be referred to as safety critical systems.

Scenario--Event or sequence of events that results in undesired consequences.

Security--See "Access Security."

Separation--The use of administrative controls and physical means to achieve independence between protection layers.

Sensor--A measurement device (instrument) or combination of devices that detect process variables or conditions (e.g., transmitters, transducer, process switches, and toxic gas detectors). The sensor boundary includes the process connection, sensor, transmitter, and signal connection to the logic solver.

Shutdown--A process by which an operating process or system is brought to a safe state.

Significant Consequence Events--An event that potentially causes significant harm to humans, damage to the environment, loss of property, or business interruption.

Site Acceptance Test (SAT)--Conducted to determine and document that a new or modified instrumented protective system meets the design basis, is installed in accordance with construction, installation, and software requirements, and is ready to start-up.

Software--Intellectual creation comprising the programs, procedures, data, rules, and any associated documentation pertaining to the operation of a programmable electronic (PE) device or system.

Software Languages--The programming languages can be classified into three categories:

- **Fixed Program Language (FPL)**--Language used to implement a dedicated function, such as measuring pressure using a pressure transmitter. The user is unable to alter the function of the device, but is able to adjust (e.g., configure) a limited number of parameters (e.g., the range of the pressure transmitter). The functional specification of an FPL device is documented in the manufacturer's data sheets, user manuals, or the like. Typical examples of FPL devices are transmitter, sequence of events controllers, analyzers, and small data logging systems.
- **Limited Variability Language (LVL)**--Language that provides the user with the ability to customize the functionality of the programmable electronic system (PES) to achieve specific requirements. LVL documentation consists of a programming guide (e.g., a user manual) that defines the basic program elements (e.g., function library) and the method for combining them into an application program. Examples of LVL include ladder logic, function block diagrams, and sequential function charts.
- **Full Variability Language**--Includes any general-purpose computer-based language that is equipped with an operating system. The operating system provides system resource allocation and a real-time multi-programming environment. Computer specialists, using high-level languages, tailor the language for a specific application, yielding a unique program for each specified application. This language is typically not used in instrumented protective system application. When this language is required, more management system rigor must be applied to reduce the potential for systematic errors.

Software Types--Software can be classified into three types:

- **Application Program**--Software specific to the intended application. In general, it contains logic sequences, permissives, limits, expressions, etc., that control the appropriate input, output, calculations, decisions necessary to meet the design basis. See “full variability” and “limited variability” under “Software Languages.”
- **Embedded Software**--Software that is part of the programmable electronic (PE) device or system supplied by the manufacturer and is not accessible for modification by the owner/operator. Embedded software is also referred to as firmware or system software.
- **Utility Software**--Software tools for the creation, modification, and documentation of application programs. These software tools are not required for the operation of the protective systems.

Software Lifecycle--Activities occurring during a period of time that starts when the software is conceived and ends when the software is permanently retired. Software, thus, follows the same lifecycle as the hardware, including specification, verification and validation procedures, security provisions, and management of change.

Spurious Operation--Failure causing equipment to take action on the process when not required. Spurious operation has an immediate impact on process uptime and potentially on process safety.

Spurious Trip--Refers to a process shutdown, or disruption, due to the spurious operation of equipment. Other terms often used include nuisance trip and false shutdown.

Spurious Trip Rate (STR)--Expected rate (number of trips per unit time) at which a process shutdown, or disruption, occurs due to the spurious operation of equipment. Other terms used include nuisance trip rate and false shutdown rate.

Standby--Equipment that is normally not running but must be ready to run or operate, when required, e.g., an emergency diesel generator.

Subsystem--A portion of a system.

Supervisory Layer--Protection layer, which implements protective alarm systems, operator monitoring and supervision, and process actions to achieve or maintain a safe state for the process to reduce the frequency of the hazardous event.

Support Systems--Human machine interfaces, communications, wiring, power supplies and other utilities which are required for the system to function.

System--Collection of devices organized to accomplish a specific function or set of functions.

Systematic Failure--Failure related in a deterministic way to a root cause, which can only be minimized by effective implementation of the protective management system.

System Software--See "Software Types."

Task Analysis--A human error analysis method which breaks down a procedure or overall task into unit tasks and determines the detailed performance required of people and equipment and the effects of environmental conditions, malfunctions, and other unexpected events on both.

Test Interval--Time period between two successive proof tests.

Tolerable Risk--Non-zero risk that is tolerable (or acceptable) to an owner/operator. This risk tolerance is often defined in terms of frequency of occurrence and consequence severity.

Trip--A process shutdown that may be due to a process demand or to a spurious failure of an instrumented protective system.

Turnaround--Maintenance activities associated with a process, unit, or plant, which require the process, unit, or plant to be taken out of normal service with all equipment taken to a shutdown or out-of-service state.

Unavailability (Instantaneous)--The probability that the equipment is not capable of performing its required function under given conditions at a given instant of time, assuming the required external resources or support systems are provided.

Unavailability (Mean)--The fraction of time that the equipment is not capable of operating in the desired manner. It is the mean of the instantaneous availability over a given time interval.

Undetected Failure--Failure detected by inspection, proof test or process demand. Synonyms include hidden, concealed, unannounced, unrevealed and covert.

Uninterruptible Power Supply (UPS)--Means by which uninterrupted power is provided to protective functions for a limited period of time in the event of plant power failure. This usually accomplished by an inverter, transfer switch and batteries.

Unmitigated Risk--See "Process Risk."

Useful Life Failure--Random failures that occur during the time period where the failure rate can be considered relatively constant, because early failures have been corrected and wear out failures have not begun.

User Approval--Management system used to determine and document that equipment and associated documentation, procedures, and training is suitable for an intended purpose, classification, operating environment, and function complexity.

Utility Software--See "Software Types."

Validation--Activity of demonstrating that the installed equipment achieves the core attributes and the requirements of the design basis.

Verification--Activity of reviewing, inspecting, checking, testing, or by other means determining and documenting whether the outcome of work processes, activities or tasks conform to specified requirements and traceable input information.

Voting--Specific configuration of equipment within a subsystem. Voting is often expressed as MooN (M out of N). "N" designates the total number of devices (or channels) implemented; "M" designates the minimum number of devices (or channels) out of N required to initiate, take, or maintain the safe state. Also called voting system or voting architecture.

Wear-out Failure--Failures identified late in the equipment's installed life, indicating that the equipment has reached the end of its useful life. The equipment failures within the device are due to aging or other mechanisms that lead to a predictable increase in the failure rate relative to the useful life failure rate. Example mechanisms include: battery consumption, mechanical or thermal cycling, gradual contact corrosion or fouling, etc.

Worst Credible Scenario--The most severe incident of all identified outcomes and their consequences that is considered plausible or reasonably believable.

B

PROTECTION LAYERS

A protection layer is a physical entity supported by a management system, which is capable of preventing a hazardous event from propagating into an undesired consequence. Figure B.1 illustrates how, starting from the inside and working outward, each protection layer builds on the next. The order of presentation also indicates the typical order of protection layer deployment as part of a risk reduction strategy.

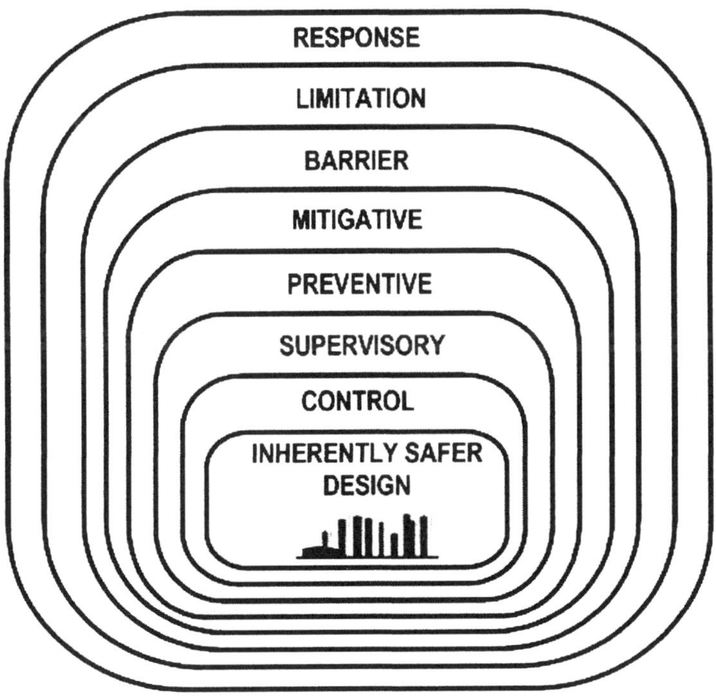

Figure B.1. Protection Layers.

The inherently safer layer is the most important, as it reduces or eliminates the inherent risk of the process design. The next five protection layers act to control hazards and risks that could not be eliminated by process design. The control, supervisory, and preventive layers reduce the hazardous event frequency by taking action on the process to achieve or maintain a safe state. Conversely, the barrier and mitigative layers are designed to minimize the consequence severity resulting from the hazardous event. The barrier and mitigative layers generally result in secondary or residual consequences, which may or may not be acceptable. In some instances, the secondary consequence must be addressed using other protection layers.

The limitation and emergency response layers take action after hazardous event occurrence. These layers are intended to reduce the consequence severity of the hazardous event, but their effectiveness in doing so is unknown. The emergency response layer musters response personnel to the incident location and minimizes exposure of surrounding plant personnel and the community using shelter-in-place or through relocation to safe areas.

Protection layers can be considered independent protection layers (IPLs), when they are designed and managed to the rigor necessary to achieve the core attributes, namely independence, functionality, integrity, reliability, auditability, access security, and management of change (see Appendix C). The necessary level of rigor is related to many issues, including:

- Site safety culture,
- Consequence severity,
- Process and IPL novelty and complexity,
- Specific core attribute,
- Inherent quality of the core attributes,
- Propensity for systematic errors, and
- Anticipated number of changes.

IPLs can be viewed as "lines of defense" against hazardous events, while the core attributes establish the performance metrics. The independence of the IPL from the initiating cause(s) and from other IPLs is very important. In general, IPL equipment should be included under a mechanical integrity process. The equipment should be subject to inspection and proof tests at a defined interval to validate the protective function and to return the equipment to the "as new" condition. All IPLs depending on operating personnel should be covered by written procedures, and identified personnel should be trained and tested on the procedures. Access to protective equipment should be controlled and proposed changes should undergo management of change review prior to implementation.

B.1 INHERENTLY SAFER DESIGN

The physical design and process chemistry are the origin of process risk. This risk can be reduced or eliminated during the process design through choices in process technology, equipment design, and operating parameters. This concept of process design is known as "inherently safer design" and is discussed in *Inherently Safer Chemical Processes: A Life Cycle Approach* (CCPS/AIChE 1996). Inherently safer design can minimize or eliminate the need for protection layers.

Inherently safer design involves the use of four key design principles:

- Minimize--use smaller amounts of a hazardous material,
- Substitute--replace a material with one that is less hazardous,
- Moderate--use less hazardous conditions, a less hazardous form of the material, facility siting, or equipment location and spacing, or
- Simplify--design facilities that make operating error less likely and are more tolerant to process excursions.

  Example--Reboiler failure is inherently safe

  Consider a hazardous event involving potential over-pressure of a distillation column due to reboiler control loop failure. An inherently safer approach would be to design the condenser with more heat removal capacity than the reboiler heat addition capacity is capable of providing. Three approaches (Dowell, et.al, 1996) could be used to accomplish this design change:

  - Adjust the ratio of the heat transfer capability of the reboiler and condenser,
  - Limit the maximum flow of heating medium to the reboiler by restricting the flow using smaller piping, or
  - Limit the maximum temperature of the heating medium using a pressure regulator and a pressure relief valve on the steam piping to limit steam pressure.

Since the distance between the process equipment and the populated areas greatly influences the process risk, facility siting and plant layout are very important inherently safer features. Consideration should be given to segregation of certain types of risk, access restriction to hazardous areas, the design and location of control rooms, and the site topography.

Risk assessment activities occur throughout the process life from earliest research and development through long-term process unit operation. Consequently, the risk reduction strategy begins during the process design using inherently safer principles in the selection of the process chemistry, process chemicals, process variables, equipment design, occupied areas, and chemical inventories. During the process design, deviations from normal operation are examined to determine whether the equipment can be designed to withstand the potential extremes. When abnormal operation is made safe through design, the process is simply more tolerant to process upset.

The use of a protection layer in lieu of inherently safer design results in devices, systems, and procedures that must be specified, implemented, audited, tracked, maintained, proof tested, etc. While enabling safe operation, these systems and procedures can cause operational problems, such as spurious trips of the process unit, if they act when they should not. Further, given the wide range of random failures and systematic errors, it is not "if" a protection layer will fail, but "when it will fail" and "what is the consequence of its failure."

Lifecycle costs should be considered, such as:

- Capital investment,
- Operating and maintenance costs,
- Spurious trip costs, and
- Post incident clean-up and rebuild costs.

Inherently safer design is most cost effective when applied in early design phases as shown in Figure B.2. The opportunities for inherently safer design diminish as engineering progresses and equipment is procured and commissioned. Any risk gaps (i.e., differences between process risk and owner/operator's risk criteria) found during later project stages or subsequent operating phases are more likely to be addressed with protection layers than inherently safer design practices. Changing the design simply becomes more difficult and expensive as the project matures to full-scale implementation.

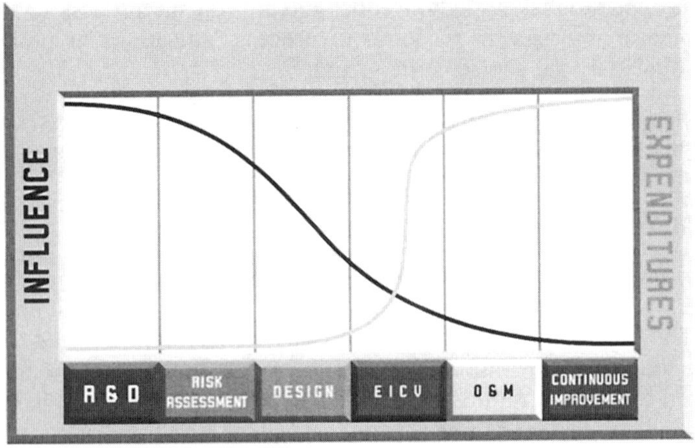

Figure B.2. Ability to Cost Effectively Influence Inherent Risks.

Inherently safer design should be considered throughout the process lifecycle. Existing processes benefit from an occasional review of inherently safer options. While the options may be limited on an existing process, some design options that were not considered good investments may become favorable. When planning an instrumentation and controls upgrade, re-visiting the process design package may identify opportunities to eliminate process hazards, thereby reducing the instrumentation and controls requirements. These are cost improvement opportunities, because decreasing the number and/or severity of process incidents improves the bottom line.

When inherently safer design relies on equipment specification, it should be clearly indicated in the equipment design basis. Inspection and auditing is required to ensure that the design feature remains throughout the equipment life. Changes to the process operation or to process equipment should be reviewed using a management of change process to ensure that inherently safe design features maintain their integrity.

B.2 CONTROL

The control layer implements standard operating procedures, basic process control systems, and process alarms to accomplish normal regulatory actions. The control layer is used to control or maintain the process within the normal operating limits. A process may be controlled in various ways, such as:

- The operator taking direct action on the process equipment according to written operating procedures, such as starting or stopping pumps or manually opening and closing valves,
- The operator taking action either directly on the process or through the operator interface in response to process indications or process alarms provided by the control system, and
- The control system may automatically control the process according to process conditions that it measures and regulates to a set point.

B.2.1 Control Function

The ability of automation to yield more consistent product quality and production rates have led over the years to significant investments in instrumentation and controls. The majority of automation in the process industry is focused on controlling the process within the normal operating limits. Control functions are implemented in what is referred to as a Basic Process Control Systems (BPCS). These systems are designed to be separate and independent from the SIS. A BPCS does not perform any safety instrumented functions. Appendix F.2 discusses the evolution of the BPCS and SIS.

The BPCS is generally designed and managed to achieve business and production goals, especially uptime, production, and quality. It responds to input signals from the process and its associated equipment, other programmable systems, and/or from an operator and generates output signals causing the process and its associated equipment to operate in the desired manner. The BPCS consists of many hardware and software components and relies heavily on communication equipment to access and display process information.

The BCPS logic solver is often referred to as the "controller" and can utilize pneumatic, hydraulic, electrical, electronic, or programmable electronic (PE) technology. A modern PE-based BPCS provides nearly seamless integration of controllers and operator displays. PE technology enables complex control algorithms, such as advanced process controls, sequencing, predictive controls, and batch reactor recipe management. The advantages of programmable electronic systems are numerous and are well documented, including:

- Ease of modification,
- Ergonomic display of complex information,
- Ability to logically display a great deal of information to the operator in a user-friendly format,
- Troubleshooting capabilities,
- Real-time data tracking and displays,

- Quick response-time to process excursions, and
- Enhanced preventive maintenance tasks.

When a controller (i.e., input/output (I/O) modules, main processors, and communication interfaces) fails, many control functions can fail simultaneously. The controller performance can be improved using redundant components throughout the system, including I/O modules and main processors. However, the high degree of interaction between the numerous components of the control layer and the open access frequently allowed to its various systems limit the integrity realistically achieved by the control layer.

Many facilities allow remote data access and collection to support troubleshooting and optimization activities performed from external locations, such as corporate or centralized engineering offices. Remote access reduces travel costs from centralized offices to facilities. It also allows the engineering staff to download large amounts of real-time data directly to their computers for analysis, saving hours of data entry. Access to real time data is a great benefit to research and development personnel, as well as managers and supervisors. This enhances the process monitoring capability and reduces the cost of obtaining this information significantly. This access increases the risk of data highway corruption, introduction of viruses into networks and systems, and unintended or unknown system failures. This access also increases complexity, making it more difficult to predict how changes in interconnected systems affect the overall system.

During the risk assessment phase, control function failures are analyzed to determine whether they lead to hazardous events. The dangerous failure rate (or hazard rate) of a control function implemented in a BPCS that is not designed and managed according to ISA 84.01/IEC 61511 may not be assumed to be lower than 10^{-5}/hr. Control functions are composed of many components that can fail such that the process deviates from normal operation. Examples of control function failure are:

- Field sensor fails,
- Inadequate control loop tuning,
- Input/outputs (I/O) fail to change states,
- Main processor ceases processing information,
- Inadequate or slow data acquisition (e.g., sluggish response under upset conditions),
- Positioner fails,
- Final control element fails,

- Loss of the operator HMI information, and
- Disrupted communication with field devices.

In the best case scenario, the failure is detected and the process maintains an acceptable operational state because of the action of other systems. At worst, the failure leads to a process incident. Between these best case/worst case extremes is the outcome of the process's response to the IPL actions. For example, a process excursion can cause an instrumented protective system (IPS) to take the process to a safe state, stopping incident propagation, yet causing downtime and lost production.

BPCS logic solvers typically incorporate internal diagnostics to detect specific component failures. Detected failures should initiate a fault alarm and a specified output action. In some cases, this action is configurable; in others it is not. When configurable, the affected process signal is taken to the configured state to achieve the desired performance, such as fail last, fail upscale, or fail downscale. When it is not configurable, the BPCS configuration must be understood so that the control scheme can take the failure states into account. From an operating standpoint, it is desirable to have the BPCS provide the highest possible uptime or reliability. Consequently, the typical practice is to fail the faulted components in the last position to allow the process to hold its last state (at least for the faulted process variable), thereby maintaining operation.

The BPCS logic solver should allow I/O segregation and a modular approach to system configuration, which can reduce initial programming and training costs. Modularity provides many options to achieve increased process reliability. Consequently, control loop interdependence should be considered during design to assure that a single function or controller failure does not escalate into a cascade of events or into a system failure.

Automation brings great flexibility and complexity, which can become a hindrance to its proper use. Automation creates a dichotomy. When the automation is good, fewer decisions are made by the operator on a regular basis, leading to knowledge erosion. When the automation is poor, the operator is more likely to ignore and override it. Therefore, the effectiveness of this layer is highly dependent on the operator's relationship with the control layer automation.

Remote monitoring capabilities can reduce manpower needs for some outlying areas, such as tank farms. It reduces personnel exposure by making site entry an exception rather than a normal daily task. This can be a mixed blessing, because operations personnel sometimes do not accept ownership of equipment with which they have little to no interaction. Additionally, this loss of local monitoring can extend the time it takes to detect small problems before they become significant ones, e.g., small flange leak forming an ice ball or a pump starting to 'sound' funny.

Operator procedures associated with the control layer should be documented according to good engineering practices. Operators are often an integral part of the control function, e.g., in batch operations, the decision to proceed with a batch sequence may be determined by the operator based on BPCS information. Operators are expected to manually line-up equipment and to maintain control of the process during start-up of equipment. Personnel should understand how the failure of their actions could lead to hazardous events.

B.2.2 Protective Function

The BPCS can also be used to implement protective functions, such as supervisory functions, when the BPCS design and installation provides sufficient independence between the initiating cause and the supervisory function. Figure B.3 illustrates how control and supervisory functions can be implemented in the BPCS using separate field sensors, controllers, and final elements.

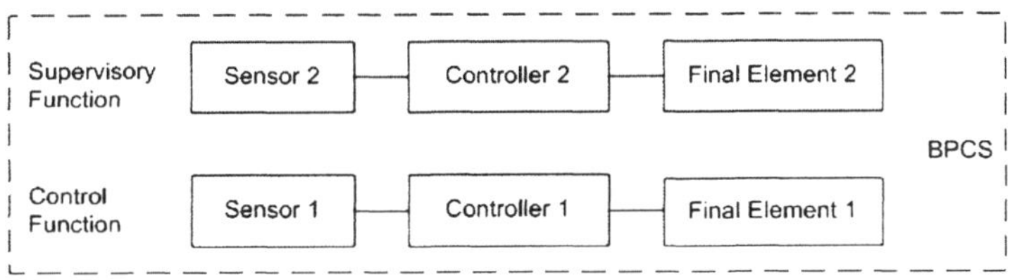

Figure B.3. Control Function and Supervisory Function Implemented in the BPCS.

As illustrated in Chapter 3.5.6, when the BPCS is used for control functions and supervisory functions, it must be managed with greater emphasis on the core attributes. Consequently, there is a limit to what can be achieved by a system that has "managed" separation. The risk reduction allocated to a safety function in the BPCS must be less than 10. For a BPCS where control and supervisory functions are implemented as shown in Figure B.3, the claimed hazard rate must be less than 10^{-6}/hr.

Common cause should be evaluated when allocating the functions to the BPCS to ensure the function is:

- Separate and independent of the initiating cause (i.e., the supervisory function is in an independent controller from the initiating cause controller), or
- Implemented in a BPCS controller with sufficient diagnostics and fault tolerance against dangerous failure to meet the required hazard rate.

Within the process industry, there are market sectors, which permit the operator to place functions within the BPCS in manual operation and to control the process through manual changes of the output signal to the final control element. When this occurs, the control function does not respond to the process conditions. When under manual operation, the control function integrity is limited by basic human reliability. If the function is also a supervisory function, the manual operation of the function should be considered equivalent to a bypass, because the output does not respond to a change in the process condition. Placing a function in manual must be covered by a procedure, which ensures that it can be performed safely.

B.3 SUPERVISORY

The supervisory layer implements protective alarm systems, operator supervision, and process actions to achieve or maintain a safe state of the process to reduce the risk of the hazardous event. When these functions are physically allocated to the BPCS, the considerations discussed in Section B.2.2 apply. Chapter 4.6.2 provides additional considerations for the physical allocation of functions to the BPCS or IPS.

It takes time for an operator to detect a problem and to take action. The amount of time required is directly related to control room distractions, operator interface design, the information and training provided to the operator, and the complexity of the required response (e.g., control action, shutdown action). The operator must be physically capable of taking the action under all required conditions (e.g., rain, snow, fire) within the available time. An operator's capability to recognize and stop the propagation of hazardous events is affected by many factors, such as:

- Process information clarity and relevance,
- Operating procedure clarity and completeness,
- Time available for operator response,
- Operator interface, including display clarity and ergonomic design,
- Task familiarity,
- Task complexity,
- Human reliability,
- Fatigue level,
- Stress level, and
- Operator supervision.

Devices in the supervisory layer must be periodically proof tested. At a minimum, the equipment should be tested prior to start-up after maintenance outages, turnarounds, or instrumented system upgrades. Historical field device performance should be reviewed periodically to determine whether the test interval is appropriate. The alarm devices and interface must be under configuration management and the alarm set point must be controlled under management of change processes. Proof testing and change management is necessary to assure that the supervisory system meets the allocated risk reduction.

B.3.1 Operator Activities

Operators can reduce the risk of hazardous events by taking action according to documented operator procedures. Table B.1 provides examples of risk reduction factors (RRF) for operator and supervisory activities. The operator should be trained and tested on the procedure, which should list the process condition(s) that clearly indicate the need to take action. The procedure should provide a list of the action(s) required by the operator when the process condition(s) are unacceptable, the time available for the operator to take such action, and the consequences if action is not taken.

IPL	Conditions	RRF
Process Related Rounds and Inspections	**Frequency of operator rounds must be sufficient to detect and prevent the hazardous event Operator is trained to recognize and respond to unacceptable out-of-range values. If a specific process variable is being monitored, the operator should record the specific value displayed by equipment independent of the initiating cause.**	**10**
Observational	**Frequency of operator rounds must be sufficient to detect and respond to the hazardous event The need to take response must be obvious to the operator through normal visual or hearing range, e.g., loud noise, high vibration, serious leaking, etc.(1)**	**10**
Review	**Independent inspection/verification and sign-off that a required operator action was performed as intended (e.g., valve line-up is confirmed as correct).**	**10**
Action	**An operator action that uses a different operator, relying on independent observation.**	**10**
Corrective Action	**An operator action taken based on a scenario where the propagation is so slow that the operator has sufficient time to gather further information (e.g., laboratory tests, product quality, and material balance) as necessary to recognize the error and to correct it.**	**10**

Note 1: At a minimum, the release quantity should be estimated based on the amount potentially released between rounds or, as in the case of simple observational, the amount released in the time it takes to detect and respond. The response should not require the operator to go into the hazard zone.

Table B.1. Examples of Operator or Supervisory Activity RRF.

Additional information on operating procedures is provided in Chapter 6. Compliance with procedures used to reduce process risk should be audited periodically to ensure that personnel are following procedures and that procedures remain up-to-date. In all cases, the operator should have sufficient time to recognize the problem, determine the solution, and take action.

If the supervisory function is a verification activity, operating procedures should identify when these independent reviews are required and provide a place for the name of the reviewer to be recorded. For example, an independent review verifies that the valves are properly lined-up, reducing the potential that the valves are in the wrong position. This activity is included in the procedure with a place for sign-off. Personnel responsible for executing the verification activity should be trained on its importance in reducing the risk of the hazardous event.

Example--High level in knock-out drum

A knockout drum receives fuel gas from an upstream processing unit. The incoming gas is relatively dry; therefore, the drum catches small amounts of liquid. The knockout drum is equipped with a level control function that automatically operates a level control valve to drain the boot on increased level. The level control valve is equipped with a hand jack. The observation of the boot level using an independent level indicator is part of routine operator rounds.

A process hazard assessed by the team is liquid carryover to the downstream equipment. The H&RA team identifies the failure of the process control loop to maintain level as a potential initiating cause. The team reviewed the rate of level rise in the knock-out drum and determined that it takes 16 to 24 hours for the level to get high enough that there is a risk of carryover. The level is currently checked once per 8-hour shift and the level is recorded on the operator's job checklist. The team determines that the operator's surveillance of the level meets the protective function requirements. This procedure and associated instrumentation and controls are denoted as protective.

B.3.2 Operator Alarms with Response

Alarms should only be used when the operator is expected to take a specified action, which is covered by operating procedure. Alarms are often cited as the first safeguard in reducing unacceptable risk. For most hazardous events, only one protective function can be claimed in the supervisory layer, irrespective of the number of indications or alarms.

An alarm can be classified as a protective function, if it meets three criteria:

1. The alarm function is independent of the initiating cause and other protective functions addressing the identified hazardous event,
2. The alarm function, including inputs and outputs, is designed to provide the allocated risk reduction, and
3. There is sufficient time for the operator to detect a problem exists, to determine what to do and to take the appropriate action necessary to return the process to the normal operating limits.

Operators often receive information from the BPCS and take action on the process through the BPCS. Operator response to an alarm requires three subsystems: 1) the alarm sensing device to the operator interface, 2) an operator trained according to a written procedure, and 3) the equipment used to take action on the process. Each subsystem must operate correctly and timely to achieve the intended outcome.

Various human factors should be considered when assessing the operator's ability to achieve the required risk reduction:

- Clarity of process information available,
- Control room manning,
- Distracting duties,
- Decision-making complexity,
- Response ergonomics,
- Time available for response,
- Written procedures,
- Stress level, and
- Training related to hazardous event response.

Time Available After Alarm[1]	Conditions [2]	RRF
Any response time	Operator action is complicated, e.g., large number of alarms generated by initiating cause or the required response is not documented in a written procedure or the operator is not trained on the written procedure.	1
<10 minutes response time	Operator must troubleshoot to determine what the appropriate action is.	1
2 to 10 minutes response time	Drilled and practiced response, also known as a "never exceed, never deviate" response. If the alarm is received, the operator must execute the safe state action without delay. Alarm is independent of the BPCS.	10
≥10 minutes response time	Operator response does not require troubleshooting or investigation prior to action. Alarm may be implemented in the BPCS or independent of the BPCS.[3]	10
≥40 minutes response time	Operator response requires minor troubleshooting or investigation prior to action. Alarm may be implemented in the BPCS or independent of the BPCS.[3]	10
24 hours response time; multiple operators can take action	Alarm should be automatically repeated at an interval necessary to ensure that each shift is notified of the process condition. Minor troubleshooting may performed prior to action. Alarm is independent of the BPCS.	100

Note 1: The operator response time should consider the time it takes to recognize the alarm, to diagnose the problem, to complete the required action on and for the process to reach the designated state. This is compared to the allocated process safety time, which considers how rapidly the process moves from the alarm condition to the hazardous event.

Note 2: The required action must be clearly indicated by the alarm, the response covered by a written procedure, and the operator trained and tested on the procedure.

Note 3: As long as independence from the initiating cause and other IPLs is demonstrated, allocation is influenced by the operator interface design and the importance of the operator response. In all cases, the operator should receive the information in a clear, unambiguous, and prioritized manner.

Table B.2. Examples of Operator Response to Alarm RRF. (adapted from ISA TR84.00.04-2005 Appendix B)

The risk reduction achieved by successful alarm response depends on the time available, which is limited by the process dynamics. The process safety time starts with the process deviation and ends with the hazardous event. The operator's response time begins with the recognition that action needs to take place and ends when the process achieves the safe state. The process safety time is compared with the time required to execute the operator procedure. The total operator response time should be less than one-half of the available process safety time. Table B.2 provides examples of RRFs for operator response to alarms.

Information quality can be seriously affected by alarm flooding occurring during emergency events if alarms are not specified and managed properly. Chapter 5.5.5 provides a brief overview of alarm management. Procedure clarity and operator training become critical when the operator is expected to make complex decisions, e.g., examining temperature and pressure related to the batch recipe and sequence step.

An important consideration is how the alarm condition physically appears to the operator. A protective alarm must stand out against other alarms that may be generated simultaneously. The interface design should follow good human factors practices. It should also take into account site specific considerations, such as existing interfaces and the operator's skills and limitations, including color blindness.

The alarm system should be capable of indicating multiple alarm conditions at the same time. A method to prioritize alarms should be established to prevent potential alarm overload, in which the hazardous event results in a stream of alarms so numerous the operator cannot mentally process the information provided. Safety alarms should be considered high priority alarms, since these alarms indicate the potential for a hazardous event posing harm to people. These alarms should be distinct from other non-safety alarms.

Alarms should be audible until they are acknowledged by the operator and should remain visible until the process condition has been corrected. Once acknowledged, the alarm can automatically reset when process conditions return to the normal operating envelope. The operator should not be able to disable a protective alarm during any operating mode where the hazardous event could occur. Changes to the protective alarm set point must also be controlled through procedures and management of change processes.

A dedicated interface should be considered for protective alarms, since this clearly and unambiguously communicates their priority. The interface can also be placed to allow other control room operators to see and respond to the area alarms. If it is not practical to implement a dedicated interface, appearance and/or sound should be utilized to draw attention to the protective alarms. When a BPCS interface is used, the BPCS management system must place greater emphasis on access security and administrative control for the hardware, data, communications, and software.

Owner/operators have found that, when dedicated displays are configured similar to traditional alarm panels using light boxes, operators develop pattern recognition and identify the root cause faster. Separation also provides better access and management of change control. Operator response integrity can be improved by displaying the required operator actions on request. Alarms placed on a separate display should produce a digital alarm message so that the message can be used in alarm statistics in measuring alarm frequency per unit time per operator, operator actions against alarms, etc.

Operator action may involve a variety of activities in the control room, as well as in the operating unit. The expected operator action should consider the time available for the response following the alarm initiation and the physical ability of the operator to complete the required action. In stressful situations, there is a high potential for operator (or procedural) error. The probability of operator error is also related to the available time for operator response. The less time that is available for diagnosis and response, the greater the likelihood that the operator will fail to correctly execute the task. There are also other work-related factors to consider, such as job responsibility (e.g., is this one of many safety or protective alarms), complexity, and stressors.

The level of automation associated with the operator action is also important. For example, does the operator press a push button initiating closure of the process isolation valves or must the operator manually close the valve? If the valves are manually closed, how long does it take to get to the valve and close it? When the operator takes the action in the field, how does the operator know that the action is not working and that it is time to escape? Field personnel taking manual actions to prevent a hazardous incident are potentially being sent into the hazard zone. They need to know when to evacuate in case the event is not preventable by their documented and trained actions.

Operator action is documented in a written procedure that highlights the protective actions and explains the consequences of deviating from these actions. The operator's capability of executing a task correctly is increased when the instructions are clear and straightforward with adequate feedback from the process, allowing the operator to determine whether the actions are having the desired effect.

A separate document binder is recommended to emphasize the significance of the procedures considered as protective functions and to provide quick reference should they need to be reviewed during an alarm event. If the alarm procedures are fully integrated into the main operating procedures, special tagging is recommended to allow rapid identification of the appropriate procedure. Consideration should also be given to providing the procedures in electronic format. Some alarm systems allow electronic procedures to be linked to alarms.

These procedures should be part of the routine operator training with periodic drills to assure the operator response integrity and to improve response time. Operators should also be tested periodically on their knowledge of and response to protective alarms. The management system should document the requirements for procedures, checklists, sign-offs, or independent audits for each activity performed in this layer.

B.3.3 Instrumented Systems

Instrumented systems may provide information to the operator who is responsible for taking the corrective action (e.g., control action or process shutdown). The instrumentation may also take specified action automatically. Instrumented supervisory functions can be implemented in a dedicated system, the BPCS, or IPS (SIS). Allocation to the BPCS is acceptable when the BPCS is designed and managed to achieve the required risk reduction. See Section B.2.2 for additional discussion on the implementation of protective functions in the BPCS. If the supervisory function is allocated to the IPS, see Section B.5 for further discussion.

B.4 PREVENTIVE

The PIS layer is consists of sensors, logic solvers, final elements, and support systems (e.g., utilities, communications) implemented to achieve or maintain a safe state of the process in response to an unacceptable process condition. PISs are sometimes divided into various classes to allow the systems to be managed separately, if desired. For example, some owner/operators distinguish between asset protective systems, environmental protective systems, and safety instrumented systems (SIS). An SIS is a special class of instrumented protective system (IPS), which is designed in accordance with ISA 84.01/IEC 61511. The SIS must be separate and independent of the BPCS to assure all of the following:

- Clear and unambiguous indication of what constitutes BPCS and SIS equipment,
- Access to the SIS is minimized to avoid inadvertent changes and spurious trips,
- Access to the BPCS is provided without compromising safety or causing spurious trips, and
- Common cause failure between the SIS and the BPCS is minimized.

<u>Example--High level in compressor knock-out drum</u>

A compressor knockout drum receives gas from an upstream processing unit. The incoming gas is usually dry, but the upstream process can "burp" occasionally sending very wet gas to the knock-out drum. The knockout drum is equipped with a level control loop that automatically drains the drum when the level increases beyond a control set point.

A process hazard assessed by the team is liquid carryover from the knock-out drum to the compressor. The team determines that the liquid could cause significant damage to the compressor, but no loss of containment. The hazard and risk analysis team identifies a level control loop failure as an initiating cause. The resulting process risk requires a risk reduction of 10 to achieve the risk criteria for asset damage.

The team allocates the protective function to the protective instrumented system (PIS). The final design includes an independent PIS that uses a dedicated level transmitter, logic solver, and final elements to initiate compressor shutdown on high drum level.

B.4.1 Instrumented Systems

The integrity level (IL) represents one of four discrete ranges used to benchmark the risk reduction reasonably achievable by each protective instrumented function (PIF) and the PIS. The integrity level is defined numerically by order of magnitude ranges, as shown in Table B.3. During the hazard and risk analysis, the allocated risk reduction is conservatively estimated, because the actual performance should exceed the required performance to ensure adequate risk reduction. In general, the risk reduction factor is assumed to fall near the low end of the range. For example, for IL 1, IL 2, IL 3, and IL 4, the risk reduction is assumed to be 10, 100, 1000, and 10,000, respectively.

Integrity Level (IL)	Hazard Rate [1] (failures/hour)	Risk Reduction Factor (RRF)	Average Probability of Failure on Demand (PFD_{AVG})
4	10^{-8} to 10^{-9}	10,000 to 100,000	10^{-4} to 10^{-5}
3	10^{-7} to 10^{-8}	1,000 to 10,000	10^{-3} to 10^{-4}
2	10^{-6} to 10^{-7}	100 to 1,000	10^{-2} to 10^{-3}
1	10^{-5} to 10^{-6}	10 to 100	10^{-1} to 10^{-2}

Note 1: Also known as frequency of failure and probability of failure as a function of time.

Table B.3. Integrity Level Relationships.

For PIS, the numerical criteria changes, depending on the PIF mode of operation, namely continuous and demand. When the PIF is expected to operate continuously to maintain safe operation, a dangerous PIF failure immediately results in a hazardous event, unless independent action is taken to prevent it. Consequently, the integrity level for a continuous mode PIF is based on its hazard rate (see table B.3). However, most PIFs operate in a dormant or standby mode, taking action only in response to a process demand. A demand mode PIF is judged by its average probability of failure on demand (see table B.3).

It is strongly recommended that a defense-in-depth strategy be used to address the process risk when an IL 4 PIF is indicated. Inherently safer design practices should be employed, where possible, to reduce the inherent risk. Then, other IPLs are used to reduce the process risk, allocating the required risk reduction across multiple IPLs. This defense-in-depth strategy has proven to be more effective than any single layer can be.

Achieving, and more importantly, maintaining an IL 4 design becomes a distinct challenge when transferred from paper to the physical world. While the judicious (and often complicated) application of available technologies implemented in redundant platforms can result in a numerically satisfactory IL 4 design, the lifecycle management of an IL 4 PIF is an arduous process. This level of performance requires a very rigorous and extensive management system to minimize systematic errors in the risk assessment, design, operation, maintenance, and management of change. It is questionable whether the core attributes of any single protection layer can be managed sufficiently to achieve IL 4.

If an IL 4 PIF is implemented, equipment user approval must demonstrate extensive experience in the specified operating environment and substantial analysis and testing. Information must be gained for equipment performing functions of similar complexity, so the hardware and software behavior can be assessed. Operating time must be of sufficient duration that random failures and systematic errors are well understood, allowing rigorous prescriptive assessment, design, operation, inspection, maintenance, and management of change practices to be defined.

B.4.1.1 BMS

Burner management systems (BMS) are addressed in various practices, such as those published by the National Fire Protection Association (NFPA) and the American Petroleum Institute (API). A burner management system (BMS) is a protective instrumented system (PIS) that protects fired equipment, such as boilers, process heaters, ovens, and furnaces. A BMS is comprised of permissives, start-up sequencing, and shutdowns protecting against a wide variety of hazardous events, including safety, environmental, and equipment damage.

The primary hazard associated with fired equipment is the accumulation of uncombusted fuel in the firebox. The ignition of the accumulated fuel can result in significant damage to the fired equipment and potential loss of containment to the surrounding area with impact to plant personnel in the immediate vicinity. The hazards associated with fired equipment can be significant, especially when manual light-off is used. Supervisory functions, such as permissives, are often implemented to facilitate safe light-off and proper start-up sequencing. ISA 84.01/IEC 61511 applies when BMS functions are classified as safety.

BMS design and management is covered by various application-specific practices, depending on the type of fired equipment:

- NFPA 85--Boiler and Combustion Systems Hazards Code
- NPFA 86--Standards for Ovens and Furnaces
- API RP 556--Instrumentation and Controls for Fired Heaters and Steam Generators

The interested reader is referred to the indicated good engineering practices for more specific information.

B.5 MITIGATIVE

The mitigative layer includes mechanical equipment, such as pressure relief devices, blowout panels, and vacuum breakers, and instrumented systems, such as life safety systems, high integrity protective systems, and reactor kill systems. The action of the mitigative layer should achieve or maintain a safe state of the process. The outcome of this action should be predictable and effective and should prevent the hazardous event under review.

These systems are referred to as mitigative because there is often a residual consequence to the equipment taking action. The residual consequence represents a hazardous event that should be analyzed during the risk assessment phase. The consequence may be acceptable, e.g., pressure relief to an appropriately sized flare system, or unacceptable, e.g., the atmospheric release of a toxic substance.

B.5.1 Mechanical Equipment

Most mechanical equipment acts in response to specified process conditions (e.g., pressure at or higher than a set point) and initiates a specific mitigative action (e.g., PRD opens). The action of this equipment is very predictable and the equipment failure modes are well understood. Lack of adequate inspection and preventive maintenance is the primary cause for mechanical equipment failure.

Table B.4 provides some examples of mechanical equipment and RRF achieved based on specified conditions. In general, a mechanical device should meet all of the following to be considered a protective function:

- Be specifically designed to mitigate the identified hazardous event,
- Be installed to minimize potential operating environment problems, e.g., the use of purging with pressure relief valves, and

- Be maintained such that periodic inspection reports demonstrate that the mechanical device works as required.

Mechanical Device	Conditions	RRF
Pressure Relief Valve	Clean Service. PRV must be sized to completely mitigate the scenario.	100
	Multiple full-load PRVs are available to mitigate scenario.	1000
	Multiple partial-load PRVs are available and sized such that more than one PRV would need to fail for the scenario to occur.	100
	Multiple partial-load PRVs are available, but more than one is required to mitigate the full load. This includes staged release PRVs.	10
	Plugging service with no protection. An unprotected PRV used in a plugging service is not considered sufficient for consideration as an IPL.	1
	Plugging service with protection. The design is based on prior history in similar services and may include the use of specially designed PRVs, inlet header purges, and close coupled rupture disks. If pluggage can be caused by polymerization during venting, these special designs are generally insufficient.	100
Vessel Rupture Disk	Must be designed to mitigate scenario.	100
Flame Arrester	Designed for the hazard (deflagration or detonation) and inspected periodically.	100
Vacuum Breaker	Designed for the hazard and inspected periodically.	100
Blowout Panels	Must be designed to mitigate scenario.	100
Overflow Line	Overflow line is designed to discharge to a safe location and is sized to address the scenario.	100

Table B.4. Examples of Mechanical Mitigation Device RRF.

Example--Mitigative layer secondary consequence

In Seveso, Italy in 1976, a partially processed batch of chemicals used to produce 2,4,5-trichlorophenol, a bactericide, decomposed exothermically. The mixture of chlorinated organics produced 2,3,7,8-tetrachlorodibenzo-p-dioxin (TCDD), which is a known unavoidable reaction byproduct. TCDD is widely considered one of the most toxic, man-made chemicals.

At the elevated temperatures, more TCDD was produced than typical, so that when the rupture disk opened to relieve the excess pressure, a significant quantity of TCDD (~2 Kg) was released to the surrounding area with the reactor contents (~3000 Kg). The rupture disk worked as designed and the reactor was essentially undamaged by the event. The risk reduction strategy was still poor, because it ignored the consequences of the atmospheric release.

Within a week of the incident, animals began to die in mass and skin lesions appeared on children. More than 80,000 animals were eventually destroyed to prevent the contamination from getting into the food chain. Approximately 37,000 people were exposed to the release with those directly exposed to the gas cloud developing choroacne, a severe skin disorder usually associated with dioxin, within a couple of months of exposure. The most contaminated area, which covers approximately 100 hectares (247 acres), was completely evacuated and is now a park.

B.5.2 Instrumented Systems

Life safety systems, high integrity protective systems, and reactor kill systems are examples of mitigative systems that rely on instrumentation and controls. The systems are generally designed according to ISA 84.01/IEC 61511. These systems may be implemented in a separate, dedicated system or in the unit PIS (SIS), taking into consideration:

- Application,
- Initiating causes,
- Other protection layers,
- Function complexity,
- Risk reduction requirements,
- Common cause, and
- Process safety time.

In all cases, these systems must be separate and independent of the BPCS and must be designed and managed to achieve the required risk reduction.

B.5.2.1 High Integrity Protective Systems (HIPS)

In conventional design, a Pressure Relief Device (PRD) is used as the primary means of over-pressure protection, and a disposal system, such as a flare or scrubber, is used to safely process the material relieved during an over-pressure event. However, there are cases where a PRD is simply ineffective:

- PRD is routed to relief system (relief valves or header) that is not sized to mitigate a single or multiple relief scenario,
- Pipeline protection where piping is underspecified by design to reduce cost or due to physical limitations, and relief devices are not adequate or cannot be routed to a safe location,
- Pipeline protection where existing manifolds and piping is used to connect to higher pressure wells and where submersible pumps are used to raise the pipeline pressure to reach long distance collection points or to increase production rates,
- Pipeline protection where existing pipeline is de-rated due to corrosion or erosion,
- Exothermic reactions that can "run away" at rates faster than the relief device is capable of reducing pressure,
- Polymerization or depositing of materials during normal operation that historically have resulted in partial or complete relief device blockage, or
- Reactive process chemicals that when relieved into common headers may interact with other materials in the header, thus plugging the header and rendering the relief device useless.

API and the American Society of Mechanical Engineers (ASME) provide criteria for the protection of vessels and pipelines from excess pressure using instrumented systems. HIPS are a special class of SIS and should be designed and managed according to ISA 84.01/IEC 61511, as well as applicable API or ASME requirements. Specific regulatory and enforcement jurisdiction requirements should also be determined. In some instances, approval of local authorities may be required.

For new vessels, ASME Code Case 2211 (1999) requires that the user ensure the maximum allowable working pressure per Section VIII, Division I, Para UG-98 of the vessel is greater than the highest pressure that can reasonably be achieved by the system. The Code Case is only applicable when it is not intended to install a PRD on the vessel. If a PRD is required for any case, the PRD must be sized for all credible overpressure scenarios.

The Code Case requires a detailed analysis to determine the maximum pressure developed due to credible abnormal events. This analysis should not take consequence severity into account when determining the risk reduction requirements for the instrumented systems. The Code Case requires that the frequency of vessel over-pressure be reduced to non-credible.

The fourth edition of API 521 allows credit for a favorable response of instrumented systems when the use of a PRD is impractical. The recommended practice refers to these systems as HIPS and states that they should provide protection equivalent to or better than a pressure relief device.

The need for HIPS may be identified during a unit hazard and risk analysis or in a focused study examining global hazardous events. Additional analysis is often performed for these events, such as process simulation, reactive calculations, dynamic simulation, and flare load calculations. See Chapter 3.2 for more discussion on global hazardous events.

The failure modes of PRDs are well known and predictable. When a spring-operated PRD opens during normal operating conditions, the effect on the process is typically a process disruption and is self-correcting when the pressure returns to normal. In contrast, the HIPS results in a process shutdown, because it is designed to take the process to a safe state. It is also designed to maintain the safe state until the operator resets it.

For runaway reaction applications, the HIPS may be preventive or mitigative. For example, the HIPS may prevent a runaway reaction by closing the feed valves when the feed rate is above the safe operating limit. Or, it may mitigate a runaway by detecting pressure and injecting a chemical that kills the reaction.

HIPS are generally expected to achieve SIL 3. This was considered by Lawley and Kletz (1975) who suggest that if an SIS is used instead of a relief valve, it should be designed to provide a reliability which is 10 times that of the latter. The reason for this is the uncertainty in the relief device failure rate data and the difference in the modes of failure; a relief device that fails to operate at the set pressure may nevertheless operate at higher pressure, whereas an SIS is more likely to fail completely.

The significant impact of HIPS on the process operation often demands that the HIPS be highly reliable. Redundant equipment, fault tolerance, diagnostics, and frequent inspection and proof testing are generally necessary for the HIPS to achieve reliability similar to a PRD. HIPS complexity requires careful consideration for common cause, common mode, and systematic failures. This complexity increases when networks of vessels are involved, risk propagates due to trips, and fast response times are required.

A HIPS may protect a network of vessels from a global hazardous event or a single vessel from abnormal operation. Examples of global hazardous events are provided in Chapter 3.2. Many global events are caused by loss of utilities, which may affect the facility, process areas, or process units. Single vessel events may involve runaway reactions, hot spot formations, or de-rated vessels or pipelines. Additional examples are provided at the beginning of this section.

In many cases, the consequence severity is difficult to assess due to the uncertainty of where the overpressure will be relieved in the vessel network. If low pressure and high pressure equipment relieve simultaneously, the backpressure may affect the low pressure system first, assuming there are no mechanical integrity issues with the equipment or the relief system.

When multiple vessels are involved, the process risk can be difficult to assess unless there is a detailed evaluation of the relief load for each identified global hazardous event. The evaluation should identify the combinations of vessel relief that potentially challenge the relief system design basis. These combinations determine the fault tolerance of the relief system design basis. It is highly recommended that the relief system be fault tolerant and include sufficient safety margin to account for uncertainty.

If the relief system cannot tolerate the relief of a vessel, the relief system cannot tolerate the failure of the vessel HIPS function. When multiple vessels simultaneously relieve, the system design and verification becomes more complex. If the relief system cannot handle any of the loads, the failure of any HIPS function results in the failure of the relief system.

The risk reduction must be achieved by the overall system of HIPS functions protecting vessels connected to the relief system. Achieving the risk reduction for the overall system places very high risk reduction requirements on the individual functions. If multiple vessels are involved, it may be extremely difficult to achieve the overall risk reduction. The design basis should place special emphasis on independence and separation of the functions, equipment redundancy, and frequent proof testing.

It is preferable that the relief system be designed to relieve simultaneous vessel loads. When the relief system is capable of safely managing multiple vessel loads, the relief system design generally lowers the requirements on the individual HIPS functions. For example, if the relief system can tolerate the relief load of any one (1) of 5 vessels, but not 2 of 5, the relief system can tolerate the failure of only 1 of the 5 HIPS functions.

The susceptibility of the overall system of HIPS functions to common cause, common mode, and systematic errors should be assessed and means implemented to reduce their occurrence when practical. When multiple HIPS functions are placed in the same logic solver, the logic solver becomes a single point of failure for the spurious shutdown of multiple vessels and for the dangerous failure of the relief system.

The propagation of risk should be considered during the development of the risk reduction strategy and the process requirements specification. The HIPS takes action on the process to prevent over-pressure, which results in isolation or venting of a piece of process equipment or a process unit. Interconnected equipment are typically affected, potentially leading to process demands on other protective systems. For example, the shutdown of a reboiler results in a loss of fractionation (stripping), causing light components to pass to the next column which may subsequently over-pressure.

Process safety time is generally critical in HIPS applications. A PRD opens in milliseconds in direct response to process pressure greater than its set pressure. In contrast, the HIPS must detect an unacceptable process condition and take action on a process. HIPS response time is generally longer than the response time for a PRD. For example, if the HIPS initiates the closure of large valves, the response time can extend to tens of seconds unless special design practices are followed. Then, the process itself may take seconds, or even minutes, to respond to the HIPS action. For example, it may take minutes for heat input to stop after a reboiler is shutdown.

B.5.2.2 Life Safety Systems

Life safety systems originated as part of NFPA 101, "Life Safety Code," which essentially dealt with fire and flammable gas protection. The semi-conductor industry adopted the term and expanded its use to toxic gas protection, as well as the detection and response to pyrophoric material releases.

Although not formally defined in industry, life safety systems are intended to either increase the level of safety for occupants or control the spread of harmful effects due to loss of containment. A life safety system should achieve or maintain a safe state. Life safety systems may restrict access to a building, such as locking a door, or initiate automated actions, such as closure of a feed supply valve, initiating a deluge system, increasing ventilation, etc.

When determining the risk reduction provided by life safety systems, consideration of process safety time is critical. The number and location of sensor(s) should be examined to determine the likelihood of detection within the process safety time. Achieving any level of certainty associated with detection requires careful design and sensor placement.

In addition to event detection, the likelihood of response success and the time required to respond and achieve the desired action should be evaluated. In the case of flammable or toxic gas release, the outcome of the action can be very difficult to estimate. Systems that take action directly on the process are easier to assess and the outcome is more predictable.

B.6 BARRIERS

Physical barriers do not have to act to reduce the loss of containment or incident frequency; they passively reduce the consequence through their physical presence. Examples include sloped dikes, insulation, berms, bunds, and explosion barriers.

In general, when the physical barrier is designed, constructed, inspected, and maintained according to good engineering practice, there is a very high probability that it will function as intended when required. Barriers should be inspected at a frequency sufficient to ensure their integrity, and their design basis should be periodically audited to verify its appropriateness given site or installation changes (e.g., tanks were supposed to operated at less than 50% capacity, but now run 85% capacity).

The hazard and risk analysis procedure should define how physical barriers are considered during the study. Many owner/operators consider the presence of these layers when evaluating the consequence severity of the hazardous event. In this way, barriers are treated similarly to the process equipment and piping. The barriers are assumed to work if they are designed and maintained according to specification. Others evaluate the consequence severity without consideration for the physical barrier. Then, they list the physical barrier as a protective function with a defined risk reduction target.

B.7 LIMITATION

The limitation layer implements fire & gas, emergency dump, emergency shutdown, hydrocarbon disposal, water/steam curtain, and deluge systems. The limitation layer is designed to reduce the incident severity associated with loss of containment. The hazardous event may arise due to a wide variety of events, including an initiating cause that was not considered credible, an unforeseeable chain of events, a general loss of equipment integrity (e.g., flange leak, pump seal leak), or the failure of a protection layer during a process demand. Rapid detection and response to leaks is important to minimize the potential consequence severity. The faster the detection and response, the smaller the potential vapor cloud, and therefore less potential energy is released if it ignites.

Limitation systems have a role in reducing risk but, in general, they are not treated as an IPL in the hazard and risk analysis due to their unknown and uncertain effectiveness. A limitation system does not achieve or maintain a safe state of the process. It acts after the event has occurred to reduce the incident severity. The preferred risk reduction strategy is to focus on the use of the control, prevention, and mitigation layers, which yield a known outcome and a specified safe state.

Fire and gas, fire protection, and emergency shutdown systems are designed to respond after a loss of containment within a designated area. These systems are implemented to lessen the effects of the release within the zone, to minimize the potential for the incident to propagate to surrounding equipment, and to bring the incident under control faster.

Fire and gas systems detect the presence of a gas (e.g., hydrocarbons, toxic gas), or the presence of fire and/or smoke. The layout of fire and gas detectors depends on the type of leak, release size, location, and air movement patterns. For flame detectors, the likely source of flame must be identified, followed by an assessment of the detector cone of vision and the physical obstructions potentially blocking its view--flame detectors have been known to cause spurious alarms due to sunlight.

Heat detectors are much slower in detecting an event than flame detectors, but cause fewer spurious events. Smoke detectors are used to protect personnel from exposure and should be used in any areas where there is a high potential for fire involving equipment or electrical systems. Infrared and ultraviolet electronic detectors can fail to detect releases due to smoke from combustion and to background electromagnetic radiation.

Designing a fire and gas system is very difficult. It is a challenge to detect the hazard with certainty and to achieve a predictable outcome from the system response. Fire detectors include fusible links, pneumatic fusible loops, and electronic flame, smoke, and heat detectors. Gas detectors are provided for ventilation systems and identified risk areas.

Detection by personnel has proven to be at least as good, if not more effective than automated detection, in facilities that are normally staffed. An operator is significantly more effective in detecting a liquid hydrocarbon leak than electronic detectors. A means for notifying personnel of the need to initiate the system is a requirement for those fire protection systems that pose risk to personnel, such as the use of carbon dioxide to inhibit combustion.

Fire and gas systems typically initiate alarms to notify personnel that evacuation is required and to activate fire protection systems. Active fire protection systems include water deluge, carbon dioxide dumping, dry chemical agents, clean agents, foam, or water. In some cases, fire and gas systems also activate an emergency shutdown system (ESD). In off-shore applications, fire and gas systems and emergency shutdown systems are integrated systems known as the emergency support systems (ESS).

The primary purpose of an ESD is to isolate the source of the loss of containment from major facility inventories, e.g., isolation from main supply pipeline or feed tanks. ESD action may be an automatically initiated response to a fire and gas system as described above; however, a manual pushbutton should be provided to allow the operator to initiate inventory isolation, as well as any other necessary ESD action. These manual pushbuttons are sometimes called "panic buttons" and may be locally or remotely initiated. ESD systems may also take other actions associated with the process unit, such as:

- Emergency depressurization,
- Secondary isolation of the affected equipment from other process units,
- Electrical equipment isolation to prevent electrical fires or ignition sources,
- Ventilation system shutdown to minimize ingress of smoke, flammable or toxic gas,
- Automatic fire protective system initiation, and
- Evacuation of personnel.

The criticality of the above functions should be considered during design and should be included in any risk reduction evaluation, as appropriate. As system complexity increases, it becomes more difficult to achieve the integrity requirements. In estimating the effectiveness of the limitation system, the following should be considered:

- Several sensors are usually present within a designated zone,

- The outcome of the event is a function of detection capability, which is related to the release size, release location, and dispersion,
- A number of hazardous events can occur within a zone and each has a potentially different outcome,
- Partial failure to function may not significantly increase the risk (e.g., even if the isolation valve failed partially closed rather than fully closed, the release rate would be decreased, lessening the incident), and
- Other means of process isolation generally exist.

Although fire and gas systems and ESD are instrumented systems, they are not SIS and are not designed in compliance with ISA 84.01/IEC 61511. They are safety equipment and should be tested, inspected, and managed as such. These systems are sometimes implemented in the same logic solver as the SIS. When this occurs, the entire system must be managed to at least the level of rigor required for the SIS.

In contrast to the control, preventive, and mitigative layers, limitation systems, especially the fire and gas systems, tend to be implemented as energize-to-trip. This is due to the undesired consequences of spurious operation. See Appendix F.5.2.2 for discussion of energize-to-trip design. For hydrocarbon releases, the equipment should be classified as intrinsic or explosion proof, since it is intended to survive the event and not serve as an ignition source. Consideration should be given to fire, blast, and flying debris protection for IPS equipment.

Limitation systems are often implemented according to good engineering practices, local regulations, and insurance requirements. When the hazard and risk analysis does consider limitation systems, the system must be designed and managed to achieve the allocated risk reduction. Most owner/operators limit the claim for these systems to a risk reduction factor of 10, because a higher risk reduction claim is difficult to achieve and maintain using conventional technology and, in many cases, is simply impractical. Higher risk reduction claims should be justified by detailed analysis, such as facility design and layout, fire and explosion study, historical equipment performance, a detailed quantitative analysis, and consequence modeling. Table B.5 provides examples of RRFs for limitation systems.

System	Conditions	RRF
Fire Detection with Water Deluge System (FDS)	Using fire or smoke detectors to automatically activate a system to prevent or control a fire, e.g., wet pipe or dry pipe systems with fusible links, deluge systems or water curtains applying water, foam, or other chemical fire suppressants.	10
Gas Monitors with Automated Deluge (GDS)	Using mist, vapor, or other detectors to automatically activate a system to "knock down" a chemical mist or vapor release, or to ventilate a building, prior to ignition of the release. For example, a deluge system or emergency ventilation system.	10

Table B.5. Examples of Limitation System RRF.

Fire and gas hardware manufacturers provide recommended installation and maintenance practices (e.g., inspection and test intervals), and these recommendations should be considered during the design and operating basis development. Following industry practices, manufacturer's recommendations, and an appropriately rigorous management system enable fire and gas systems and ESD to achieve good engineering practice.

> <u>Example--Limitation layer importance</u>
>
> In Mexico City, Mexico in 1984, an 8" line ruptured at a liquefied petroleum gas terminal which had a storage capacity of 16,000 m3 in 6 spheres and 48 horizontal cylinders. A loss of pressure was indicated in the control room and in the pipeline pumping station. The operators tried to identify the cause of the pressure loss as the gas cloud grew to encompass a large area. The release took 5 to 10 minutes to reach the flare.
>
> Plant workers attempting to return to the control room and to man the fire pumps were found dead and badly burned. Gas detectors were not installed to detect the release, so when low pressure was detected, plant workers were sent to the field to investigate. It only took minutes for the cloud to reach the flare, which was an insufficient time for workers to troubleshoot hazards in the field and take corrective action. At some point during the event, the ESD button was pressed, but it proved inadequate because it simply isolated the facility from the main supply. It did not address the inventory of gas that would be released due to the interconnectivity of the vessels. An integrated gas detection and ESD would have reduced the consequence severity.
>
> The spacing between the spheres and cylinders was also insufficient, demonstrating the importance of plant layout. The vessels were

simultaneously affected by the hazardous event. The vapor cloud ignition created a ground fire that led to the subsequent sympathetic explosions with the eventual loss of 4 spheres and 44 cylinders. Fragments of vessels became missiles and were thrown off-site causing fires to spread through the surrounding community. The facility was completely destroyed and more than 500 people were killed and 4,500 people were injured.

B.8 RESPONSE

The response layer includes various emergency alarms that notify personnel and/or the community to shelter-in-place or to evacuate to safe zones and muster emergency response personnel Requirements for emergency response began appearing in European regulations in the 1970s and in United States regulations in the 1980s. Now, most governments require that owner/operators develop emergency response plans to coordinate owner/operator activities with those of local emergency responders, fire departments, medical facilities, and police.

The international clamor for emergency preparedness regulations was in direct response to four incidents that shocked the world. In two cases, there was a near total loss of plant assets and in all four cases, there was significant community impact. These incidents were in Flixborough, United Kingdom (1974), Seveso, Italy (1976), Bhopal India (1984), and Pasadena, Texas (1989). Summaries of these incidents can be found in *Lee's Loss Prevention in the Process Industries*. The importance of good plant and community response plans cannot be overstated.

From an instrumentation and controls perspective, emergency response often relies on communication equipment and alarm systems to warn people inside and outside of the facility of the hazardous situation and the need for shelter-in-place or prompt evacuation. The alarm systems and communication equipment should be independent of systems that could be affected by the process incident. They also should be frequently tested. Many companies test emergency response alarms at weekly.

C

CORE ATTRIBUTES

Independence, functionality, integrity, reliability, auditability, access security, and management of change are fundamental characteristics of an independent protection layer (IPL). These characteristics are considered core attributes, which can be assessed throughout the lifecycle to determine the degree to which the attributes have been achieved. The core attributes are interrelated, such that a lack of rigor associated with implementation of one attribute can influence the effectiveness of another attribute. Consequently, the specific means used to address each attribute must consider its potential affect on other attributes.

For example, management of change is typically supported with access security provisions. Controlling access minimizes the risk of unapproved changes. Access security and management of change processes provide an audit trail to determine who obtained access and whether their activities were approved. Controlling changes ensures that functionality, integrity, reliability, and independence are maintained throughout the equipment life.

General work processes, assessment criteria, documentation, and required approvals should be specified in internal practices to ensure that the core attributes are given appropriate consideration. These practices may incorporate what is known about the IPL from its design and management limitations. The IPL cannot achieve better risk reduction than its design and management allow.

A distinguishing characteristic of each IPL is the level of rigor used to define, apply and assess the core attributes. The level of rigor appropriate for a particular IPL may be established by government requirements, good engineering practices, insurance requirements, or manufacturer's recommendations. The following sections focus on the application of the core attributes to instrumented protective systems (IPS). In some cases, this application is contrasted with the attributes of the basic process control system (BPCS). This is not intended to indicate how a BPCS should be designed or to place any requirements on the BPCS design. It is merely provided to contrast the rigor associated with the IPS attributes with that of the BPCS.

C.1 INDEPENDENCE

For a protection layer to be considered independent, its performance should not be affected by the occurrence of the initiating cause, its consequences, or by the failure of another protective function used to reduce the risk of the same hazardous event. The correct operation of the protection layer should not be conditional on any other layer and its separation from other layers should be unambiguous.

The interdependencies in the design and management of the physical systems used to control and shutdown equipment must be understood, so single points of failure are addressed appropriately. Dependencies, or common cause, should be identified and means implemented to reduce their probability of occurrence is sufficiently low compared to the overall hazard rate or risk reduction requirements

Independence within the management system should be demonstrated for any administrative process used for risk reduction or used for reduction of systematic error. For example, how will an independent verification of the manual operation be incorporated into the procedure and audited for compliance? The independent assessment should also consider who performs and manages the inspection, preventive maintenance, and maintenance activities for each protection layer. Lack of independence necessitates focus on procedures, training, and approval processes to ensure the layer is managed appropriately.

Independence of the physical systems is easy to demonstrate and maintain when the protection layer is physically separate from the initiating cause and from any other protection layer used to reduce the risk of a hazardous event. Figure C.1 shows an example of a physically separate and diverse BPCS and PIS.

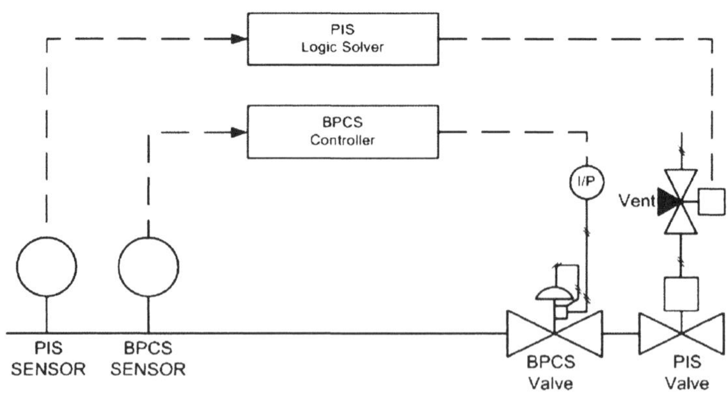

Figure C.1. PIS and BPCS are Separate and Independent

Figure C.2 provides an example where the BPCS and PIS are not independent. When protection layers are combined in this way, the system must be managed to the highest level of rigor required by the layers. The system must also achieve the overall hazard rate or risk reduction required from the two layers. For example, if the PIS is a safety instrumented system (SIS), the system represented in Figure C.2 must be designed and managed as an SIS.

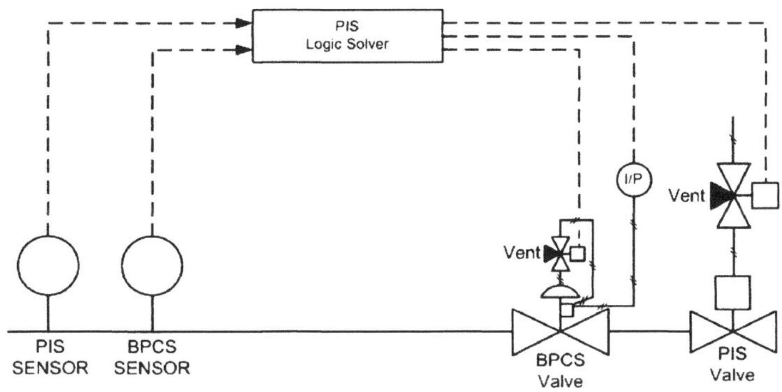

Figure C.2. PIS and BPCS are not Independent or Separate.

C.2 FUNCTIONALITY

Functionality is defined by the required operation of the protection layer within the overall risk reduction strategy. The protection layer must be capable of operating as expected during actual service conditions and during all process operating modes where the hazardous event can occur. Operating modes include any planned process operating state, such as:

- Normal,
- Start-up,
- Shutdown,
- Maintenance,
- Instrument testing,
- Preparation for maintenance, or
- Equipment bypassed or out of service.

Gaining confidence in the protection layer operation is why inspection, preventive maintenance, and proof testing activities are so important. The protection layer must be capable of responding effectively within the time required by stopping the propagation of the initiating cause, even in the presence of other protection layer failures. This requirement along with the core attribute of independence generally results in the use of separate equipment and management systems for each protection layer.

The above establishes the minimum functionality requirements; however, most owner/operators have greater requirements for IPS functionality. IPSs are expected to be easy to use, be simple to implement, have a low cost of implementation, require little training for a diverse set of users, have a low cost to upgrade and to maintain, and have the ability to communicate to a wide range of other systems.

Modular designs are often required to simplify the operation of each function within the IPS. Modular design minimizes the connections between functions, reducing coupling. Each protective function should be clearly specified to achieve a desired outcome, establishing what is commonly referred to as the "cause and effect" relationship between the initiating cause and the desired effect. An IPS implements instrumented protective functions (IPFs), which detect abnormal or unacceptable process conditions and take action to achieve or maintain a safe state of the process. An effective action may include sequential or immediate shutdown.

The reduction of the system to its individual functions allows function classification and provides traceability between the design and management of the function and the required risk reduction. The system should also be examined holistically to understand the potential interactions between the hardware, software, and communication components of the system.

Personnel responsible for maintaining these systems must understand how these components work within the complex system. An incorrect change in one component may adversely affect the system, leading to a loss of functionality. For this reason, access security must be provided at the system level and management of change processes must assess system impact, as well as impact to the function.

C.3 INTEGRITY

A protection layer's integrity is related to the risk reduction that can reasonably be expected given the protection layer's design and management. *Guidelines for Safe Automation of Chemical Processes* (CCPS 1993) referred to integrity as "dependability." Integrity is limited by the rigor of the management system used to identify and correct equipment failures and systematic errors. Random and systematic events can prevent the protection layer from operating as intended when a process demand occurs.

During the risk assessment phase, the required integrity is defined for each protective function using qualitative, semi-quantitative, or quantitative methods as discussed in Chapter 3. Each protective function is allocated to a protection layer whose core attributes are managed sufficiently to achieve the required integrity. The term used to describe the integrity requirement varies with the hazard and risk analysis method, and include availability, risk reduction factor (RRF), hazard reduction factor (HRF), probability of failure at a function of time (PFD(t)), average probability of failure on demand (PFD_{AVG}), and frequency reduction factor (FRF). This guidelines book uses the term "risk reduction" for consistency purposes, but other terms are equally acceptable in expressing the concept.

C.4 RELIABILITY

People often associate reliable with dependable, trustworthy, and consistent performance (Green and Bourne 1972). A reliable protection layer operates as intended when required (in low demand mode) and continuously (in continuous mode), does not require frequent repair and maintenance, and does not cause an inadvertent process disruption or shutdown.

Reliability is an attribute of a protection layer related to its equipment operating as intended, under stated conditions, for a specified time period. The attribute is affected by the protection layer design and measured during the operational and mechanical integrity phase. Reliability and availability are the two main probabilistic parameters used to verify that the protection layer operates as specified in the operating environment.

A variety of measures for estimating the attribute reliability have been suggested. Most revolve around some expression of success or failure. For example, consider a given number of devices that have experienced a given number of failures in a specified time period. As the number of failures within the device pool increase, the protection layer is perceived as less reliable and eventually as unreliable. As the number of failures in a given device pool decrease, the protection layer is considered more reliable and eventually as dependable. Another measure is the system availability (ratio of the up time by the period of time under consideration) or the failure rate (ratio of the number of observed failure by the cumulated running time).

The impact of equipment unreliability can be seen in near miss and incident investigation reports, trip logs and reports, and maintenance reports. If an IPF fails to operate when required, it has experienced the worst form of unreliability; one that may have, or could have, resulted in an incident. This facet of reliability is addressed through the integrity requirements, which are discussed in Section C.3.

Spurious operation results when IPS equipment fails in a way that the IPF partially or completely takes action on the process in accordance with its specification. A spurious trip has an immediate impact on the process operation and does not necessarily lead to a safe state. Spurious operation can be dangerous, especially when the function partially acts. Potential hazardous events caused by partial or complete IPF action should be analyzed during the risk assessment phase.

Management sees the direct impact of IPS reliability when IPS equipment initiates an unnecessary process shutdown. The IPS design should consider the false, nuisance, or spurious trip rate, which may be defined by a quantitative performance metric. The IPS design is then adjusted to minimize the spurious trip rate, while achieving the required integrity. Long-term performance is monitored to determine if the reliability requirements are met.

C.5 AUDITABILITY

The protective management system defines the work processes used to develop and maintain the required information, documentation, and procedures. Layers of oversight are built into the management system based on many factors, including consequence severity, required risk reduction, potential risk, and task complexity. Auditability is intended to make risk management more transparent, evidence based, and demonstrable. It defines the way in which an owner/operator ensures that personnel are doing what they are supposed to do and IPSs are performing as intended. It relates to the ability to demonstrate the adequacy of and adherence to the design, maintenance, inspection, testing, and operation practices used to achieve the core attributes.

Assessments, verifications, inspections, and audits provide the opportunity to self-examine internal processes and to identify ways to become more effective. A variety of techniques may be used, from informal qualitative reviews to formal detailed audits. These techniques are the internal checks and balances, which provide evidence that appropriate efforts are being made to address process risk.

Auditability is often limited by the identified, observable characteristics required by the management system. The "audit trail" or "paper trail" should be traceable through the lifecycle from the hazard and risk analysis findings to reports on the installed and operational IPS. Increasing auditability rigor results in higher quality protection layer design and management. This ultimately impacts plant operation in a positive manner with better performance and improved accountability.

C.6 ACCESS SECURITY

Significant business disruptions have resulted from viruses, Trojans, worms, and similar computer and communication network attacks. According to a study conducted in 2005 by the Computer Security Institute, 90 percent of the 650 participating companies detected computer security breaches during the previous 12 months. Seventy-four percent acknowledged financial losses due to these security breaches with 273 respondents. The reported losses exceeded $265 million.

Despite the implementation of firewalls, there are significant numbers of unauthorized "inquiries" and penetrations into instrumented systems each year. According to the Industry Security Incident Database (ISID), 22 participating worldwide companies in the oil & gas, chemical, power, food, and water industries reported 83 confirmed and 35 likely, but unconfirmed intrusions; plus another handful in the unknown or unlikely category.

During a demonstration in 2005, three cyber security researchers from the Idaho National Laboratory hacked into a "secure" BPCS and successfully penetrated two firewalls. The researchers obtained access to a proprietary data communications highway, falsified the operator's interface, and started a pump without the operator being alerted. It took less than 20 minutes for the researchers to access the BPCS and start the pump.

Security experts characterize attackers as:

- GRP I – *Crackers*, persons who modify scripts and programs to compromise computer performance,
- GRP II – *Terrorists*, organizations intent on creating shock and awe, and
- GRP III – *Nation States*, organizations who are intent on creating serious, perhaps unrecoverable, financial destruction.

The Flattening of the Line cyber-security risk graphic (Figure C.3) illustrates the likelihood and resultant consequences each group is likely to target. A fourth group of potential attackers are the *Insiders* who are typically disgruntled personnel with easy access to passwords, systems, and assets.

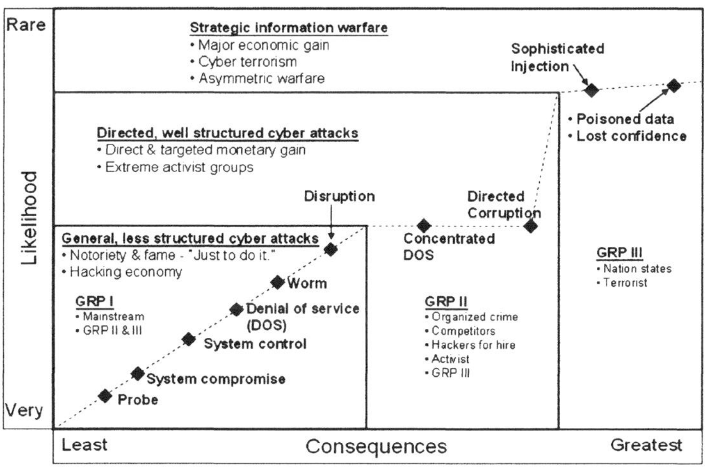

Figure C.3. Cyber-security risks. (Nelson 2006)

Consequently, access security is a safety-critical issue, requiring careful planning and discipline. The first question to consider when determining physical security requirements is "What is the reason for security?" Given that the range of responses includes minimizing inadvertent human error to preventing sabotage, the requirements can vary significantly.

Access security is achieved with a two pronged approach: 1) administratively with policies, practices and procedures to monitor, control, and audit and 2) physically with independence equipment supported by barriers, such as keys, locks, and passwords. Data, software, and hardware must be managed at the level of rigor required for the protection layer to achieve the performance expectations.

Access security should be implemented for any activity that potentially affects a core attribute of the system. Engineering activities often include logic, set point, and functional changes, while maintenance activities may involve installation, commissioning, repair/replacement, configuration management, and bypassing. Access security should be implemented for operations activities, such as bypassing, set point changes, alarm response, event reporting, and documentation.

Many control systems can now be accessed by other computer systems from remote locations, as discussed in Appendix B.2.1. While most BPCS controllers use proprietary real-time operating systems, the controllers are not immune to attack from internal and external sources, as discussed previously.

Security risks are aggravated in modern BPCSs, where the operator interface uses Intel/Microsoft technology. This software platform can provide the operator with access to standard business software, such as company e-mail, Microsoft Office software and other company-specific applications. While this capability may provide some nominal benefit, it also increases the risk that viruses, Trojans, worms, etc. can be introduced into the operator interface or that hackers can successfully falsify or "lock-up" the interface.

Remote access and the presence of other software increases the probability that corrupt information or viruses can be transmitted to the BPCS operator interface, engineering work stations, and potentially to other interconnected systems. When open access is provided, the interconnected systems become vulnerable to the outside world. Passwords, firewalls and encryption are used to provide basic protection against hackers. Additional means should be used to isolate and filter network information so that it can not affect any protection layer.

Some PE logic solvers provide the ability to define user access levels. A tiered access system is created where a maintenance technician or operator can view the current state of a parameter, but is unable to alter that parameter. User authentication can be multi-level with the implementation of name, location, and process identifiers to ensure that personnel are accessing the correct information.

Tiered access systems and passwords are generally considered appropriate for controlling access to the BPCS, but are not sufficient to control access to IPS, especially SIS. Access controls and monitoring should tighten as the integrity requirement increases. Effective security uses a combination of methods to ensure access control of the hardware and software and to prevent unintentional or unauthorized changes to protection layers. In some cases, more rigorous means should be considered, such as controlling physical access to the interface by locking it in a room or cabinet, disconnecting the engineering interface when it is not in use, or placing the interface in an area where it can be continuously monitored.

Access to IPS field equipment is typically managed using administrative controls supported by design features, which provide notification or reporting capability. Even the best designed, most highly inspected and tested system can be defeated by the inappropriate use of a bypass switch or the inadvertent change of a set point. Where practical, a bypass alarm should be used to notify the operator when a bypass is enabled.

Administrative controls may include management of change, approved procedures, required notifications, permits and their approvals, or any combination of these. Access security is enhanced when IPS equipment is distinct in appearance and labeling. It should be readily apparent to casual observers when someone is working on IPS equipment. For example, field devices can be painted or tagged with a color coding scheme to alert personnel and to trigger a response to double check actions accordingly.

Preventing intentionally malicious sabotage requires significantly more intensive and formal oversight and monitoring. Greater security typically results in a less user-friendly system with a high level of procedural bureaucracy. Consequently, each operating facility and process should define the degree of security needed and the measures necessary to achieve and maintain the core attributes.

C.7 MANAGEMENT OF CHANGE

Successfully managing changes to process chemicals, technology, equipment, and facilities is a significant attribute of safe and reliable operation. Properly managing and documenting modifications is necessary to sustain the core attributes throughout the protection layer lifecycle. The management system should employ a management of change (MOC) or change management process to review, document and approve changes prior to implementation.

MOC establishes a formal process where risk arising from proposed changes is identified and managed properly. MOC ensures that changes made to the process unit and its associated protection layers do not negatively or inadvertently compromise the other attributes. Changes that are considered "replacement in kind," such as the substitution of one item for an identical one, are not reviewed under the MOC process.

Changes affecting the management system implementation should be evaluated under MOC. Resource changes can affect the ability to maintain the rigor of the management system or to fulfill its requirements. Changes such as staffing, reporting authority, or site organization may result in the need to modify work processes or the design basis. For example, reductions in maintenance staffing may require design basis modification to reduce maintenance requirements.

D

UNDERSTANDING FAILURE

Control systems are designed to keep the process within specified process parameters considered acceptable for normal and safe operation. When the process exceeds the normal operating limit, it may propagate into a hazardous event with potential consequence to human life, the environment, and plant assets. The risk assessment phase identifies hazardous events that pose significant consequences and defines the means used to prevent their occurrence. The identified risk may be reduced using an instrumented system implemented in the preventive layer, such as a protective instrumented function (PIF).

> Example--Loss of vessel level
>
> Consider the case where a loss of vessel level allows high pressure gas to pass to downstream equipment that is not rated for the gas pressure. Assume a PIF is specified to reduce the risk of this event. The PIF detects low level and responds by closing an outlet block valve to prevent the gas blow-by.
>
> Three redundant level transmitters are used to detect a low level condition. Other independent level devices are used by the basic process control system to monitor and control the vessel level. When any two of the three (2oo3) PIF level transmitters detect low level, the PIF closes the outlet block valve. The PIF can still work if one level transmitter fails dangerously; however, if two transmitters fail dangerously, the PIF will fail to close the valve, allowing the level to be lost in the vessel with potential catastrophic consequences.
>
> The dangerous failure of two level transmitters may be unlikely if the devices are designed, installed, operated, inspected, and maintained according to the requirements of the design and operating basis. When these requirements are not met, multiple simultaneous failures may occur. This is known as a common cause failure.

During a hazard and risk analysis, risk reduction is allocated to protective functions used to reduce the process risk below the owner/operator risk criteria. When the protective function is implemented as a protective instrumented function (PIF), the allocated risk reduction is related to its integrity level (IL). The risk reduction and IL establish a benchmark for the design and management practices used throughout the PIF life. This benchmark serves two purposes: 1) it defines the minimum performance that should be achieved with regard to random failures; and

2) it determines the rigor of the protective management system required to reduce the potential for systematic errors.

The required risk reduction establishes a minimum level of management system rigor that must be provided to reduce the potential for systematic errors to a sufficiently low probability. Systematic errors are caused, or indirectly induced, by human error or unforeseeable complex process conditions. Systematic failures are not random events and must be addressed by the management system, using quality management processes to minimize systemic error. Most systematic errors are not easily included in the verification calculation.

The risk reduction can be used to establish a benchmark for the design and management practices used for any instrumented protective function (IPF). These requirements can be met with an IPF that is unreliable due to frequent fault alarms and spurious operation or with one that is reliable. Spurious operation of an IPF causes an unnecessary process disruption by taking action on a process that is operating normally. The costs of the process disruption should be considered, so an appropriate target spurious trip rate can be established for each IPF. The design basis should achieve the target risk reduction and spurious trip rate.

The risk reduction establishes three criteria: 1) equipment should be user approved for the operating environment; 2) the subsystems should have the necessary fault tolerance against dangerous failure for the claimed risk reduction; and 3) the average probability to fail on demand (PFD_{AVG}) for a demand mode IPF or to a hazard rate for a continuous mode IPF should be met.

User approval assures that a sufficient amount of information is known about the equipment, so that it can be properly implemented to achieve the desired performance. Fault tolerance prevents a single component failure from causing IPF failure. Finally, the performance calculation determines whether the planned design can theoretically achieve the desired integrity and reliability, taking into account six design parameters:

- Mean time to failure (MTTF),
- Voting architecture,
- Diagnostic coverage (DC),
- Testing interval (TI),
- Mean time to repair (MTTR), and
- Common cause failure.

These parameters are addressed in the design basis and affect the IPF performance. While the risk reduction is the benchmark for the IPF integrity, it is also one of seven core attributes (see Appendix C) that the IPF must achieve. The design basis considers all seven attributes when selecting these parameters, so they are discussed in Chapter 4 rather than in this appendix.

Random failures are generally considered in the verification calculation. Random failures occur when stress causes a fault to develop in a component of a device. Random failures occur due to various random events and can be assumed statistically to occur at a constant failure rate. Random failures are easily modeled using probabilistic math, allowing the performance to be estimated. Random events can induce CCF in the form of single points of failure or the failure of redundant devices. The nature of the event determines the techniques necessary to reduce its potential and the analytical method used to predict it.

Consequently, to fully understand the lifecycle requirements, it is necessary to understand the types of failures and their potential effects on the instrumented protective system (IPS). Once failures are understood, it is possible to identify optimal solutions that meet the process unit's integrity, operability, functionality, maintainability, and reliability requirements. Any personnel assigned responsibility for verifying the risk reduction should understand each failure type (i.e., random, systematic, and common cause) and the strategies that can be used against it. There are many books available on the subject of reliability engineering, so this appendix provides only a brief overview of the mathematics for calculation of the risk reduction and spurious trip rate.

D.1 CAUTION-IT'S A BENCHMARK

For many years, the process industry used of prescriptive design and maintenance practices. These practices were intended to be sufficiently conservative to assure that the risk reduction could be met in a wide variety of process applications.

Internal practices are frequently criticized for what is perceived as "over-design," because the practices seem to go beyond the minimum requirements of ISA 84.01/IEC 61511 or other good engineering practices. However, it should be recognized that any practice documents what is broadly acceptable to participating stakeholders. Minimum practices may not be sufficient to adequately address risk.

Further, prescriptive designs generally consider only a single safe state action. When the protective function requires multiple actions, the prescriptive practice may result in "under-design." A specific application may necessitate more rigor, including additional fault tolerance, diagnostics, testing, etc. For more guidance related to ISA 84.01/IEC 61511, refer to ISA TR84.00.04.

Project and plant personnel are under pressure to "optimize" (or value engineer) anything and everything. This "optimization" is generally intended to make the design as simple and cost effective as possible. However, when taken to extreme, optimization can reduce the design safety margin to a minimum and can result in the heavy reliance on devices whose failure mechanisms are not well known. As less and less safety margin is built into the design, greater importance is placed on the precision of the predictive calculation. Thus, the device's failure rate can become a critical design parameter rather than simply a performance benchmark. Excessive focus on minimal cost can result in the IPF being designed and operated with less tolerance for failure.

While this appendix presents a discussion of how to classify and manage failures, it must be recognized that the calculations are not a means to an end. Safe operation is achieved by designing and implementing IPSs that take into account a wide variety of site specific criteria. The primary advantage of internal practices is that, rather than turning every IPS design into an exercise in optimization, the risk management philosophy is clearly laid out with a sufficient safety margin to address uncertainty. The risk reduction is then verified using estimates of the random dangerous failure rates of the devices that comprise the IPS. Long term performance is then monitored and tracked to determine if the actual failures exceed expectations.

D.2 A "Bathtub" Viewpoint

The failure behavior of a population of hypothetical devices over their lifecycle (i.e., from production to disposal) is commonly represented by the "bathtub" curve, as shown in Figure D.1A. The bathtub curve for an individual device is developed by counting the failures that occur in a population of identical (or sufficiently similar) devices over a certain period of time. When the devices are in their useful life period, the collected data set is used to calculate the failure rate using statistical techniques. The plot shows how the overall failure rate changes with time. As indicated by Smith (1997), the bathtub curve is in reality composed by three overlapping curves, one for each of three regions, as shown in Figure D.1B.

The initial failure rate of the hypothetical device is driven by its burn-in, infant mortality, or early failure rate, which declines rapidly. The middle flat section of the curve represents the "useful life" of the device and is typically characterized by a constant failure rate. The IPF performance calculation is based on this middle section. The last part represents the "wear out" or "end-of-life" failure rate and is characterized by an increasing failure frequency.

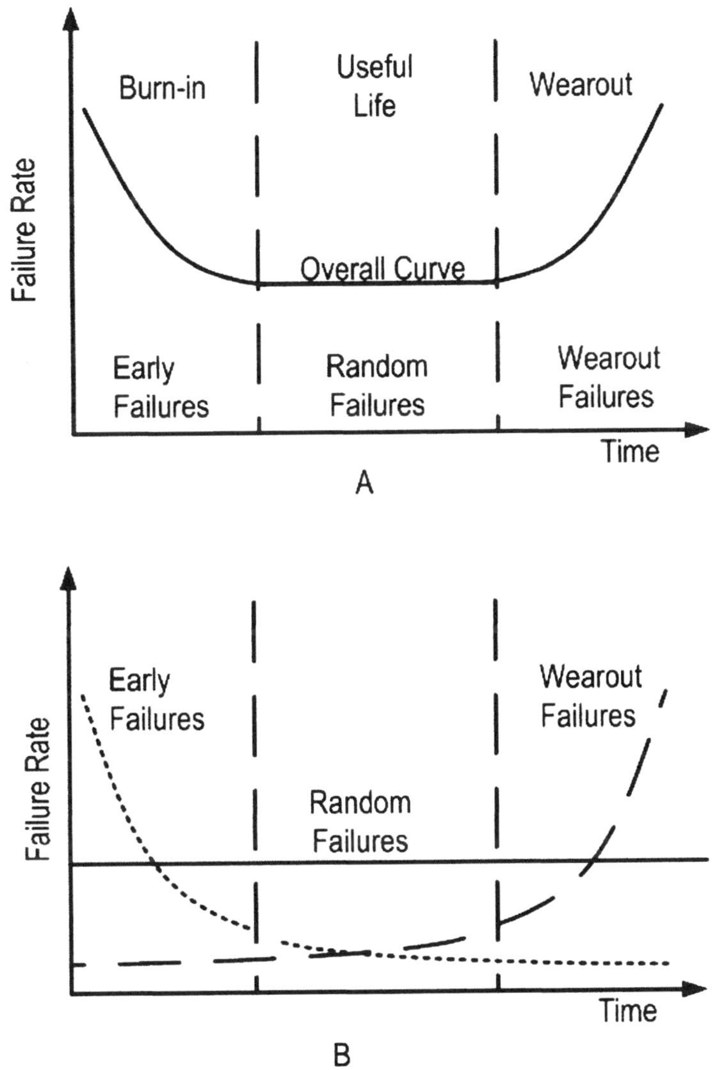

Figure D.1. Overall Bathtub Curve (A) and Components of the Bathtub Curve (B).

Early failures occur during a component's initial life and are caused by manufacturing, assembly, test, installation, and commissioning errors. Manufacturing flaws in the components that comprise the device may cause device failures during the burn-in period. Manufacturers affect the shape of this curve when they perform burn-in and function testing prior to release from production.

Early failures also occur as a result of device handling and installation. Devices can be damaged during shipment, unpacking, storage, transport to work site, and installation. Many early failures are caused by rough device handling, improper pre-installation storage, poor installation practices, and sloppy construction practices.

Materials used for installation activities, such as paint, pipe dope, insulation, and small pieces of a wielding rod, have been shown to cause devices to fail by getting into places where they are not supposed to be. Water and moisture intrusion during installation and commissioning can seriously damage IPS equipment.

Rigorous inspection, commissioning and validation activities are necessary to identify and correct these failures. However, not all flawed components fail immediately, and some cases of early failure occur during the useful life of the device. This is indicated by the extension of the burn-in curve to the useful life and wear out region.

The transition between the burn-in and useful life (Figure D.1A) occurs when the failure rate becomes constant. Failures during the useful life are caused mainly by random events that increase the stress on the device beyond its physical limits. When the wear-out failures become dominant, the overall failure rate increases. This indicates the end of the device's useful life.

The wear out period is characterized by an increased slope related to the device technology. Programmable devices tend to have a very sharp increase in the failure rate due to the large number of aging components. Electromechanical devices tend to have a more gradual increase in failure rate as they age, and the slope of their failure rate curve is heavily affected by how well their mechanical components are inspected and maintained. Preventive maintenance extends the useful life of electromechanical devices. A lack of inspection and preventive maintenance has been cited as a primary cause of early failure.

As with infant mortality, not all wear-out failures occur after the end of the useful life; these failures can happen during any stage of the device lifecycle and as time passes its rate of occurrence increases. The device should also be replaced when maintenance records demonstrate that it has reached its wear-out phase. When infant mortality and end-of-life issues are addressed within the device installation, commissioning and maintenance plans, it is assumed that the device failure rate is constant. The useful life section of Figure D.1A illustrates the predictable failure rate that is the goal of any effective inspection and preventive maintenance program.

For IPF analysis, it is very common to assume a constant failure rate; however, it has been demonstrated that this region does not exist for many devices. Some devices, such as block valves, are characterized by dominant failure mechanisms which are a function of time (i.e., lower failure rate at early stages of useful life, high failure rate at wear-out stages). Other statistical distributions should be considered, such as Weibull, log normal, or the bathtub curve, when the failure rate varies significantly with time.

The random failure rate is characteristic to the device when it is operated in accordance with its specification. The failure rate is estimated based on specific operational and maintenance conditions implicit in collected data. Reducing inspection and maintenance rigor will affect a device's failure rate. For example, mechanical equipment has a higher likelihood of failure when left in one position for extended time. Periodic movement of mechanical equipment reduces the likelihood of some failure mechanisms.

Operating the device outside of the manufacturer's specification for the device may damage or stress it, thereby causing early failure. Unexpected process application impact, such as chemical attack, corrosion, and deposition, or external environmental impact, such as electromagnetic interference, vibration, and heat, can also shorten its useful life.

D.3 FAILURE TYPES

An IPF uses many different devices to execute the protective functions intended to reduce the risk of identified hazardous events. The IPF performance depends on the IPF device characteristics (e.g., individual failure rates), the properties of the system itself, and the interactions among its components (e.g., voting architecture, common cause failures). An implicit assumption made during IPF design and its performance verification is that the devices are in their useful life period and are replaced when their performance indicates unacceptable wear-out failure.

The IPF probability of failure is a function of the random dangerous failure rate of its devices (e.g., field sensors, final elements, logic solvers, and support systems) and the design basis parameters (e.g., redundancy, architecture, proof test interval, and diagnostic coverage). The IPF may also operate spuriously causing unnecessary process trips. Spurious and dangerous failures are considered critical failures, because these failures result in the loss of the equipment's ability to operate as specified.

Other systematic, non-random contributions also affect the observed performance. Systematic and random faults can cause the failure of an individual device or the simultaneous failure of multiple devices. The individual device failure is easily assessed using probabilistic techniques that are described in many publications, such as Smith (1997), Mannan (2005), and ISA TR84.00.02.

Device failures are tracked and analyzed as part of the protective management system. The goal of a failure tracking program is to identify root causes and data trends and to develop means to reduce the potential for failure recurrence. The analysis may identify systematic errors and common cause failures. When a common cause failure occurs, the root causes tend to be installation- and/or application-specific.

Systematic errors occur because a human error broke through the administrative controls and quality assurance processes that were supposed to detect and correct it. Therefore, systematic errors tend to be perceived as a local personnel issue. Regardless of the nature of the failure, dangerous failure reports and identified trends should be communicated to appropriate personnel, because when personnel better understand how devices can fail, the better prepared they are to prevent failures.

Failures are managed within the ISA 84.01/IEC 61511 work process using different strategies depending on whether they are random or systematic. Consequently, the following sections briefly discuss random, systematic, and common cause failures, so the strategies for their management can be better understood.

D.3.1 Random failures

Random failures are the result of hardware degradation mechanisms, which can be accelerated by stress factors due to the operating environment. Stress factors are caused by many types of events, such as:

- Normal or abnormal process conditions,
- Presence of an adverse microenvironment (e.g., corrosion produced by chemical impurities or metallurgy),
- Presence of solids or other materials which randomly deposit in IPF devices,
- Exposure to electrostatic discharge,
- Operating for long periods at the extreme of the device's environmental specification, and
- Excessive device wear and tear.

Low frequency atmospheric events (e.g. snow in Houston, Texas, USA) can also be considered random events. The random failure does not follow any pattern, but instead occurs randomly during the device's life. The user approval process should ensure that a device's random failures and failure modes are well understood prior to approval.

Random failures can be identified by internal device diagnostics, external diagnostics, inspection, and proof tests. Redundant, fault tolerant subsystems are often used to reduce the probability that a single failure will cause the IPF to fail to operate correctly. Redundant subsystems also provide the opportunity for external diagnostics, where a diagnostic algorithm is executed at a specified interval to detect device failures.

The inspection and proof test interval is generally chosen based on maintenance history, manufacturer's recommendations, good engineering practice, insurance requirements, regulatory requirements, and what is necessary to achieve the required performance. As the test interval gets longer, there is an increased probability that multiple devices within a subsystem have malfunctioned prior to fault detection.

D.3.2 Systematic failures

Systematic failure is related in a deterministic way to a root cause, which can only be minimized or eliminated by changes in the design basis, installation practices, software systems, or operating basis. Systematic errors which have resulted in process safety incidents are:

- Risk assessment errors,
- Design errors,
- Specification errors,
- Unexpected operating environment impact,
- Installation and commissioning errors,
- Operator errors,
- Maintenance errors, and
- Change management errors.

While random hardware failures are caused mainly by physical degradation mechanisms, systematic failures are the direct consequence of IPF complexity. Every device is subject to failure due to design, specification, operating, maintenance, and installation errors. These mistakes immediately put the devices on the path to failure.

As IPF complexity increases, the potential for systematic errors increases due to the combination of failures. Additionally, the probability of detecting these errors decreases. Each device has many known opportunities for systematic error. With any new technology, there is the potential for many unknown (or as yet unidentified) failures. When issues associated with interconnectivity, communication, and support systems are added to the analysis, there are generally a large number of potential systematic failures.

The complex nature of systematic failures often makes them difficult to analyze probabilistically. Collected data typically includes some systematic failures, which contribute to the observed failure rate. Figures D.2A and D.2B illustrate how the observed failure rate is higher due to the presence of systematic events that are exhibited randomly. These failures are illustrated in Figure D.2B using additional failure rate lines.

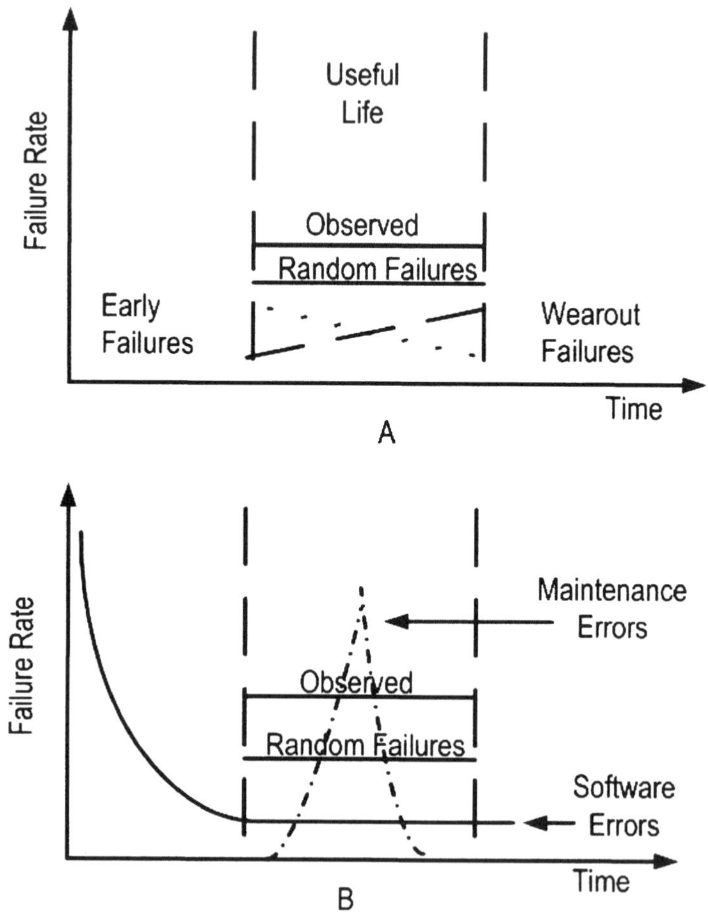

Figure D.2. Hypothetical Device's Random Failure Rate is Constant (A), However Other Non-Random Failure Sources Affect the Observed Failure Rate of the Device (B).

Since the intent of the quantitative analysis is to predict IPF performance, the systematic failures should be tracked and their inherent presence considered when estimating the random failure rate. As more information is collected, trends can be identified and used to minimize random and systematic failures in new or modified designs. For example, it may take multiple failure reports before it is recognized that the instrument air quality is causing the equipment failure.

Only a limited number of device failures and failure paths can be tested. When the failure patterns are not detected by the limited testing that is practically achievable, failure can happen every time the specific set of conditions occurs (e.g., every time there is lightning in the area). This potential failure becomes an intrinsic part of the IPF. Systematic errors are a major source of common cause failure, having the potential to disable redundant devices.

Systematic failures include many types of errors, such as:

- Manufacturing defects, e.g., software and hardware errors built into the device by the manufacturer,
- Specification mistakes, e.g. incorrect design basis and inaccurate software specification,
- Implementation errors, e.g., improper installation, incorrect programming, interface problems, and not following the safety manual for the SIS devices, and
- Operation and maintenance, e.g., poor inspection, incomplete testing and improper bypassing.

Systematic errors related to manufacturing defects can be reduced through the use of diverse redundancy. Diversity typically involves the use of different technologies or different manufacturers. While manufacturer errors can be addressed by diversity, this increases the IPF complexity. Incorrect specification, implementation, operation, and maintenance constitute root causes that are not solved by diversity and can actually increase if unnecessary complexity is introduced into the IPS design. The need for diversity should be balanced with the need for error free operation and maintenance.

The perceived improvements gained by diversity are based on the assumption that different devices exhibit different failures and failure modes. In other words, it is less probable for all of them to fail simultaneously, if they are different. However, diversity only reduces the potential for common mode failures. Many common cause failures are not addressed by diversity.

According to the data from a sample of 34 incidents published by the Health and Safety Executive (HSE 2003) there is no single lifecycle phase to blame for IPS non-performance. Mistakes are made throughout the lifecycle. Internal guidelines and procedures should define work process to support the day-to-day work processes for project engineering and on-going plant operation and maintenance. Guidelines and procedures serve as a training tool and ensure consistent execution of required activities.

Systematic errors are best addressed through with a protective management system, which emphasizes continuous improvement through performance monitoring. A rigorous system is necessary to decrease systematic errors and enhance safe and reliable operation. Each verification, assessment, audit, and validation recommended by this guidelines book is aimed at reducing the probability of systematic error to a sufficiently low probability. As errors or failures are detected, their occurrence should be investigated, so that lessons can be learned and communicated to affected personnel.

Example--Poor specification

Consider an engineering phase error where the presence of chloride in the operating environment is not listed on the specification form. The equipment is purchased with inappropriate materials of construction. While the purchased equipment fulfills user approval requirements, the incorrect specification results in the premature failure of the device.

Example--Temperature detection lag

Three thermocouples voting 2oo3 are installed in the same thermowell. The designer feels that this is acceptable, because the specification indicates that the process is clean. However, in reality, the process produces a polymer which rapidly fouls the thermowells. The polymer insulates the thermocouples from the actual process condition. The detection lag in large enough that the thermocouples are unable to detect high temperature within the process safety time.

Example--Improper configuration management

A recent embedded software upgrade resulted in a change in the equipment configuration instructions. However, the procedure was not updated, resulting in the equipment being configured improperly.

D.3.3 Common Cause Failures

Common cause failure (CCF) is a term used to describe random and systematic events that cause multiple devices, systems, or layers to fail simultaneously. Another term is “common mode” failure, which describes the simultaneous failure of two devices in the same mode. Common mode failure is related to the use of identical devices in the redundant subsystem. For example, two redundant differential pressure sensors can be simultaneously disabled due to loss of signal (common mode failure) originated from diaphragm damage (failure cause) caused by water hammer. Common mode failure is a subset of common cause failure.

All common cause failures have the potential to reduce the IPF performance; however, they are addressed in different ways depending on the nature of the failure (e.g., systematic or random). Throughout the ISA 84.01/IEC 61511 lifecycle, it is recommended that devices, systems, or protection layers be assessed for independence and the potential for common cause failure. Independence and common cause are often interrelated. A lack of independence means that there is a potential for a common cause failure. Likewise, an identified common cause indicates a lack of independence and therefore some dependency.

Diversity is often suggested as a means to eliminate common cause failure. However, common cause can impact identical and diverse devices. For example, the process application or external environmental condition can affect different technologies simultaneously when the conditions trigger each device's failure mechanisms. These devices may eventually fail due to different reasons, but the abnormal process condition is root cause that started the failure propagation. The use of different technologies (i.e., diversity) does reduce the potential for common mode failure. Diversity reduces the potential for dependent failure by minimizing common mode failure, but does not eliminate the potential for common cause failure.

The approach taken to manage CCF is specific to the nature of the failure. Two types of CCF are addressed: 1) single points of failure where one malfunctioning device causes an IPF failure; and 2) single events that lead to multiple failures in a redundant subsystem. Figure D.3 and Table D.1 refer to these as "single point of failure" and "redundant," respectively.

Single points of failure can occur due to systematic (A) or random (B) events. Systematic failures (A) occur when human errors result in the violation or invalidation of design and operating basis assumptions (e.g., process assumed to be clean but in reality is not). Random failures (B) can occur throughout the useful life of a device. These failures are managed using redundancy, diagnostics, and proof testing.

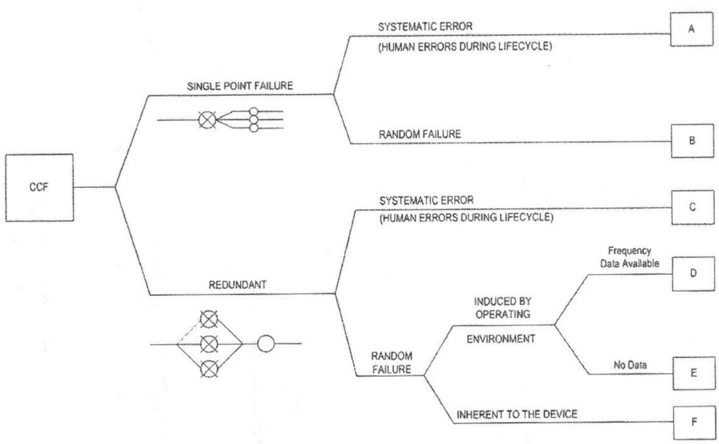

Figure D.3. Taxonomy of Common Cause Factors (CCF) and Methods for Their Analysis. (see Table D.1)

CCF TYPE METHOD	COMMENTS	EXAMPLES
A **Lifecycle management**	**Caused by mistakes and puts devices immediately on failure path.** **Assumptions are violated and invalidated.** **Failures unknown due to lack of technical knowledge.** **Difficult to model.**	**Software error poor installation, incorrect configuration lack of inspection, maintenance leaves device in bypass.**
B **Analytical** **(Single event in fault tree)**	**Device fails due to stressors in the operating environment infant mortality, or wear out.**	**Typical random failure.**
C **Lifecycle management**	**Caused by mistakes and puts devices immediately on failure path.** **Assumptions are violated and invalidated.** **Failures were unknown due to lack of technical knowledge.** **Difficult to model.**	**Wrong design, software bug, defective device due to manufacturer error.** **Wrong specification leading to failure in dirty environment believed to be clean.**
D **Analytical** **(Single event in fault tree)**	**Caused by the installation and device interconnection.** **Explicit failure data available.**	**Single power supply, single instrument air supply, shared digital communication.**
E **Analytical** **(β-factor)**	**Caused by the operating environment, installation, device interconnection, and maintenance practices.** **No explicit failure data available therefore β-factor estimated on prior-use information and experience for similar devices.**	**Environmental stress on the device (e.g., abnormal corrosion, vibration, pressure, temperature, solids, erosion, etc.) due to transient or abnormal operation.**
F **Analytical** **(β-factor)**	**Inherent part of device design.** **β-factor estimated on prior use information and experience.**	**Device failure due to manufacturing errors, such as defective raw materials, poor quality production processes, or software bugs.**

Table D.1. Methods Used to Address the Different Types of Common Cause Failures. (see Figure D.3)

As with single points of failure, redundant subsystems can fail due to systematic errors (C) in the device manufacture, specification, design, installation, and maintenance. These errors typically happen due to lack of knowledge, information, and training and are generally unknown to personnel. Test procedures may not identify these errors, since they are not expected. Systematic errors are difficult to test even if easily identified. For example, it is possible that the valve actuator is incorrectly specified, but how do you test to determine that the valve actuator will not close under emergency process conditions?

Specification errors must be caught during the design and engineering phases using independent verification. Checklists can be used to identify CCF. The list of questions guides the engineer through the design aspects examining opportunities for CCF. Installation, commissioning, and maintenance errors are reduced by independent checks, verifications, and audits.

As shown in Figure D.3, random failures of redundant subsystems (D, E, and F) can be caused either by conditions that are inherent to the device (F) or inherent to the system (D and E). Random failures inherent to the device (F) are generally manufacturing defects which may include hardware and/or software failures.

These failures are estimated using the beta factor method. The operating environment, the installation, and the interconnection to other systems (D and E) affect the device operation. This system-induced random failure can be divided into two categories depending on the availability of failure frequency data. If data is available (D) the failure can be modeled explicitly as an event. For example, if fault tree analysis is the selected analytical methodology, this type of CCF is treated as a basic event with its own failure rate. If data is not available, the CCF can be addressed using the beta factor method (E).

The beta factor accounts for random events that cause a dangerous failure in the operating environment. The value of the beta factor is selected based on engineering judgment. Many owner/operators use a beta factor between 0.1% and 5.0% when good engineering practices are applied in the design, installation, inspection, and maintenance practices. The beta factor can be substantially higher if good engineering practices are not followed.

D.4 FAILURE CLASSIFICATION

All devices eventually fail. A fault occurs somewhere within the device's components and propagates into a critical failure, resulting in the device's inability to operate as specified. A device failure is observed by its effects on the device's operation. The device's failure mode is the symptom, condition, or effect by which the failure is observed, e.g., the signal does not change with the process variable. The user approval process (Appendix F) relies heavily on gaining sufficient understanding of the device's failure modes in the operating environment.

Failure mode and effects analysis (FMEA) is a qualitative analysis method used to analyze the effects of identified device failure modes on the device operation, so the design basis can take these modes into account. The effects (OREDA 1992) may include:

- Dangerous failure (e.g., failure-to-operate),
- Spurious operation (e.g., premature function),
- Degraded failures (e.g., out-of-tolerance), and
- Incipient conditions (e.g., damaged electrical insulation).

Critical failures cause the loss of the device's ability to operate as specified, resulting in either a safe or dangerous failure. Critical failures can occur suddenly or gradually over time. Pass-fail criteria are established for each device to determine when a failure is critical. If a device fails to operate as specified in the design basis, the failure is considered dangerous. The risk reduction capability of a protective function is related to its dangerous failure rate. If the device has spuriously operated, the failure is considered safe. Critical failures are discussed in more detail in Section D.5.

Failures can also develop gradually over time. Devices may experience partial failures, resulting in reduced, increased or erratic operation. Degraded operation may cause the device to act earlier or later than required but does not prevent its operation within the process safety time.

Incipient failures are conditions typically identified during inspection and preventive maintenance activities. An incipient failure does not currently effect the device operation. However, if correction action is not taken, the failure could propagate into a degraded or critical failure. For example, the screen is missing from the vent port of a solenoid-operated valve. There is no obstruction currently in the vent port, but if the screen were not replaced, debris could accumulate in the port, resulting in a degraded or critical failure.

D.4.1 Safe and Dangerous Failures

Critical failures can be further classified as safe or dangerous. A safe failure causes the device to take the safe state action defined in the design basis. A dangerous failure causes the device to fail to operate according to its specification.

> Example--Electronic pressure transmitter
>
> Table D.2 provides failure causes for an electronic pressure transmitter. Each cause results in an erroneous signal, which may be identified as degraded or critical, depending on the device specification and pass-fail criteria. Critical failure may be further classified based on its effect on the equipment operation. A transmitter failing high would be a safe failure, if the function normally takes the safe state on high process variable. The

transmitter failing high would be a dangerous failure in the event of a low process variable measurement.

There are various models for describing a device's critical failures. Some devices are non-repairable and are replaced when failure is detected. Other devices are repairable and are inspected and maintained in a manner that achieves a constant failure rate throughout their useful life. In some cases, a "device" may actually be a complex system of repairable /replaceable components. Such complex devices may not be easily described by any one model.

Failure	Mode	Effect
Electronic Failure	Signal frozen	Dangerous
Isolation Valve Closed	Signal frozen	Dangerous
Left in the Test Mode	Signal frozen	Dangerous
Partially Obstructed or Plugged Tap	Signal slow to respond	Dangerous Degraded
Fully Obstructed or Plugged Tap	Signal frozen	Dangerous
Impulse Line Crimped	Signal slow to respond	Dangerous/safe Degraded
Sensor Deformation	Signal high or low	Dangerous/safe Degraded
Out of Adjustment	Signal high or low	Dangerous/safe Degraded
Build Up of Fluid In Impulse Line	Signal high	Dangerous/safe Degraded
Power Supply	Signal low	Dangerous/safe
Loss of Seal Fluid	Signal low	Dangerous/safe Degraded
Impulse Line Leak	Signal low	Dangerous/safe Degraded

Table D.2. Example Failures, Modes and Effects for an Electronic Pressure Transmitter.

The device's critical failures are described by the mean time to failure (MTTF), which is the average time to the first device failure. The MTTF is related to the expected operating life. For a repairable device, the critical failure rate can be determined based on the average time between critical failures, or the mean time between failure (MTBF). Since the MTBF is the time between one failure and the next, it includes the MTTF and the mean time to repair (MTTR).

$$MTBF = MTTF + MTTR \qquad D.1$$

When the MTTR is small compared to the MTTF, the MTTF represents the average time that the device is in the operational condition.

The MTBF assumes that the device is repairable and can be returned to service in an "as good as new" condition. To achieve this, the inspection, preventive maintenance, and proof testing activities must be performed at the frequency necessary to ensure that the device operates throughout its life according to its design specification within the acceptable design tolerance. Maintenance activities extend the useful device life, increase device reliability, and identify when device performance has degraded to the point where replacement is required.

Reducing the frequency of maintenance activities may allow more incipient and degraded failures to progress to critical failures, resulting in earlier failure and a smaller MTTF. A reported MTTF is always related to the operating environment and the specific maintenance practices used to maintain the device. Historical data for similar devices or systems can serve as a useful MTTF estimate.

Many published databases only provide the critical failure rate, λ_{CRIT}. The ratio of safe and dangerous failures is then estimated based on the device's tendency to fail to the spurious or dangerous in the operating environment. For example, if the device is as likely to fail to the safe state as to fail to the dangerous state, it is assumed that the total failures are 50% dangerous and 50% spurious. This equates to a 50% safe failure fraction. If the λ_{CRIT} is 0.01 failures/year, the dangerous failure rate, λ_D, is 0.005 failures/year and the safe failure rate, λ_{SP}, is 0.005 failures/year.

D.4.2 Detected and Undetected Failures

Each safe and dangerous failure can be further classified depending on the ability to detect the failure using diagnostics. Diagnostics may be provided using hardware and software within the device itself or implemented externally to identify and report specific failure modes. The diagnostic coverage (DC) is determined by the fraction of the dangerous failures that can be detected by the operation of diagnostics that report faults to the operator and take a specified action on fault detection.

Considering the failure classification and failure detection, the following failure rates are estimated:

Dangerous failure rate, λ_D:

$$\lambda_D = 1/\text{Mean Time Between Failure Dangerous } (MTBF^D) \quad \text{D.2}$$

Dangerous detected failure rate, λ_{DD}:

$$\lambda_{DD} = DC \times \lambda_D \quad \text{D.3}$$

Dangerous undetected failure rate, λ_{DU}:

$$\lambda_{DU} = (1\text{-}DC) \times \lambda_D \quad \text{D.4}$$

Safe failure rate, λ_S:

$$\lambda_S = 1/\text{Mean Time Between Failure Safe } (MTBF^S) \quad \text{D.5}$$

Spurious failure rate, λ_{SP}:

$$\lambda_{SP} = \lambda_S + \lambda_{DD} \quad \text{D.6}$$

Critical failure rate, λ_{CRIT} :

$$\lambda_{CRIT} = \lambda_D + \lambda_S \quad \text{D.7}$$

$$\lambda_{CRIT} = \lambda_{DU} + \lambda_{SP} \quad \text{D.8}$$

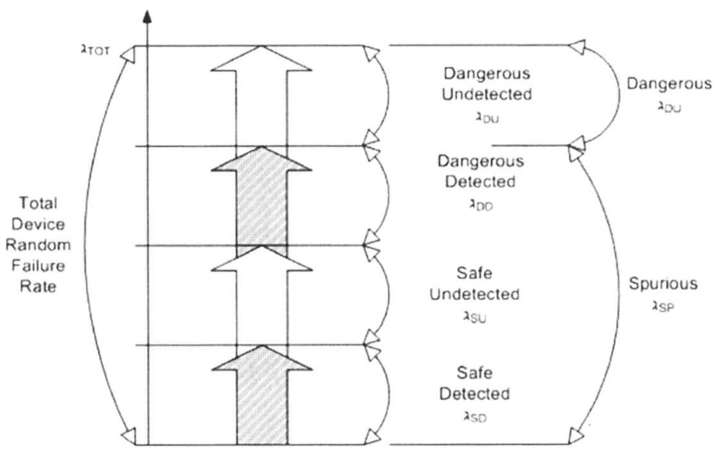

Figure D.4. Components of the Total Random Failure Rate.

The significant difference between detected and undetected failure is the time that the device remains in the failed state before detection. For devices with continuous on-line diagnostics, the failure is detected at the diagnostic interval (DI) and is repaired and returned to fully operational condition mean time to repair (MTTR). The diagnostic interval is also known as the mean time to detect. The DI is generally significantly smaller than the MTTR, so DI is often neglected in the analysis.

For a device without on-line diagnostics, the failure is detected by proof test or demand. Assuming that the proof test is sufficient to detect the failure, the time the device stays in the failed state is set by the test interval (TI) plus the MTTR. In contrast to the DI, the TI is generally much larger than the MTTR, so it will dominate the analysis. In terms of risk, the longer the time to detect a failed device, the higher the likelihood of a hazardous event.

D.5 PF Performance Metrics

The verification of the risk reduction sometimes involves a predictive calculation. The performance may be defined in terms of the hazard rate for continuous mode IPFs or the average probability of failure on demand (PFD_{AVG}) for demand mode IPFs. Some IPFs may be better modeled using the instantaneous probability of failure PFD(t). These models or calculations should be viewed as tools to benchmark and compare different options. The calculations should never be perceived as a precise measure of IPF performance.

The numbers used to benchmark IPF performance should never be construed to indicate that there is an acceptable level of dangerous failure. Equipment must be designed, constructed, installed and maintained to minimize the risk of significant consequence events, especially those involving highly hazardous chemicals. IPS equipment must not be run to failure, but maintained in the "as good as new" condition. This requires an on-going mechanical integrity program, which assures the equipment's long-term integrity.

The protective management system must seek to identify and correct systematic errors, which reduce the IPF's actual performance. While the calculation does not explicitly address human error, the limitations of the human systems in supporting the required performance should be considered when estimating the achievable failure rate.

A significant assumption in the calculation is that inspection, preventive maintenance, and proof testing are performed at a rate sufficient to maintain a constant failure rate. The calculation does not implicitly or explicitly account for how changes in the design, operating, and maintenance practices affect the achievable failure rate.

The calculation may indicate that a change in testing from once per year to once in 10 years is acceptable. However, the change in testing will result in less opportunity to detect incipient and degraded failures, allowing more critical failures to occur. Less frequent movement of mechanical components and linkages may result in a higher potential for the component to freeze in position or lock-up.

As the test interval is extended, it becomes more likely that the test will find a critically failed device, rather than a degraded one. Breakdown maintenance is not acceptable for the IPS. Consequently, when dangerous failures are found, an investigation should determine the root cause and should identify means to reduce recurrence, as necessary, including an increase in the inspection, preventive maintenance, and/or proof test frequency or rigor.

D.5.1 Failure Rate

The calculations assume that the device faults are repaired after detection and the device is returned to service in the "good as new" condition. The dangerous failure rate can be determined from the mean time to failure dangerous ($MTTF^D$):

$$\lambda_D = \frac{1}{MTTF^D} \qquad \text{D.9}$$

D.5.2 Instantaneous Probability of Failure

For IPFs, the probability of failure can be derived for the transition of a device from the operating state to the failed state, as shown in Figure D.5. The device is capable of performing its design function in state 1. When the device fails, it transitions to state 2, where the device is no longer capable of achieving its design function. The transition between state 1 and state 2 is calculated from a probability density function f(x) that describes the probability of device failure as a function of time. It is commonly assumed that the device is in its useful life, so that the dangerous failure rate, λ_D, is constant with time (see Section D.3).

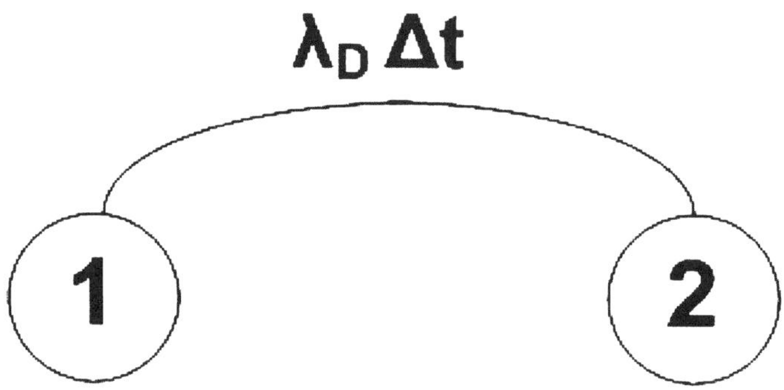

Figure D.5. States of a Device.

The probability of the device being in operating state, state 1, at time t is given by:

$$P_1(t+\Delta t) = P_1(t) - \lambda_D P_1(t)\Delta t \qquad \text{D.10}$$

Rearranging,

$$\frac{P_1(t+\Delta t) - P_1(t)}{\Delta t} = -\lambda_D P_1(t) \qquad \text{D.11}$$

The derivation results in the following:

$$P_1(t) = e^{-\lambda_D t} \qquad \text{D.12}$$

The probability of being in the failed state at time t, where P2(t) = 1-P1(t).

$$P_2(t) = PFD(t) = 1 - e^{-\lambda_D t} \qquad \text{D.13}$$

The instantaneous PFD(t) is a snapshot of the failure probability of the device taken at time t. The initial state is at time = 0, where the probability of failure is zero. At time = ∞, the probability of failure is 1. As shown in Figure D.6, PFD(t) increases exponentially with time. When the device is proof tested at the test interval, TI, the PFD(t) is reduced to its initial value. This involves two implicit assumptions: 1) all device failures are detected by inspection and proof test and 2) the device is repaired and returned to service "as good as new" condition. The effect of the proof test is illustrated by the saw tooth shape shown in Figure D.6.

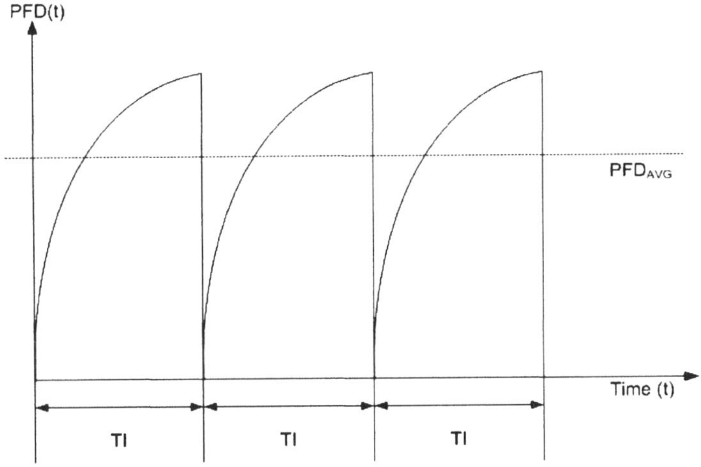

Figure D.6. Typical Saw Tooth Shape for the PFD(t).

Many technical references use the exponential series expansion of equation D.13 as follows:

$$PFD(t) = \lambda_D t - \frac{(\lambda_D t)^2}{2} + \frac{(\lambda_D t)^3}{6} - \ldots \qquad \text{D.14}$$

Since most dangerous failures are "rare events," the squared, cubed, etc. terms are generally small and can be neglected. The "rare event" assumption is considered valid when $\lambda_D t$ is less than 0.1 and results in the equation describing the instantaneous probability of failure at time t:

$$PFD(t) = \lambda_D t \qquad \text{D.15}$$

A proof test can be divided into a series of tests conducted at different test intervals. All tests should be executed during validation and the scheduled, off-line proof test. The partial test reduces the theoretical PFD(t) by the degree of test coverage. The effect of partial testing is illustrated in Figure D.7, which provides the PFD(t) on a logarithmic scale as a function of time.

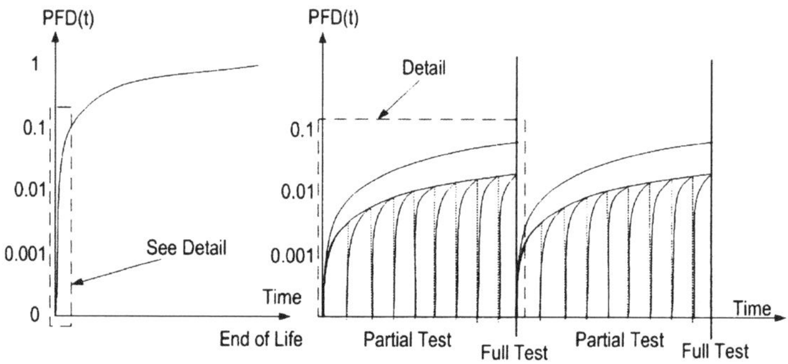

Figure D.7. Effect of Partial Testing on PFD(t).

Each test often detects only specific failure modes. The ratio of the dangerous detected failures to the total dangerous failures is the diagnostic coverage (DC). The PFD(t) is determined for each partial test using an appropriate diagnostic coverage and the partial test interval. Care should be taken to ensure that the total diagnostic coverage provided by the tests does not exceed a value of 1. For example, if the partial test (PT) detects 70% of the failures, the full test (FT) detects the 70% and the remaining 30%. Equation D.15 is expanded as follows:

$$PFD(t) = (1\text{-}0.7)\,\lambda_D TI_{FT} + (0.7)\,\lambda_D TI_{PT} \qquad \text{D.16}$$

D.5.3 Average Probability of Failure on Demand

The PFD_{AVG} is used for demand mode IPFs. It is the average probability that the device fails to respond to a process demand and is calculated by averaging the instantaneous PFD (t) over a defined time interval. As shown in Figure D.6, the PFD_{AVG} is the average of several snapshots taken at different times.

For demand mode IPFs, the time interval is assumed to be the proof test interval (TI). It could also be other time intervals, such as the diagnostic interval or partial test interval. The PFD_{AVG} is determined by the integration of PFD(t) from 0 to TI, as shown in equation D.17 when MTTR<<TI.

$$PFD_{AVG} = \frac{1}{TI}\int_0^{TI}\left(1 - e^{-\lambda_D t}\right)dt \qquad \text{D.17}$$

The integration can be simplified using the exponential series expansion (equation D.14) and the rare event assumption (D.15).

$$PFD_{AVG} = \frac{1}{TI}\int_0^{TI}\left[\lambda_D t - \frac{(\lambda_D t)^2}{2} + \frac{(\lambda_D t)^3}{6} - \frac{(\lambda_D t)^4}{24} + \ldots\right]dt \qquad \text{D.18}$$

$$PFD_{AVG} = \frac{\lambda_D TI}{2} \qquad \text{D.19}$$

When a dangerous failure is detected, the device is either repaired or replaced. This activity results in a time period where the device is not available to perform its activity. This time period is the MTTR, which is the average repair time given site maintenance practice. The derivation of D.19 assumed that the MTTR is significantly less than the TI, which is true for any IPS designed and managed according to this guidelines book.

Equation D.20 adds the contribution of MTTR to the PFD_{AVG}:

$$PFD_{AVG} = \frac{\lambda_D TI}{2} + \lambda_D MTTR \qquad \text{D.20}$$

The λ_D MTTR represents the period when the device is unavailable because it is under repair or replacement. As shown in equation D.20, this period is assumed to occur only once in the $MTTF^D$ of the device. However, repair periods also occur in response to safe and detected faults. Devices should be removed from the approved equipment list (see Appendix F and Chapter 4) when they are not performing as required. Otherwise, the design and installation practices should be modified to reduce the likelihood of device failure.

Equation D.19 represents the PFD_{AVG} for a single device assuming a constant failure rate. Since all things degrade with time, a constant failure rate can only be achieved when the device is subjected to a minimum level of inspection, preventive maintenance, and proof testing. The amount of inspection and preventive maintenance is affected by many factors, such as the device manufacture, the operating environment, and support system quality.

When N devices are installed, voting subsystems can be created, where M represents the number of devices that must agree out of the total N. MooN (M out of N) voting subsystems can be modeled using Markov modeling or fault tree analysis. Markov modeling and fault tree analysis yield different numerical results for architectures where M≠N. Both approaches are considered acceptable for the PFD_{AVG} calculation. The error associated with the data assumptions is generally significantly larger than the error introduced by the equation.

The derivations shown in this section assume an exponential failure distribution, which may not be true for all devices. Other failure distributions may apply such as Weibull, log normal, Poisson, or binomial. Given the number of assumptions required for the analysis, the calculations should not be the only form of evidence used to justify deviation from what is considered good engineering practice within a particular market sector.

Markov modeling was developed by a Russian mathematician, Andrei Markov, around 1900. Markov models are based on "average after logic" and assume that initially all IPF subsystems are working; therefore the PFD at time 0 is zero. With time, each device can fail bringing the IPF to degraded intermediate states; when enough devices fail, the IPF fails dangerously. For consideration of common cause failures, see Sections D.3.3 and D.7.4. The logical relationship between the devices is developed as an equation for PFD(t), which is integrated to yield the PFD_{AVG}.

Assuming the device failures are independent, i.e., no common cause failure, the PFD_{AVG} for an MooN subsystem, where R=N-M+1:

$$PFD_{AVG} = \frac{N!}{(N-R)!\,R!} \times \left[\frac{(\lambda_D)^R}{(1-DC)TI + DC(DI)}\right] \times \left[\frac{[((1-DC)TI + DC(DI)) + MTTR]^{R+1} - (MTTR)^{R+1}}{R+1}\right] \qquad \text{D.21}$$

Fault tree analysis was developed at Bell Laboratory in the early 1960s for missile launch control reliability for the Polaris project. Fault tree analysis is generally based on the "average before logic" approach, which simplifies the mathematics considerably. In this method, the PFD_{AVG} is determined for each device and the resulting PFD_{AVG} is combined using Boolean algebra. Common cause failures are addressed according to Sections D.3.3 and D.7.4. Assuming the device failures are independent, i.e., no common cause failure, the PFD_{AVG} for an MooN subsystem, where R=N-M+1:

$$PFD_{AVG} = \frac{N!}{R!(M-1)!} \qquad \text{D.22}$$

$$\left[\frac{(1-DC)\lambda_D TI}{2} + \frac{DC\lambda_D DI}{2} + \lambda_D MTTR\right]^R$$

D.5.4 Beta Factor

A common cause failure is the occurrence of an event that results in the failure of a subsystem or the entire system. Common cause failures can occur due to many types of events, such as manufacturing defects in redundant devices, aging components, operating environment stresses, common process connections, and common support systems.

These failures are referred to as dependent failures and are most commonly modeled using the beta factor method. The probability of common cause failure (PFDβ) is estimated as a percentage of the highest failure rate among the devices in the subsystem, β x λ_D. For a repairable device, the PFDβ is calculated as:

$$PFD_\beta = \frac{\beta(1-DC)\lambda_D TI}{2} + \frac{\beta DC\lambda_D DI}{2} + \beta\lambda_D MTTR \qquad \text{D.23}$$

The value of the beta factor is selected based on prior-use data in the operating environment. The prior use data can be for a specific device technology or for a sufficiently similar device. Many owner/operators use a beta factor between 0.1 and 5.0% when the devices are user approved for the application and good engineering practices are applied in the design and installation to minimize CCF. The beta factor can be substantially higher, if good engineering practices are not followed. Refer to ISA TR84.00.02 and ISA TR84.00.04 for more discussion of common cause failure and the beta factor.

D.6 Spurious Trip Rate

The spurious failure rate, λ_{SP}, can be determined from the mean time to failure spurious ($MTTF^{SP}$):

$$\lambda_{SP} = \frac{1}{MTTF^{SP}} \qquad \text{D.24}$$

For a simplex device, the spurious trip rate (STR) is the device spurious failure rate, λ_{SP}:

$$STR = \lambda_{SP} \qquad \text{D.25}$$

The spurious trip rate is typically calculated without consideration for the support systems such as instrument air and power supplies. The failure of the support systems should be monitored and tracked separately from the instrumentation and controls.

When the subsystem is fault tolerant for spurious failure, the STR is calculated by examining the frequency of failure of one device and the probability of failure of the second device in the time it takes to detect and repair the first device failure. When continuous diagnostics are provided, this diagnostic interval is typically much smaller than the MTTR, so the time to detect and repair can be assumed to be MTTR. For an MooN subsystem of identical devices operating under active diagnostics, the spurious trip rate can be determined by:

$$STR_{MooN} = \frac{(N-1)!}{(N-M)!\,(M-1)!} \times N\,\lambda_{SP}\left[\lambda_{SP} MTTR\right]^{M-1} \qquad \text{D.26}$$

The time to detect can be assumed to be TI, when the safe failure is detected by proof test only. For an MooN subsystem of identical devices with no active diagnostics, the spurious trip rate can be determined by:

$$STR_{MooN} = \frac{(N-1)!}{(N-M)!\,(M-1)!} \times N\,\lambda_{SP}\left[\lambda_{SP} TI\right]^{M-1} \qquad \text{D.27}$$

D.7 EXAMPLE APPLICATION

A protective instrumented function (PIF) consists of a series of subsystems, including input devices, logic solvers, final elements, and required support systems, such as power sources and instrument air. A PIF may be implemented in a dedicated logic solver or with other PIF in a shared logic solver. The integrated system is known as a protective instrumented system (PIS). Each subsystem that comprises the PIF is characterized by a PFD_{AVG} for a demand mode PIF and the hazard rate for a continuous mode PIF. Each subsystem also has a spurious trip rate (STR).

As an example, the hazard rate, PFD_{AVG}, and STR are determined below for the PIS shown in Figure D.8 with a test interval of 1 year. Example PIS devices are shown in Table D.3 with corresponding values for the safe and dangerous failures. The values are for illustration purposes only and are not representative of any specific application of device technology.

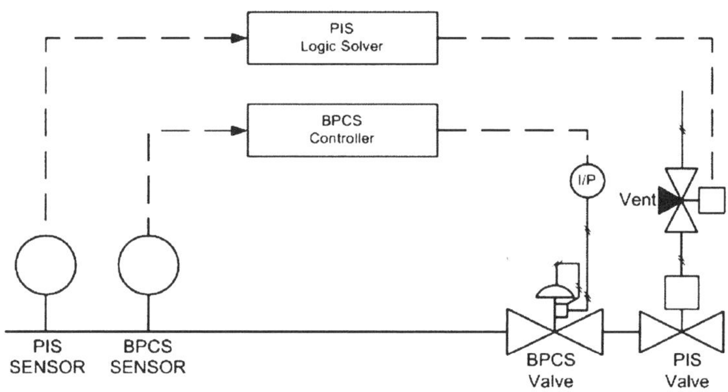

Figure D.8. Example Architecture Illustrating an Independent Control System and Protective Instrumented System (PIS).

Device	Dangerous Failure Rate (1/Years)	Spurious Failure Rate (1/Years)	$MTTF^D$ Years	$MTTF^{SP}$ Years
Pressure Transmitter	6.67E-03	1.25E-02	150	80
Non-programmable Trip Amplifier	1.38E-03	5.00E-03	725	200
Solenoid Operated Valve (de-energize to trip)	1.67E-02	3.33E-02	60	30
Ball Valve (fail to close)	1.67E-02	6.67E-03	60	150

Table D.3. Failure Rate Data Used in the Examples.

To obtain the frequency of dangerous failure, the individual device failure rates are summed as shown in Table D.4.

Device	Dangerous Failure Rate (1/Years)
Pressure Transmitter	6.67E-03
Trip Amplifier	1.38E-03
Solenoid Operated Valve	1.67E-02
Ball Valve--Fail Closed	1.67E-02
Hazard Rate	4.14E-02

Table D.4. Hazard Rate Results for Figure D.8 Architecture.

To obtain the PFD_{AVG}, the simplified equation (eq. D.19) can be used. The results are shown in Table D.5.

Device	PFD_{AVG}
Pressure Transmitter	3.34E-03
Trip Amplifier	6.90E-04
Solenoid Operated Valve	8.34E-03
Ball Valve--Fail Closed	8.34E-03
PFD_{AVG}	2.07E-02

Table D.5. PFD_{AVG} Results for Figure D.8 Architecture.

The spurious trip rate for the system is given by the sum of the device spurious trip rates as shown in Table D.6.

Device	STR (1/years)
Pressure Transmitter	1.25E-02
Trip Amplifier	5.00E-03
Solenoid Operated Valve	3.33E-02
Ball Valve--Fail Closed	6.67E-03
Spurious Trip Rate	5.75E-02
$MTTF^{SP}$	17 years

Table D.6. STR Results for Figure D.8 Architecture.

E

PROCESS EQUIPMENT RELIABILITY DATABASE

The Center for Chemical Process Safety (CCPS), an American Institute of Chemical Engineers (AIChE) industry technology alliance, has been facilitating an industry effort whose mission is:

> *Operation of an equipment reliability database, making available high quality, valid, and useful data to the process industry enabling analyses to support availability, reliability, and equipment design improvements, maintenance strategies, and life cycle cost determination.*

Multiple companies from the oil, chemical, and industrial gas industries, as well as consultants, insurance companies, and equipment manufacturers joined together to achieve this mission. CCPS is providing an infrastructure to facilitate development and sharing of technical information with respect to an industry process equipment reliability database (PERD).

To accomplish its objective, the CCPS has created a forum for participating companies to support and direct this effort (participation in the project does not require membership in CCPS). The participant companies, through representatives on a Steering Committee, are actively involved in the design, implementation, startup, and operation of the database. General meetings are conducted three times per year with smaller working groups meeting monthly and as needed for specific sub projects.

The PERD project technology allows the collection of data using a definable and reproducible taxonomy. PERD taxonomies provide a fundamentally sound structure to capture data in a format that allows immediate analysis, facilitating the understanding of what the raw data means or represents. The taxonomy defines data, failure modes, causes, mechanisms, and data field formats needed to calculate key performance metrics. Benefits accrue from being able to aggregate data among multiple plants and companies without degrading its quality. The taxonomies represent the data structure incorporated into the database software developed as part of the PERD initiative.

The strategic thrust of the PERD initiative is to:

- Develop equipment taxonomies to establish the technical foundation for population to a meaningful level of detail,

- Leverage the equipment taxonomies to develop software that supports statistical analysis,
- Develop an infrastructure so that companies can collect and transfer data to PERD for sharing among the participants,
- Provide a working forum to assist companies in their effort to provide data to an operational database, and
- Educate new participants on how to accomplish the above.

Those companies that embrace the PERD thought process and methodologies and then proceed to implement them are in a position to obtain and use "proven in use" data. Benefits of such data are that it:

- Supports risk decision making,
- Improves mechanical integrity via quality processes,
- Verifies safety integrity levels, and
- Reduces systems engineering costs.

The flow chart in Figure E.1 illustrates how participants can gain increasing benefits by taking the time to understand the technical foundation and using the tools and aids that have been developed as part of the initiative.

Progress to date includes the development and publication of two guidelines books and several taxonomies, many having been made available to the general public. In addition, PERD first generation software has been developed that includes plants, spring operated relief valves and in line centrifugal compressors. The entire work process has been validated via a pilot project that has seen multiple companies take their respective data and map/translate it into a PERD compatible format. The data was then imported, aggregated, and made anonymous for subsequent circulation.

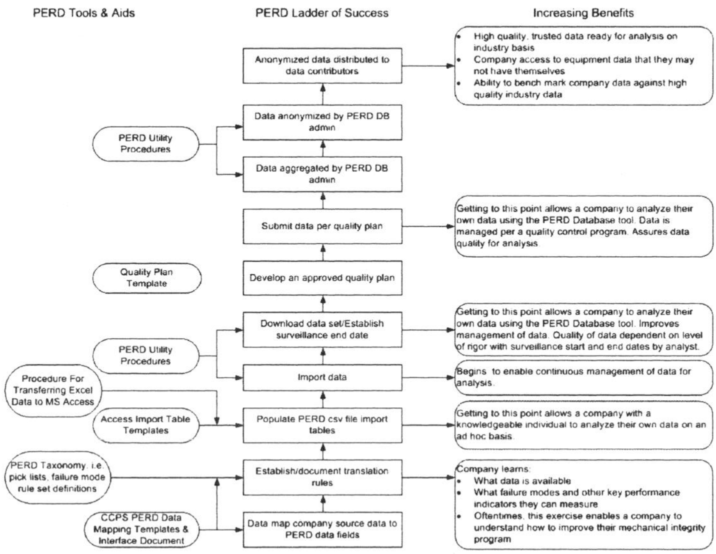

Figure E.1. PERD Process.

Future plans include the development of 2nd generation software intended to be much more robust and to also include instrument loops as part of the database. This addition will support tracking the performance of safety instrumented functions. While this work is underway, another focus is aimed at relief valve data to fully realize operation of the database, enabling fulfillment of the PERD mission. The pace of advancement is at the discretion of the participating companies as a function of available budget and "sweat equity."

F

USER APPROVED EQUIPMENT AND PRACTICES

F.1 USER APPROVED

This appendix is distributed "as is." All warranties, either expressed or implied, are disclaimed as to quality, performance, or merchantability, whether expressed, implied or statutory, hidden defects, or fitness for any particular purpose. The author makes no representations about the suitability of the information in this appendix for any purpose. Reader bears the entire risk relating to the use of this appendix.

A paragraph similar to the above is contained in many end-user license agreements accepted during the installation of user software and with the hardware manufacturer's terms and conditions. The underlying concept is that the "user" of the product bears the risk associated with determining whether a product is fit for a particular purpose.

The manufacturer limits its liability associated with the operating environment by providing environmental specifications which must be met. Other limitations are generally addressed in the installation and maintenance manual (or in the safety manual for safety equipment). Product manuals provide warnings concerning the installation, inspection, configuration, preventive maintenance, and proof test requirements. These agreements, terms, and warnings establish the boundary for the manufacturer's claims. Deviation from the manufacturer's recommended practices generally voids all warranties, either expressed or implied.

> Example--Requirements for instrument air quality
>
> Solenoid operated valve (SOV) manufacturers state in the SOV installation and maintenance manual that the SOV should be supplied with clean, dry, and filtered (<25 micron) instrument air. Any claims made by a manufacturer regarding the SOV generally assume that the instrument air meets these requirements. In the process industry, instrument air quality is rarely this good, with particulates and moisture being major problems at some facilities due to climate and environmental conditions. The contribution of instrument air quality to the failure rate of the SOV is significant and should be considered when estimating the failure rate for the SOV as a protective function device.

No manufacturer's claim, third-party analysis, or compliance report reduces the user's responsibility for determining the product is fit for its purpose in a particular application. This user responsibility was first acknowledged in Guidelines for Safe Automation of Chemical Process (CCPS/AIChE 1993) and later in ISA 84.01-1996. Both used the term "user approved" as "hardware, software, procedures, etc. that the user has evaluated and determined to be acceptable for the application."

This guidelines book defines user approved as the "hardware, software, procedures, etc., which the user has determined and documented to be sufficient to implement the required function such that it achieves the core attributes of the protection layer." The seven core attributes are independence, functionality, integrity, reliability, auditability, access security, and management of change. The equipment boundary includes the hardware, process connections, utilities, interfaces, communications, and software elements necessary for the equipment to perform its design intent.

A user approval process should be established to determine and document that equipment is suitable for the application, classification, operating environment, and function complexity. The capability assessment should systematically evaluate the ability of the equipment to operate according to the design basis in the intended operating environment under normal and abnormal process conditions. The evidence used to support user approval for protective equipment is discussed in the following sections. Similar work processes and evaluation rigor are used for all IPSs. For safety instrumented systems (SIS) implementing safety instrumented functions (SIF), ISA 84.01/IEC 61511 provides some specific requirements based on the assigned safety integrity level (SIL). These requirements are discussed in the relevant sections below.

The user approval process examines equipment using analysis and testing (see Section F.1.2) and demonstrated performance in a similar operating environment (see Section F.1.3). The user approval process justifies that equipment:

- Has been subjected to analysis and testing (see Section F.1.2),
- Has been demonstrated to perform its intended function properly (see Section F.1.3),
- Has been proven to provide the required risk reduction,
- Has operation and maintenance procedures to support continued functionality,
- Has access security features to limit access of critical features to only authorized personnel,
- Is under configuration management,
- Is monitored and tracked for performance, and
- Meets applicable good engineering practices.

This process should result in an Approved Equipment List (see Chapter 4.8.2), which is supplemented as required, by installation, configuration, inspection, preventive maintenance and proof test details. Special configuration is often required for SIS (or PIS) applications. In ISA 84.01/IEC 61511, these details and configuration requirements are referred to collectively as the "safety manual."

User approval requires a close relationship between the manufacturer and the owner/operator as a product moves from early development through obsolescence. Technology evolution results in some equipment moving quickly toward obsolescence. Rapid technology evolution has shortened the lifespan of some programmable products to less than 10 years. Such evolution results in the manufacturer selling products that have some of the desired hardware and software features with a migration path for future upgrades. In some cases, significant changes in the product may be required to correct problems or to provide requested functionality.

The owner/operator should have a process to control replacement of equipment other than "replacement in kind," which is the identical substitution of one item for another. When the manufacturer offers a new version of an approved product or recommends a replacement, these new products should go through the user approval process. For new versions of approved products, this process can be simplified when the manufacturer provides documentation explaining changes to the products and highlighting any configuration changes.

Configuration management can be challenging, but is essential to achieve the core attributes. Small changes to manufactured products occur regularly, and many of these changes do not affect how the equipment works in the application. Significant changes, such as changes to equipment materials of construction, configuration instructions, installation requirements, or maintenance recommendations, should result in a review of the equipment design and operating basis.

Success requires that the owner/operator identify IPS specialists, who serve as an information resource for the company. These specialists should be highly familiar with the application requirements, receive and review advisory notices and product information from the manufacturer, and make decisions regarding the addition or removal of equipment from the Approved Equipment List. It is recommended that the product specialist be independent of any project team seeking approval for new equipment.

Field performance and lessons learned in applying equipment should be shared with the manufacturer to assist in future product development and in the dissemination of information regarding critical failures. Monitoring field performance and sharing failure reports with project and plant personnel is important for performance improvement. See Chapter 6.9 for a discussion on monitoring IPF performance.

F.1.1 Operating environment

The operating environment can be identified by drawing an imaginary bubble around the device as installed in the process. The operating environment includes a variety of items that affect device operation, such as:

- External environmental conditions,
- Process operating conditions,
- Communication robustness,
- Process and system interconnections,
- Power, grounding and surge protection provisions, and
- Support system quality, e.g., instrument air or power.

The long-term management of the device has a significant affect on its performance. The following items should also be considered when selecting a device:

- Operator interface design and management,
- Access security provisions,
- Hardware and software configuration management,
- Inspection and preventive maintenance practices, and
- Proof test requirements.

F.1.2 Analysis and Testing

Equipment should be subjected to analysis and testing to evaluate the device's design, manufacture, and validation procedures, as well as the manufacturer's quality and change management systems. The rigor of the analysis and testing is related to the equipment complexity and the maturity of its technology. When completed, there should be an understanding of how the equipment functions and fails. This analysis considers the operating environment and results in the following:

- Description of the analysis boundary,
- Identification of dangerous failures and failure modes,
- Documentation of means to detect dangerous failures by diagnostics or testing,
- Description of assumed diagnostic coverage and any user requirements necessary to achieve the diagnostic coverage, e.g., configuration requirements and the need for external diagnostics, and

- Statement of the probability of safe and dangerous failures with any assumptions regarding maintenance activities and test intervals.

Analysis and testing should also include evaluation of the equipment's conformance to applicable codes, standards, and practices. Meeting the area classification (e.g., hazardous environments) and following applicable electrical codes is important for all instrumented system applications. An improperly powered system is a potential ignition source if a loss of containment occurs.

Some owner/operators develop installation details for each type of technology. These owner/operators then analyze the planned installation and define its minimum requirements based on the core attributes. This provides a consistent operator and maintenance interface and allows the analysis and testing for equipment approval to be limited to the equipment and its ability to work within the approved installation.

Manufacturers should analyze and test SIS products for compliance with IEC 61508. This generic functional safety standard applies to safety-related systems used in many different industry sectors, including process, rail, machinery, and medical. IEC 61508 Part 2 provides hardware requirements while IEC 61508 Part 3 provides the software requirements.

Since IEC 61508 applies to many sectors, there are products meeting its requirements, which are not sufficiently robust for the operating environment in many process industry facilities. A product developed for use on a manufacturing floor may not be appropriate for use in a rail transit system or a refinery SIS. Alone, IEC 61508 compliance is not sufficient evidence for user approval of a product for an SIS application.

An IEC 61508 evaluation may yield a "certification" if the product is evaluated by a recognized approvals body. When the product meets the relevant requirements, it is said to meet an SIL Claim Limit or to be fit for use in a specified SIL. In the United States, certifications are issued by Nationally Recognized Testing Laboratories (NRTL), such as those listed on the US-OSHA website (www.osha.gov). Other governments establish certification requirements by policy or regulation.

When a product manufacturer claims compliance to IEC 61508, the manufacturer supplies a safety manual providing the criteria for product implementation. In some cases, there may be relatively few special requirements. In other cases, there may be many detailed requirements, such as specific configurations, minimum fault tolerance, externally configured diagnostics, provision for frequent inspection and preventive maintenance, maximum proof test intervals, and specific installation details. Deviation from the safety manual may invalidate the SIL Claim Limit.

Equipment does not require "certification" to be user approved for an SIS application. Many products, especially field devices, are successfully selected and implemented based on analysis and testing by various parties, including manufacturers, third-party assessors, and owner/operators. There is a limit to a manufacturer's ability to test the product in various applications or to identify failure modes within a specific operating environment. Accelerated life tests can be conducted on some types of equipment for the purpose of identifying wear-out mechanisms. Environmental specifications can be substantiated by environmental tests. However, all tests generally examine each variable independently. Potential synergistic effects of operating at the extremes of multiple environmental or operational conditions are not considered. Consequently, manufacturer analysis and testing should be supplemented by owner/operator testing in a similar operating environment.

Manufacturers sometimes provide field reports based on the observed performance at various owner/operator sites. The quality of field reports is highly variable, depending on the method used to collect, analyze, maintain and report the information. In some cases, field reports may only represent early failures when product users are reporting during warranty periods. The reports may not represent useful life or wear-out failures, since these often occur after warranty periods. In other cases, the manufacturer may track a specific segment of user applications and follow appropriate statistical techniques in data analysis and reporting.

More often, manufacturers supply analysis and testing results and documented predictive calculations of their product's frequency of failure, probability to fail on demand, and spurious trip rate based on the product's "shelf-state" design and manufacture. For any new technology, the analysis and testing is generally based on very limited field operating experience. Thus, the manufacturer evaluation represents the highest performance that can be expected from the equipment when it is implemented in accordance with its safety manual.

Many important failure mechanisms are typically excluded from the manufacturer's equipment boundary, such as failure mechanisms associated with the process connection, manner of installation, process operating conditions, external environment, human interface power supplies, and communication interfaces. Failures resulting from external influences can be significantly greater than failures resulting from product manufacture. Failures in the interfaces between the equipment, the process, and other IPSs should be considered when determining a equipment's suitability for addition to the approved manufacturers list.

F.1.3 Prior Use History

Previous experience in a similar operating environment provides valuable input into the user approval process. IPF equipment is installed to address significant consequence events. Prior to any equipment being used in an IPF application, there should be a demonstrated history of successful operation in a similar operating environment. Understanding how the equipment fails in the operating environment is necessary to achieve a high degree of certainty that the design, inspection, testing, maintenance, and operational practices are sufficient. In IEC 61508, this is called "proven in use;" while in ISA 84.01/IEC 61511, it is called "prior use."

Equipment should not be selected until there is a clear understanding of how the equipment fails, how the failure is detected and repaired, what inspection and preventive maintenance activities are necessary to maintain the equipment in an "as good as new" condition. Identified failures should be taken into account when defining the requirements for design, inspection, maintenance, proof test, and operation.

Field device performance is generally dictated by the operating environment and the owner/operator's inspection and maintenance practices. For example, a sampling of data for pressure transmitters from various manufacturers report theoretical mean time to failure dangerous ($MTTF^D$) values that are three to ten times better than owner/operator prior use data. When poorly applied and maintained, the pressure transmitter $MTTF^D$ can be even worse than the prior use data suggests.

When specified appropriately, non-PE logic solvers, such as relays and trip amplifiers, can achieve among the best $MTTF^D$ possible for a single device. With redundancy and adequate proof testing, non-PE logic solvers can achieve any required $MTTF^D$. The actual performance is predominantly affected by systematic errors in their installation or change management.

PE logic solver performance is highly affected by the presence of diagnostics used to detect failures of its many components. SIS applications require that the PE logic solver be safety configured to detect known failure modes. PE logic solvers tend to incur more frequent spurious operation than reported by analysis. This is largely due to unknown failure mechanisms, the large number of aging components, maintenance error, or systematic error. Poorly executed application software can result in process safety incidents or spurious operation. Managing and properly implementing changes to the embedded, utility, and application software is crucial to the PE logic solver performance. PE logic solvers can achieve any required $MTTF^D$ with sufficient diagnostics, redundancy, and proof testing.

Maintenance records are a valuable information source. The plant maintenance systems can be used to track equipment performance. Periodically, maintenance records and investigation reports should be reviewed to determine how equipment performance can be improved through better design, specification, installation, and mechanical integrity practices. Appendix E provides a discussion of the CCPS/AIChE Process Equipment Reliability Database (PERD) effort.

There are specific requirements and limitations regarding the prior use evaluation for SIS in ISA 84.01/IEC 61511 clauses 11.4 and 11.5.3 through 11.5.6. The requirements vary depending on the equipment type (sensor, logic solver, final element) and whether the equipment uses programmable elements. If the equipment is programmable, the requirements also vary depending on whether fixed programmable language or limited variability language is used to configure the equipment. See Chapter 5.6 for more discussion on programming languages and their varying requirements. These requirements are also discussed in ISA TR84.00.04 Annex L.

For SIS, ISA 84.01/IEC 61511 requires that equipment be selected based on expected performance in the operating environment. The equipment should be monitored in an operating environment similar to the intended application. The equipment performance can be observed in process control and other non-SIS applications. In many cases, owner/operators gather information on the equipment performance at other owner/operator sites.

Operating time can be gathered during alpha and beta testing where owner/operators work in close association with the manufacturer. Alpha testing is conducted to demonstrate the basic functionality of the equipment and general compatibility with the service. These are carefully controlled tests performed under the guidance of a technical support person from the manufacturer. The alpha test represents the first step toward user approved status, because it demonstrates that minimal functionality is provided. Alpha tests are followed by beta tests, which demonstrate longer term functionality and higher levels of integration with other systems. Beta tests are usually conducted using multiple installations to gather information on equipment failures in a variety of operating environments. Alpha and beta testing results alone are generally not sufficient for approving equipment for IPF applications.

Manufacturers often provide references to other owner/operators who use their product. These references can often provide valuable insight into product application and use. In addition, the manufacturer user's group can provide valuable networking opportunities to gather information concerning product performance. The owner/operator may choose to rely on the manufacturer and other owner/operator's field experience or may choose to supplement this evidence with more direct experience. This may include bench testing and field trials in process control or low consequence services. Generally, the more unfamiliar or complex the technology is, the more time that's required to understand how it works and how it fails.

Proven performance takes on special significance for IPS, because most operate as standby, or dormant, systems. Most IPSs do not modulate, but instead take action only when a process demand occurs. Consequently, equipment can fail dangerously and this failure can remain undetected until a process demand occurs or until the equipment is proof tested. If detected by demand, an incident may occur.

Conversely, the basic process control system (BPCS) equipment is expected to operate frequently, if not continuously, to maintain the process within the normal operating envelope. Degradation in performance or equipment failure is often detected by the operator through process monitoring. If the process excursion can not brought into control by the BPCS or operator, an independent IPS may be implemented to stop the hazardous event propagation. If the IPS is failed, the event propagation continues.

One of the strongest arguments for separation of the IPS and the BPCS is the common practice of using the latest technology in control applications. Many owner/operators who strive to use the newest control technology often find that the products are not as well developed as originally thought. New technology implementation can turn into an unplanned product development project. This can be a hazardous practice for IPS. The value of proven operating history cannot be underestimated.

F.2 EVOLUTION OF PLANT AUTOMATION

Within the process industry, control functions are used to achieve production and product quality targets, reduce manpower requirements, reduce human error rate, and improve process uptime. Control systems execute one or more control functions, which cause the process to operate within the normal operating limits. Control functions can be executed manually or automatically. Operators supervise the process using the control system interface, respond to its indications and alarms, and, when necessary, take action on the process through it.

The control system is implemented as part of a BPCS, which is separate and independent of the SIS. The BPCS may execute control and safety functions when it is designed and managed to achieve the assumed risk reduction or hazard rate. A safety instrumented function may not be implemented in the BPCS.

Use of the BPCS to perform a safety function is highly restricted, since a dangerous failure somewhere within the system may lead to the loss of control and potentially to a hazardous event. Control functions are often configured to continue plant operation on detected failure rather than failing to the safe state. Further, many systems allow the operator to place various functions in manual when control problems are identified. In these cases, the BPCS performance is limited to the operator capability. The dangerous failure rate of a BPCS that places a demand on a protection layer cannot be assumed to be better than 10^{-5} per hour (see ISA 84.01/IEC 61511 clause 8.2.2).

The BPCS is generally designed and managed using a less restrictive user approval process than the SIS, allowing more rapid deployment of new technologies. Validation, access security and management of change requirements are also typically more flexible and performed on an "as needed" basis, within the bulk of the process industry.

Some industry sectors do have rigid administrative controls and validation processes for the BPCS, such as those falling under food and drug requirements. However, most BPCS are not rigorously designed and managed. Many allow nearly open connection of the BPCS with other data systems internal or external to the facility. The system is therefore vulnerable to unknown impact from these other systems and the humans responsible for them.

The use of the BPCS to perform a safety function should be approved by a hazard and risk analysis team, which considers the initiating cause and any potential common cause or common mode failures within the system. The risk reduction factor for a BPCS used as a protection layer must be assumed to be below 10 (see ISA 84.01/IEC 61511 clause 9.4.2).

When the SIS is independent of the BPCS, it generally operates as a dormant system that takes action only in response to operation outside the normal operating envelope. These process demands are often caused by failures within the BPCS. The SIS is designed and managed to ISA 84.01/IEC 61511 to achieve a specified safety integrity level (SIL). Most SISs are designed to fail to the safe state on loss of power or other support systems.

Independence is a fundamental principle in the design of the SIS, regardless of the BPCS capability. Separation should be clear and unambiguous. The systems should be sufficiently independent such that one system can suffer a complete system collapse, failing all of its equipment, yet the other system continues normal functioning. If this criterion cannot be met, any shared components must be designed and managed as an SIS.

Maintaining the rigor of ISA 84.01/IEC 61511 increases the cost of BPCS ownership and significantly restricts flexibility. However, the ISA 84.01/IEC 61511 lifecycle and its associated quality management system can add significant benefit to the BPCS, because better managed systems tend to operate more reliably.

As technology has evolved, emphasis continues to be placed on maintaining the independence and separation of the BPCS and SIS functions to reduce system complexity and the potential for systematic errors. While integrated technology often decreases the resource needs, it typically demands resources with specific skills, knowledge and experience. The potential for human error increases when separation is not provided because the system components are accessed more frequently.

The majority of instrumentation and controls used in the process industry are associated with the BPCS, with the ratio of BPCS to SIS I/O being heavily weighted toward BPCS. When the BPCS and SIS are combined into a single SIS, there is an increased need to access the SIS to perform maintenance or modification of the non-SIS (BPCS) I/O. This increases the likelihood of inadvertent and unintentional changes. The management system rigor for the entire system must be increased to compensate.

Finally, when the BPCS and SIS are combined into a single SIS, the logic solver likely operates in a continuous mode, because a dangerous failure within the logic solver may cause a simultaneous loss of control and safety functions. The added system complexity makes many important aspects of the lifecycle more difficult, such as the assessment of independent protection layers in the risk assessment (see Chapter 3), the development of the fault detection and response strategy (see Chapter 4), and the design verification (see Chapter 4 and Appendix

D). However, in some specialized applications, such as turbine controls and shutdown, the greater complexity is frequently justified.

The following provides a brief overview of the evolution of BPCS and SIS technology and discusses some of the advantages and disadvantages of their use.

F.2.1 Basic Process Control System

In the early days of process automation, the basic process control system (BPCS) consisted of pneumatic transmitters and controllers that operated the control valve by adjusting its output air signal to the positioner on the valve. These pneumatic controllers were used to perform relatively simple control functions.

Over time, the BPCS evolved into programmable logic controllers (PLC) and distributed control systems (DCS), which are based on programmable electronic (PE) technology. The PE technology centralized the control logic, allowing more control functions to be executed in a single platform. This processing capability allowed the implementation of statistical process control, predictive control algorithms, and other advanced control techniques, resulting in tremendous productivity and quality improvements.

The integrity of the BPCS hardware has steadily improved over the years with increased internal diagnostics allowing these systems to be configured to obtain the desired process reliability. The capability to implement redundant components throughout the system, including input/output modules and main processors, has further enhanced reliability. Extensive internal and external diagnostics have also been incorporated into the field device design, providing more rapid fault detection.

Today, most process units are highly dependent on automated control systems. Operators rely on the BPCS and the operator interface for process information during normal operation, for alarms during process excursions, and for troubleshooting process control problems. The BPCS is now so specialized that typically only a few site personnel are knowledgeable in the control system design details. These personnel are also intermittently optimizing the BPCS to gain improvements in quality, production, and cost. BPCS technology provides significant benefits with its capabilities and flexibility, but it also introduces new and more complex failures. This creates an environment where, if administrative controls are not in place, the BPCS exists in an almost endless state of flux where control loops are routinely placed in manual operation, alarms are disabled or reset by operators based on personal choice, and process control specialists implement the newest in control algorithms while the process unit is in operation.

F.2.2 Safety Instrumented System

At a minimum, a safety instrumented system (SIS) consists of a sensor, a logic solver, a final element, and a support system. The SIS includes a combination of hardware and software elements that work in unison to detect process hazards and to take defined actions to achieve or maintain a safe state. Historically, SISs were implemented using process switches, hardwired electrical systems, and final elements, such as motor control circuits or solenoid-operated block valves. Since the SIS was physically separate and diverse from the BPCS, independence of the SIS and the BPCS was easily evaluated and demonstrated. The BPCS and SIS were typically designed and maintained by diverse personnel and departments. The two systems shared few, if any, components, technology, or personnel support.

When Programmable Electronic System (PES) technology became available, there was hesitation concerning its use in safety applications. Near misses and incidents were reported, where input and output points stuck in position without warning or the main processor locked up and stopped processing. These failures were unacceptable and could not be tolerated when safe operation was at risk. Over the years of use in control applications, owner/operators learned the nature of PES failure and how to detect failure through automated diagnostics. PESs are "safety-configured" to detect known failure modes. Safety configuration includes input, output and main processor diagnostics. Inputs and outputs are pulsed to detect stuck points and a watchdog timer circuit is used to detect a halted main processor.

Improvements in safety PES performance have been gained through implementation of extensive self-diagnostics, allowing the safety PES to identify, isolate, and alarm its faults. For SIS applications, the diagnostics are configured to take the failed component to its safe state on dangerous detected failure. For example, the diagnostics on a single channel safety PE logic solver are configured to change its outputs to the safe state. This philosophy supports safe operation, but results in impact to plant uptime. For high integrity applications, diagnostics alone are insufficient, so redundant channels are used. When SIS reliability is important, safety PES designs are modified to incorporate redundant channels with appropriate voting and enhanced diagnostics (e.g., comparison of channels).

When sufficient experience had been gained, complex functions migrated into safety PES (e.g., safety PE logic solver or safety-configured PLC), while simple functions remained in electrical systems (e.g., non-programmable relays or trip amplifiers). Electrical systems are very cost effective and easily implemented for low input/output (I/O) systems. Electrical systems are independent, physically separate, and diverse from the control system. A safety PES is generally more cost effective than a comparable electrical system when the I/O requirements are large. The analysis, design and validation become more complex as the system gets larger. Modularity becomes more difficult to ensure in larger systems, opening up the number of functions requiring proof test following change.

The safety PES brings the ability to implement more complex and extensive logic with greater flexibility for modification. With these benefits comes an increased potential for systematic faults due to human error and for common mode faults in which the random hardware failure of the safety PES can result in the loss of multiple safety functions. Application software incurs an additional security risk, since changes are so easy to make. A dedicated safety PES engineering interface should be used to facilitate administrative control of SIS logic and to reduce the potential impact of routine BPCS changes on the SIS. Software modifications are controlled through a software management of change (MOC) procedure with version tracking. Access security features are implemented, such as:

- Restricted access to the engineering interface (e.g., computer in locked cabinet),
- Password protection at the programming interface, and
- Restricted hardware access to prevent unauthorized download of a new or revised application program to the safety PES.

To reduce the potential for data corruption, communication between the BPCS and SIS is also highly limited. In many cases, read-only communication is practiced, with the BPCS only allowed to read information from the SIS. If communication from the BPCS to the SIS is absolutely necessary, the communication technique is evaluated for potential failures and steps are taken to reduce the potential for communication failure to cause SIS failure.

F.2.3 Future Technology

BPCS field installation and commissioning keeps getting easier. The evolution from mainframes to integrated servers is being mirrored in the process units as systems are migrating from large centralized control systems to small digital plant architectures. Moving the equipment back into the process unit substantially reduces the installation costs, but makes the system more susceptible to exposure to hazardous environments and events.

The amount of required wiring continues to decrease. Communication buses, such as Modbus, Fieldbus, Profibus, and Ethernet, are positioned to become the replacement for standard field wiring in control system applications. While this technology can significantly reduce the costs associated with the installation of such systems, the potential for common cause failure is significantly higher, since multiple concurrent signals can be lost. For this reason, communication buses, even those determined to be complaint to IEC 61508, have not yet received widespread acceptance within the process industry for SIS.

The cost of redundant systems has continued to decline, making it possible to cost effectively implement the BPCS and SIS in similar hardware. However, each system should have independent and separate I/O modules and main processors, unless it is intended to design and manage the entire system as an SIS. Communication from the BPCS to the SIS should remain as limited as possible to reduce the potential for data corruption or bad writes. Potential communication failures should be assessed to determine how the failure affects the SIS operation and the impact should be addressed during the design.

Some systems provide common configuration or programming tools for the BPCS and SIS. Although a common tool reduces resource and training requirements, one tool can mean one access point to both systems. This presents a significant challenge to administratively control access and changes to the SIS. Separate engineering interfaces should be provided for the BPCS and SIS to reduce systematic errors and to support access security.

Most owner/operators continue the practice of implementing separate, and often diverse, platforms for the BPCS and SIS, following the well-proven, defense in depth strategy that supports both safety and reliability. With a physically separate BPCS controller and SIS logic solver, independence is easier to assess and manage over the process equipment lifetime. Independence allows the owner/operator to implement different management systems for the BPCS and the SIS; the BPCS management system may be more flexible and less rigorous than the SIS management system.

F.3 LOGIC SOLVER CONSIDERATIONS

A logic solver is selected for its capability to meet the core attributes of the protection layer and to execute a specified function or set of functions. The selection process should consider the following:

- Manufacturer constraints and safety manual requirements,
- Embedded software, if applicable,
- Area classification,
- Building constraints,
- Design basis,
- Function modularity,
- Configuration management support,
- Access security features,
- Technology needs, e.g., I/O types, scan rates, logic complexity,
- Utility requirements,
- Wiring requirements, and
- Communication capability.

F.3.1 Technologies

The logic solver should be user approved for the defined protection layer and the required risk reduction. The logic solver can use many different technologies, including:

- Pneumatic,
- Hydraulic,
- Electrical, e.g., non-PE relays or trip amplifiers,
- Electronic, e.g., solid state logic, and
- Programmable electronic systems, e.g., PE logic solver or PLC.

Pneumatic and hydraulic logic solvers are rarely used in new IPS, but may be present in older, existing IPS. Chapter 2.6.5 discusses the evaluation of the adequacy of existing IPSs against current good engineering practices and what may be required to "grandfather" these installations.

Most modern IPS are electrical, electronic, or programmable electronic systems (E/E/PES). ISA 84.01/IEC 61511 specifically addresses the implementation of E/E/PES in safety instrumented system (SIS) applications in the process industry. These requirements are discussed for each technology.

F.3.2 Electrical Systems

Electrical systems employ well-known, proven technology, which often relies on electromechanical equipment, such as relays, trip amplifiers, and timers. Relays accept discrete inputs and generate discrete outputs. Trip amplifiers accept analog inputs and generate discrete outputs.

Electrical systems also have very predictable failure mechanisms demonstrated by decades of field application. Electrical systems have significantly less obsolescence issues and a much longer useful life than programmable systems. The technology has been used for many years throughout the process industry. Many maintenance technicians have extensive experience with this technology, can understand ladder logic diagrams, and are very comfortable troubleshooting the logic design. Maintenance and testing of electrical systems generally does not require specialized training or equipment.

Electrical systems can be designed to achieve any required integrity and reliability and can provide good, predictable performance, as long as they are well-maintained and tested regularly. For example, the mean time to failure dangerous ($MTTF^D$) for a non-programmable trip amplifier is in the range of 500 to 750 years, while the mean time to failure safe ($MTTF^{SP}$) in the range of 150 to 225 years. Or, the $MTTF^D$ for a relay is in the range of 100 to 1000 years, while the $MTTF^{SP}$ in the range of 100 to 500 years. High integrity and reliability can be achieved with fault tolerant architectures.

Electrical systems provide physical separation, as well as diversity, from the programmable electronics associated with most control systems. Further, no special design considerations are needed to ensure minimal common cause failure exists between IPFs. Each electrical system component is selected to have a specific safe state position. Elementary diagrams are generally used to describe electrical system logic.

The configuration of an electrical system is more difficult to change, thus long term administrative control is easier. Electrical system modifications are easier to control than software changes in a programmable system. A work permit is required to work on an electrical system and wiring diagrams need to be developed. This supports change control, because work permits provide the audit trail that can be tracked to determine when someone has accessed or is attempting to access the system. To achieve similar control of software systems, self-control and discipline is often supplemented with a dedicated engineering interface located in a secure area.

Electrical systems generally have higher signal to noise ratios and withstand surges better than electronic and programmable electronic systems. Electrical systems also have less restrictive environmental specifications. Consequently, electrical systems generally do not require the degree of protective design features that are necessary for the other technologies. An electrical system can be distributed throughout the plant and have low capital and installation costs for less than 10 IPFs. Small systems are also easy to troubleshoot, validate, and maintain. The systems can be designed and implemented in a modular manner to protect individual pieces of equipment. The on/off status can be determined by simply checking the contact position.

Further, while it is possible to design large relay systems, the lifecycle cost is generally higher than comparable programmable systems. Electrical systems are not self-documenting like software, so it is possible to jumper (or bypass) a function without detection. Large electrical systems can also become quite complex, but this is somewhat balanced by the "tactile nature" of electrical systems in which the physical wiring can be traced, though not always easily.

With electrical systems, diagnostics of field device failures can be provided by other means. The field device's signal can be shared with the BPCS where diagnostics can be performed and faults annunciated. Any diagnostic logic should be covered under the validation, access security, and configuration management.

F.3.3 Electronic Systems

Electronic systems include solid state relays, timers, and pulsed electronic logic. Solid state logic refers to transistor elements, such as complimentary metal oxide semiconductor (CMOS), resistor-transistor logic (RTL), transistor- transistor logic (TTL), and high noise immunity logic (HNIL). These elements perform logic through direct wiring similar to electrical systems. These systems are generally suitable if the desired functionality is fixed and unchangeable. Diagnostic technologies may be incorporated into the system design to cover the unpredictable failures of solid state technology. As with most systems, performance limitations can be overcome through the use of redundancy and fault detection.

F.3.4 Programmable Electronic System

Programmable electronic systems (PES) are used throughout the process industry. PES are easy to configure and extremely flexible. They have the ability to perform many types of functions, including complex calculations, data trend analysis, and matrix logic.

The flexibility of the PES can also be a drawback, as there is a tendency to push the software toward ever increasing complexity. Software is prone to systematic errors and industry's current capability to predict these errors is highly limited. Careful software development, documentation, review, exhaustive testing, configuration management, and management of change practices are necessary to have reasonable confidence that the software is correct. Application software should be developed according to a software specification (see Chapter 5).

A PES has a smaller footprint than an electrical system for large number of protective functions. This is achieved because many functions can be implemented in a single microprocessor. This concentration of logic is the primary disadvantage of PES technology, because the correct operation of many functions is often dependent on the continued operation of a single PES. This leaves many functions susceptible to common cause failure due to random failure and systematic error. A common cause failure within the PES can potentially lead to the loss of multiple functions simultaneously.

For SIS applications, the PES is safety-configured to detect known failures modes in the I/O modules and main processors. At a minimum, PES diagnostics should incorporate stuck-on/stuck-off detection for the I/O modules and a diagnostic circuit for the main processors. The safety configuration may be externally- or internally- configured and may be provided by the manufacturer or added as part of the system integration.

Some PE equipment require installation and configuration of extensive external diagnostics to achieve SIS requirements. The configuration requirements should be defined in the equipment manual provided by the manufacturer or in internal technical practices. The wiring and programming for these systems can be quite complex, increasing the potential for systematic errors and spurious trips during maintenance.

Fault tolerance and diagnostics are often employed to obtain higher integrity and/or reliability. The $MTTF^{D}$ and $MTTF^{SP}$ reported for safety PESs varies widely and depends on many factors, including the technology, architecture, diagnostics, configuration, proof test interval, and degree of integration. A PES with an SIL 3 Claim Limit has an $MTTF^{D}$ in the range of 2,500 years to 50,000 years, while the $MTTF^{SP}$ is in the range of 10 years to 1000 years. The variance in the $MTTF^{SP}$ illustrates how an SIL 3 Claim Limit PES can have a high integrity with a large $MTTF^{D}$, yet be considered unreliable with a low $MTTF^{SP}$.

For safety PES, IEC 61511 requires channel fault tolerance based on the safe failure fraction, which is the fraction of the total failures ($\lambda_S + \lambda_D$), which are classified as either safe λ_S or dangerous detected λ_{DD}. Benign failures, which do not affect the capability of the logic solver to perform its design function, should not be considered in the calculation. The safe failure fraction assumes that the process can be maintained in or taken to the safe state in response to the dangerous detected failure.

Dangerous detected failures do not become safe simply because they are detected. They become safe when failure detection results in the faulted channel taking the safe state. If a dangerous detected failure is configured to maintain the equipment on-line, the dangerous detected failures should not be classified as "safe" in the safe failure fraction.

When the dangerous detected is configured to take the channel to the safe state, the calculation is as follows:

$$Safe\ Failure\ Fraction = \frac{\lambda_S + \lambda_{DD}}{\lambda_S + \lambda_D} = \frac{\lambda_{SP}}{\lambda_{CRIT}} \qquad \text{F.1}$$

The safe failure fraction is an important parameter for achieving compliance with IEC 61508 and ISA 84.01-2004/IEC 61511, yet, it is extremely difficult to independently verify. The number of components used to execute functions and perform internal diagnostics is generally large, and their interaction is highly complex. Due to the complexity of the assessment, it is often necessary to rely on the manufacturer's documentation for this information.

PES scan time is important in some applications, especially in rotating equipment, where the process safety time is often specified as 100 ms or less. When fast scan times are required, consideration should be given to the use of a dedicated logic solver for the application. This limits the I/O to only that deemed necessary, minimizing scan time. The operating system has an inherent minimum scan time to be considered, in addition to the scan time required for the user programmed functions. When diagnostics rely on software algorithms, scan times can be highly influenced. Good programming techniques reduce scan time and improve software troubleshooting and maintenance.

F.3.5 Logic Solver Separation

In older installations, separation was assured because the PIS (or SIS) was typically implemented as an electrical system using electromechanical equipment housed in an access controlled (often locked) cabinet. Control functions were executed by pneumatic controllers. The BPCS and PIS were physically separate and were diverse in many aspects, including manufacturer, technology, installation, support systems, and even maintenance support.

The interaction between the BPCS and PIS is now much more complex. Field devices are often shared as discussed in Section F.4, and there may be extensive communication between the systems as discussed in Section F.6. However, experienced engineers and many good engineering practices continue to recommend implementing the PIS in a physically separate logic solver from the control functions. Some of the reasons for this include:

- Restricts communication to ensure that corrupted control system data does not corrupt the PIS,
- Limits the amount of logic to validate when software changes are made,
- Improves PIS application software integrity by ensuring that on-line changes to the control system software do not alter the PIS application software,
- Makes the PIS distinct, supporting access security and management of change, and
- Reduces the chance of human error (e.g., PIS typically require fewer changes than the control system).

SIFs cannot be implemented in the BPCS. When control functions and SIFs are implemented with a single system, all functions--control and safety--must be managed with the rigor of the SIS management system. The requirements of ISA 84.01/IEC 61511 apply to the entire system, including documentation, user approval, validation, access security, and management of change. If the control functions can place demands on SIFs executed in the same system, the system must meet the overall hazard rate requirements.

A major justification for separation is reduced long-term administrative costs. When layers are combined the management system of the highest layer applies. Means should be provided to restrict access, to limit communication to other systems and to control system changes. Generally the cost of separation is significantly less than the administrative cost to maintain the required rigor. The administrative rigor must be maintained for the life of the system, including the provision for necessary resources to verify and audit compliance.

Adequate separation is achieved by administrative controls and physical means. Physical separation is provided at the system level by executing the functions in separate and, often diverse logic solvers. Access security and management of change is enhanced by physically separate systems. When the BPCS is physically separate from the PIS, the need to access the SIS is reduced and the BPCS can be managed under a less rigorous management system.

Separation ensures that the BPCS and PIS are not dependent on each other to operate. It also provides a clear and unambiguous distinction between the BPCS and PIS, which supports long-term access security and management of change. Separation also ensures that when maintenance and testing is conducted on one system the other remains available.

A common cause analysis should be performed on the BPCS and PIS to identify shared equipment. This analysis should consider the equipment necessary for operation of each function, such as the sensors, final elements, input/output (I/O) modules, main processors, utility, and application software, support systems, and communications. Functional separation is achieved when it is possible for all non-PIS components to fail dangerously and the process equipment still be taken to the safe state.

F.4 Field Device Considerations

A field device is selected for its capability to meet the core attributes of the protection layer and to execute a specified function or set of functions. Considerations when selecting field devices include the following:

- Approved Equipment List,
- Mean Time to Failure Dangerous ($MTTF^{D}$),
- Mean Time to Failure Spurious ($MTTF^{SP}$),
- Mean Time to Repair (MTTR),
- Embedded diagnostics,
- Internal design philosophy,
- Accuracy/precision requirements,
- Response time requirements,
- Manufacturer constraints,
- Area classification,
- Environmental constraints,
- Operating and maintenance requirements,
- Access security, and
- Management of change.

Owner/operators should assess the performance of field devices in the operating environment. In the process industry, only a small percentage of the instrumentation and controls are associated with IPS; the majority are used for the BPCS. Since the functional requirements are often similar for control functions and protective functions, control functions provide a valuable opportunity to demonstrate equipment performance in the operating environment. Operation and maintenance information concerning equipment performance in control applications supports user approval.

- In general, most owner/operators choose to use similar field devices in both applications due to:
- Familiarity of plant personnel with equipment operation and maintenance,

- Field experience in control applications provides assurance equipment can operate as intended,
- Unsatisfactory operation of a field device can be rapidly identified,
- Common spares do not necessitate additional storehouse stock (note: equipment should be user approved for both services), and
- Minimizes confusion when off-hours replacement is required (will Brand X work if we are out of Brand Y?).

Over the years, many types of field devices have been used to execute IPSs, such as:

- Discrete switches (e.g., pressure switch),
- Transmitters (e.g., differential pressure transmitter),
- Analyzers (e.g., oxygen),
- Circuit breakers,
- Solenoid-operated valves,
- Spring-return block valves, and
- Motor operated valves.

F.4.1 Separation

Separation of field devices used for the BPCS from those used for PIS (or SIS) is generally necessary to achieve the core attributes. When equipment is shared for control and protective functions, the design should be executed to ensure that the PIS has the ultimate control of the equipment. PIS equipment can be shared with the BPCS using a properly rated signal isolator or other user approved communication. A failure of the means used to share the equipment should not result in a PIS failure. No failure within the BPCS should disable the ability of the PIS to obtain information from its sensors or to achieve or maintain the safe state with its final elements.

The common cause and independence analysis becomes more difficult as the systems become more interconnected. Shared equipment must be managed as PIS. This increases the long-term operating and maintenance cost associated with the equipment. The equipment is expected to achieve optimum performance as a control element, but its design, documentation, configuration, and management is driven by PIS requirements. This conflict of purpose is a source of systematic error, when lifecycle decisions are made based on control system or production needs rather than on PIS requirements.

When a single piece of equipment is expected to perform both control and safety, the equipment is likely operating in a continuous mode, where its failure directly results in a hazardous event. Control system equipment whose failure can cause a hazardous event should not also be used as the only means for detecting or responding to the event. For example, a flow transmitter may fail high resulting in high flow into a vessel. This flow transmitter should not also be used as the only sensor for a protective (or safety) alarm or for a PIS (or SIS).

For SIL 1 and SIL 2 SIS, the initiating cause equipment can be shared in a fault tolerant for dangerous failure (e.g., 1oo2 or 2oo3) configuration with dedicated SIS equipment, when the following conditions are met:

- Operating history demonstrates a sufficiently low frequency of dangerous failure of the shared equipment,
- Design basis requirements are met, e.g., technology, integrity, criticality, leakage, configuration, reliability,
- The equipment is classified and managed as SIS, and
- The equipment is proof tested, as necessary, without loss of process control.

Control system equipment whose failure can place a demand on the SIS may not be shared in a non-fault tolerant configuration. For SIL 3 SIS, control system equipment should not be used to meet the fault tolerance requirements. An SIL 3 SIS should be designed with fault tolerance = 1, using redundant dedicated equipment from input sensor through final element.

F.4.2 Inputs

Inputs may include process measurements, local or remote manual shutdown switches, and digital outputs from other independent systems. Equipment should be selected to provide the desired performance with the degree of accuracy and repeatability necessary to detect unacceptable process conditions.

> Example--Level measurement is in error when ambient temperature is high
>
> During an audit, the operator reported that the level alarm was in bypass every year during the hottest month of the year. When asked why, the operator reported that it reported a false high. An investigation was recommended by the audit. The investigation found that the differential pressure transmitter had long capillary leads that lay on the concrete in the sun. Thermal expansion resulted in a false high level. The leads were rerouted, eliminating the error.

Equipment should be chosen that provide the desired serviceability. When looking at the IPF over its operating life, equipment will require removal, repair, replacement, calibration, and testing. Whether performed on-line or off-line, these activities must be performed safely. Consequently, consideration should be given to isolating, depressurizing, purging, venting, or draining the system, as necessary.

F.4.2.1 Discrete Switches

Discrete switches are off/on digital devices that can often be direct wired to the final elements. Discrete switches provide little opportunity for continuous diagnostics, since the signal does not vary with the process condition. The equipment state only changes when the process condition is reached. This is acceptable in cases where discrete changes are to be detected, such as a normally empty vessel retaining liquid. Periodic proof testing is required to ensure the switch is working and that the set point has not drifted.

Process switches should be provided with local or control room indication. For example a temperature switch should be installed with a dial thermometer near the sensing bulb.

In energize-to-trip applications, circuit continuity checks should be provided using end-of-line monitoring to detect broken wires and short circuits. The detection of these failures only ensures the circuit integrity, not the equipment operation. Periodic proof tests of the switch must still be conducted. See Section F.5.2.2 for more information on energize-to-trip applications.

F.4.2.2 Transmitters

Many owner/operators use transmitters for process variable measurement rather than discrete switches. The transmitter generates an analog signal that allows the use of diagnostics in the IPS design. Signals outside the 4-20 mA range (i.e., over-range and under-range) should be alarmed. When redundant transmitters are implemented, the deviation between the analog signals can be used to alarm faults, such as a stuck signal (i.e., no change of state).

A minor process disruption (e.g., a blip) can cause a temporary deviation between the signals. To minimize false deviation alarms related to such events, the deviation alarm is often specified as having to occur for a period of time, e.g., a 10% deviation for three minutes or 5% deviation for 10 seconds. The percent deviation and time period often require adjustment after start-up to account for process application variability. Changes to these setting should be controlled using configuration management. When the change is approved, the affected documentation should be updated. The percent deviation and time period should be minimized to obtain the best detection of failure with an acceptable number of alarms.

Transmitters can also be used to trend process variables, providing the operator with real-time information and thereby improving reliability. Since the analog signal varies with the process condition, transmitters allow the incorporation of calculated trip set-points, such as kinetic equations, as well as predictive conditions based on current feed rates or feed ratios.

Programmable electronic (PE) technology is available for switches and transmitters. PE equipment provides increased accuracy, easier configuration, and greater internal diagnostics. The disadvantage is that they are programmable electronic, increasing the complexity of the user approval process. PE equipment should be write-protected to prevent inadvertent or unauthorized changes to the configuration. Version changes tend to be more frequent with PE equipment, requiring more resources for the monitoring and update of the Approved Equipment List.

F.4.2.3 Analyzers

Analyzers allow the measurement of specific process constituents, such as hydrocarbons, hydrogen sulfide, and oxygen. Analyzers present unique problems due to the sample loop, which often results in a long sample time, increasing the response time to minutes. Sample loops should be kept as short as possible to increase flow through the sample loop reducing sample dead time. Analyzer types can also contribute to the dead time, particularly for chromatographs which have a long analysis time (usually in the minutes range). Flow switches are often used to detect "dead" sample loops and initiate a low flow alarm to notify personnel that maintenance is required. For SIS, a "dead" sample loop should result in an analyzer vote to trip, so redundancy is required if reliability is important.

Analyzer contamination is minimized using in-line filters. While the filters are generally necessary, they are a source of additional maintenance activities due to pluggage and condensation. Frequent inspection of sample loops and filters is recommended. To prevent the blockage of lines with water or, in cold climates, with ice, sample loops and analyzer houses may require heating.

F.4.2.3.1 ***Valve position***

Valve position may be monitored using limit switches or position transmitters. Valve position may also be used to alarm that the valve is not in the commanded state. When both the open and close position is monitored, disagreement in position status is generally alarmed. In rare cases, valve position is used to initiate a shutdown.

> Example--Valve interlock
>
> A valve interlock may be used to prevent inappropriate line-up. Prior to opening a valve, its closed position is confirmed using limit switches. Then, the valve is opened and its open position is confirmed. If the valve is not confirmed as being in the commanded state, the sequence is halted until the valve position is corrected.
>
> Each confirmation step is validating the action of the commanded state with the limit switch indicating the valve open or closed status. However, the confirmation does not test or validate the valve's capability of closing, but rather failures associated with the actuator, packing, and stem (for rising steam valves).

F.4.3 Final Elements

The most common final elements are valves and motor control circuits. Final elements should be monitored so that a "trip failed" or "command disagree" alarm can be generated if a trip is initiated and the final element does not achieve the safe state. Limit switches are typically installed on valve to provide indication of closed (or not closed) and open (or not open). Run/stop status is provided for motor control circuits.

F.4.3.1 General Valve Considerations

Valves should be selected based on proven performance in a similar operating environment. The following should be considered when selecting valves:

- Operating environment (e.g., plugging, corrosion),
- Material suitability considering the internal and external environment,
- Safe state (e.g., fail open or fail closed),
- Failure position on loss of support systems,
- Leakage (e.g., tight shutoff and acceptable leak rate),
- Shutoff differential pressure in both directions of flow,
- Actuator size (e.g., closure in the presence of maximum pressure generated under emergency conditions),
- Valve installation configuration whereby the process flow aids the movement of the valve to the desired trip position,
- Fire resistance, including valve body and actuator, and
- Opening or closure speed necessary for process safety time.

F.4.3.2 Spring Return Valves

Many IPS and emergency shutdown system applications rely on spring return actuators designed to take the valve to the desired safe state condition on loss of motive force, such as the pneumatic or hydraulic supply. The utility supply to the valve actuator should be controlled by a dedicated solenoid operated valve or valves (SOV). In de-energize to trip service, the motive source passes through the SOV to the valve actuator. When the SOV is de-energized, the supply source is blocked and the valve actuator is vented through the SOV.

The valve safe state is dependent on the process application as discussed in Chapter 4.5.13. The safe state is generally fail-closed on process lines that provide energy to the process, such as isolating feed reactants that generate runaway reactions under abnormal situations. Fail-open valves are generally used on process lines that remove energy from the process, such as venting or adding coolant to the process.

Moisture or particulates in the motive source can build up in tubing and within the SOV body, leading to an inability to vent the valve actuator when required. For good performance, the utility supply should be filtered, and quality should be monitored. All tubing and fittings should be protected from weather, especially freezing, and should be properly supported to prevent crimping or bending. In extremely cold climates, heating may be required to prevent the SOV from freezing up.

The SOV vent port should be mounted so that water and debris cannot fall into it. Vent ports have been found plugged with insulation and paint. Vent ports should be inspected after area construction work is complete to ensure that they are clear. Vent screens should be maintained to minimize the chance of debris getting into the port. These screens are often called "bug screens," because they prevent wasps, dirt daubers, spiders, and other bugs from taking up residence in the port, thus plugging it. A clear vent port is essential to the SOV's successful operation, thus the vent screen and vent port should be inspected periodically.

Valve closure speed is directly related to the Cv of the circuit from the connection to the valve actuator to the vent discharge. The SOV CV should be specified to minimize its contribution to the overall circuit. For very short valve closure times, quick exhaust vents may be required to increase valve closure speed. Consideration should be given to using larger tubing to increase the speed of valve closure in critical applications. Tubing should also be kept as short and straight as possible. If valve closure speed is important, it should be field tested to ensure that the installation can achieve the defined requirements.

SOV reliability is important, because an SOV coil burn-out can result in a spurious activation of the final element. Improvement in the IPF reliability can often be obtained using redundant SOV architectures, such as 2oo2, which are tested as necessary to achieve the required integrity. In recent years, low wattage SOVs have been increasingly used in the process industry due to their longer coil life and better reliability. Low wattage SOVs are often pilot assisted and attention should be paid to the instrument air quality and specification to ensure long-term performance.

F.4.3.3 Air-to-Move Valves

Some applications requiring particularly large (>20 inches) valves use air-to-move valves that are designed to fail in the last position when air pressure is removed. These valves are not fail-safe and require special pneumatic system design to meet the fault tolerance requirements.

For SIS, a dedicated air receiver is required for each air-to-move valve with a check valve between the main supply and the air receiver to prevent backflow into the main air header when the header pressure drops. The air receiver should be monitored with a pressure sensor, usually a switch, to alarm low air pressure. The air receiver is sized to provide sufficient pneumatic supply to open and close the valve a pre-determined number of times. Good engineering practice requires that the air receiver be capable of moving the valve full-stroke at least three times (e.g., close, open, close).

F.4.3.4 Motor Operated Valves

Motor operated valves (MOVs) are primarily used for emergency isolation and venting systems. MOVs fail in the last position when power is removed. Therefore, if the valve must attain a fully open or closed position to achieve the safe state, the dangerous failure of the power supply must be considered in the performance calculations.

MOVs should be provided with local/remote status indication. Since power is required for operation, MOVs should be powered by two independent power supplies using an auto-transfer switch. Line monitoring or other means should be also used to initiate an alarm on loss of circuit continuity. See Section F.5.2 for more discussion concerning power supply design and management.

Another consideration is the travel time of the valve. Large MOVs can require up to one minute for full travel, so process safety time should be evaluated prior to the specification of motor operated valves. See Chapter 5.5.6 for more discussion on establishing the requirements for response time. In certain cases, such as emergency shutdown systems, MOVs are the only feasible solution, given the size of the process lines and the shutoff pressures generated by the process dynamics. In such applications, additional protection layers should be considered.

F.4.3.5 Motor Control Circuits

Motor control circuits are used to shutdown motor-driven pumps and compressors. The motor starter is used as the final element to disconnect power from these motors. Depending on the voltage and amperage requirements, the starters may be energize or de-energize to disconnect power.

The IPS controls the starter through the start/stop circuit, typically accomplished using interposing relays. Motor starters should be provided with a means to indicate run/stop status.

For energize to trip applications, line monitoring or other means should be used to detect and alarm the loss of circuit continuity. See Section F.5.2 for more discussion concerning power supply design and management.

F.5 UTILITIES

Many types of utilities, as defined in ANSI/ISA-7.0.01-1996, are used to support the operation of protective functions. These include electric power, steam, nitrogen, hydraulic oil, instrument air, etc. If these utilities are required for the protective function to perform its design intent, the utility is part of the function boundary. Its design and management should achieve the core attributes and its frequency of dangerous failure should be considered during verification.

F.5.1 Instrument air

Instrument air is used for many different purposes within instrumented systems, such as purging junction boxes and actuating pneumatic valves. As a result, the impact of poor quality instrument air should not be underestimated. High quality instrument air is critical to achieve good performance, so it should be filtered, dry, and continuously monitored to ensure adequate pressure is maintained.

IPS should be designed to fail-safe on loss of instrument air. Non-fail safe designs, where instrument air is required to achieve the safe state, should include a back-up air supply using an air reservoir or other pneumatic source. The back-up pneumatic supplies should be monitored and loss of pressure should be alarmed. The back-up pneumatic source should be sized to provide an adequate amount of air to achieve and maintain the safe state for as long as required.

Instrument air design should consider:

- Dedicated system isolated from the process equipment and restricted in use, e.g., pneumatic powered tools should not be connected to the instrument air supply used for IPF equipment. If pneumatic tools are connected, the starting of these tools may cause pneumatically operated valves go to their safe state position,
- Required system pressure,
- Alarms for loss of instrument air pressure,
- Development of a procedure for responding to loss of instrument air pressure, such as switching to a backup nitrogen system. Whenever nitrogen is used as the back-up system, its use should be alarmed and any areas where nitrogen might automatically replace breathable air should have oxygen sensors with warning lights,
- Dryer systems load capacity,
- Clean and dry air sufficiently for general instrumentation use,
- Number of headers,
- Appropriate valving to support isolation/bypass,
- Section or zone redundancy to ensure continued availability when maintenance is required, and
- Means and techniques to be employed to prevent common cause failure.

The requirements for hydraulic systems are similar. Hydraulic power availability is also highly dependent on providing suitable pressure and volume, as well as maintaining appropriate fluid properties and preventing contamination.

Similar analysis should be performed for other utility systems which are used to implement IPF.

F.5.2 Power

Operation and maintenance define the functional criteria for the electrical distribution system. Power requirements for each IPS are determined during power system design. Detailed knowledge of electrical design and local electrical codes is necessary to ensure electrical systems are compliant and function properly.

IPS power design is critical to achieving safe and reliable operation. Failures within the electric power distribution and signal grounding systems are significant contributors to poor IPS performance. When existing power distribution systems are used, feasibility studies should conducted to determine whether the existing infrastructure can support the new system.

Safe and reliable operation requires that attention be paid to the power supply, electrical grounding, and signal protection design and maintenance. Systematic errors related to the power distribution can significantly impact the installed performance. For a reliable IPS, the design should eliminate single points of failure, apply user approved equipment, and minimize likelihood of human error.

IPS circuits should be provided with redundant power supplies, including at least one uninterruptible power supply (UPS) or other approved power back-up. The redundant power supplies should be arranged to allow the removal of one power supply for maintenance while the other remains on-line providing IPS power.

Power system design should specify the requirements for line power, UPS, and/or emergency power for equipment, as well as power sourcing and circuit protection. The design should consider the requirements for power supply distribution and isolation to motor control circuits, individual loops, relay circuits, SOVs, etc. When user approved PES are used, the design should specify chassis power distribution.

Owner/operator should define wiring and grounding practices, including cable separation and shield grouping to minimize interference. Special consideration should be given to cabling dealing with variable frequency drives, as these are especially prone to noise generation. Any special manufacturer's requirements regarding the grounding of the IPS, BPCS, and any other interconnected systems should be reviewed and addressed.

Clearly defined expectations allow detail design engineering to proceed efficiently along a defined path. Consistent wiring practices should be used throughout the design to minimize human error. Any unusual applications, which cannot be satisfied using standardized methods, should be identified and justification for the chosen solutions should be included in the design basis.

System input/output (I/O) should have separate power distribution and be independently fuse protected to minimize common cause faults. Fuses must be coordinated and sized to handle the surge when power is initially applied.

Power systems should be protected against overload and short circuits, lightning, transients (e.g., spikes, surges), under-voltages, and over-voltages. Refer to IEC 61326, *Electrical Equipment for Measurement, Control and Laboratory Use--EMC Requirements, for more guidance concerning electromagnetic compatibility* (EMC).

F.5.2.1 De-energize-to-trip Systems

Most IPFs are designed as de-energized-to-trip (DTT), e.g., loss of power results in a fail-safe condition. The IPF is designed to operate in a normally energized state. For example, a high level switch has closed contacts when the level is less than the set point. When the level exceeds the set point, the contacts open, thereby de-energizing the circuit.

While DTT is considered a fail-safe design, there are also potential dangerous failures of the power supply. For example, one company unknowingly had a low power supply output voltage that prevented transmitters from reaching their full output signal even though the process variable had reached a high pressure trip value. This resulted in the device failing to operate in response to a process demand.

Power supply disruptions of even a short duration can cause the spurious action of a DTT circuit, resulting in the safe state. When a DTT circuit experiences a loss of power, there is no sequenced shutdown, because the outputs go to the safe state immediately on loss of power. If the power supply disruption is at the system level, all system outputs are de-energized simultaneously, resulting in sudden shutdown of multiple pieces of process equipment. This can stress equipment and has been shown to cause equipment damage and other hazardous events. Consequently, power supply reliability is critical to safe operation.

UPS are often used to improve power reliability. The UPS sustains power for a specified period of time to allow the process to ride out transient disruptions and to provide a sequenced shutdown, if necessary. For more information on UPS, see Section F.5.2.3.

F.5.2.2 Energize-to-trip Systems

Where possible, DTT should be used. Following good design practices can yield a highly reliable DTT system. In a few specialized applications, IPFs may be designed as energized-to-trip (ETT). In these cases, the IPF is designed to operate in the normally de-energized state. For example, a high level switch has open contacts when the level is less than the set point. When the level exceeds the set point, the contacts close, thereby energizing the circuit.

When ETT is used, the power supply is critical and should be classified as protective. Since the dangerous failure of the power supply causes the IPF to fail dangerously, the probability of dangerous failure should be considered during verification. Designing an ETT system to meet good engineering practice and high integrity requirements is more complex and generally more costly than an equivalent DTT system.

An ETT system does not fail to the safe state on loss of power, so its dominant failure mode is not to the safe state. Consequently, the use of ETT should be restricted to applications where it is necessary for safety reasons. There are process applications where the process risk derived from a spurious trip is significant. A hazard and risk analysis team should review and approve the use of ETT to ensure that it is only used where it is justified based on process risk. In cases where ETT is justified, an alternate means to safely shutdown the process equipment in the event of power loss should be specified. Further, a manual means independent of the ETT system should be provided to initiate process equipment shutdown for safety integrity level (SIL) 3 applications.

> Example--Cracked gas compressor
>
> Consider a Cracked Gas Compressor in an Ethylene unit. A spurious trip of the compressor has far reaching impact throughout the facility in terms of cascade trips of both upstream and downstream equipment resulting in heavy flaring and potential safety issues, along with the inherent risks associated with an unscheduled re-start of the unit. On the other hand, failure to shut the machine down can result in damage to the compressor, which is a significant asset protection risk, but has little safety consequences.

Minimizing common cause failure requires detailed attention to input and output signal separation, power distribution, fusing, and grounding on terminations. At a minimum, ETT circuits should be provided with diagnostics, such as line monitoring, to detect circuit continuity failures. Some PES provide 'supervisory' outputs, which check for line continuity through the output loop, enabling the detection of broken wires or power loss. Detected faults should initiate a fault alarm.

While line monitoring verifies circuit integrity, it does not determine whether the circuit can withstand full current load. Fuse and wiring connections can fail when full load is applied to a supposedly good circuit. ETT systems must be regularly inspected and maintained.

To provide protection in the event of loss of power, battery backed DC power and/or an uninterruptible power supply should be specified to provide sufficient power for the period of time necessary to sequence the process unit shutdown.

These systems should be tested frequently to ensure that the power system transfers (typically turnaround test), the batteries are holding a charge (annual test) and the emergency generator kicks on (monthly test).

Consideration should be given to routing cable support systems around fire risk areas, but, if this is not possible, fire protection should be provided for cable trays. For redundant circuits, signal path separation through cable, conduit, field junction boxes, and cable trays should be specified where practical.

As discussed previously, the dominant failure of ETT systems is not to the safe state. Consequently for SISs, any portion of an SIF that is operating as ETT should be fault tolerant against power failure. ETT final elements should be specified to latch or hold the equipment in the safe state until manually reset. This is necessary to prevent the equipment from returning to the normal condition should power be lost after a trip is initiated.

An analysis should be conducted to determine whether there is sufficient time between the process demand and the potential process hazard for the operator to detect the problem and to initiate the manual shutdown (see Chapter 5.5.2 for process safety time). Human factors associated with the operator interface should also be examined to define the best means of alarming the loss of power and the process demands that would require the execution of the manual shutdown.

F.5.2.2.1 ***Uninterruptible Power Supply (UPS)***

A power failure will result in loss of power to the process equipment and the instrumentation and control systems. For DTT applications, loss of the power supply results in a sudden process shutdown, which may pose significant risk. For ETT applications, loss of power results in an IPS dangerous failure, since an ETT IPS requires power to achieve the safe state.

Consequently, UPSs should be specified to provide enough time to execute sequenced shutdown. The required time is determined by the process dynamics, considering various process issues, such as volatilization, condensation, and the time required for the process to return to the safe operating envelope.

A UPS is intended to sustain power through a brief power transit or to allow emergency and backup systems time to come on-line. Periodically, UPS should be load-tested to demonstrate that they can hold the load for the required time period.

Except for very small processes, it is not feasible to provide sufficient back-up power, e.g., generator, to keep process equipment running. Therefore, the consequences of power outages, both full and partial, should be evaluated as a part of the risk assessment phase. Chapter 3 discusses the analysis of global hazardous events, such as loss of power.

The logic solver, its key peripherals, inputs, outputs, and diagnostic circuits should be provided with sufficient UPS to allow the operator to maintain view and control of the process for a specified time. Maintaining power to the operator interface gives the operator the opportunity to verify that process shutdown has occurred correctly or to take manual action, if necessary.

> Example--Reaction continues after shutdown
>
> If the process continues to react in an exothermic manner for 2 hours after a shutdown due to the reactor volume and residence time, it is necessary to be able to monitor at least some of the control system variables and be able to respond with additional manual intervention, such as a water quench, if necessary for two hours.

Many types of UPS technologies are used to provide power through distribution systems. Power supply systems consist of redundant, alternating current feeds, batteries, battery charger, and a static inverter which provides a source of AC power. Important features are as follows:

- Two reliable power feeds from separate sources (e.g., separate power feeds provided by different substations using different power sources),
- Transfer switch for selection of AC power feeds,
- Manual bypass switch with a make-before-break switch for UPS maintenance,
- Means to detect failure in UPS circuitry initiating an alarm,
- High speed voltage surge arrestors installed at the power distribution panel,
- Lightning arrestors at various points throughout the power distribution system,
- Means for equipment isolation for maintenance and testing,
- Circuit breakers and fuses properly sized and coordinated to prevent circuit overload, and
- System meets electrical area classification.

F.6 WIRING PRACTICES

Wiring practices should be addressed and agreed to as soon as practical in the design. Wiring systems associated with IPS should be clearly labeled as protective. This includes labeling cables, conduit, junction boxes, marshalling racks, panels, and enclosures.

Wiring should meet manufacturer's recommendations for the selected IPS equipment. Additionally, the wiring should meet relevant national, regional, and/or local electrical codes and practices, such as ATEX, NEC, NFPA, etc. Each circuit should be independently fuse protected, self-protecting, or current limited.

Wiring generally has excellent integrity and reliability when correctly installed, inspected, and maintained. Most problems with wiring systems occur because of improper grounding, environmental exposure, poor wiring terminations and inadequate maintenance.

Grounding is critical for personnel protection and proper equipment operation. Manufacturer's grounding requirements should be reviewed and thoroughly understood prior to energizing equipment. All grounding must meet the appropriate electrical codes.

When considering the use of existing wire and cabling in retrofit applications, the requirements for the new system should be understood, such as wire size, and shielding requirements. The use of long runs of existing cabling may be tempting on an IPS retrofit, but the economic benefit may be short-term if the IPS does not function properly.

The decision to use above or below ground conduit and cable systems should be made early on, so that routing paths can be evaluated for possible single point failures. Options should be reviewed early to minimize potential problems and to ensure that the final design is carefully considered and meets defined performance requirements.

Overhead tray systems routed through high fire risk areas (such as pump alleys) may require fireproofing, while other routes may not. Additional structural supports and levels may be required due to tray loading and to maintain minimum voltage separation distances. Below ground duct bank and cable vault systems may be inherently safer when considering fire and equipment damage, but are obviously much less flexible in routing choices given underground obstacles.

Twisted shielded pair cables are used in industry for analog and low voltage digital signals. Wiring should be protected from various environmental elements, such as chemical exposure and moisture. Wiring should be electrically isolated and appropriately shielded to prevent signal noise and interference. At a minimum, wiring should be segregated by voltage level, e.g., 120 VAC, 4-20 mA analog, and thermocouple. Consideration should be given to standardized design and wiring practices to best optimize functionality and cost.

> <u>Example--Choosing between 4-20 mA or thermocouple signals</u>
>
> If a facility contains relatively few thermocouple inputs, a decision may be made to install temperature transmitters and route these signals back via 4-20 mA cables, effectively removing the need for the dedicated thermocouple wiring system. On the other hand, if thermocouple wiring systems already exist, the decision may be made to utilize these and install temperature transmitters only in cases where higher measurement accuracy is desired. Either decision can impact the initial design and cost estimate of the project.

F.7 COMMUNICATIONS AND INTERCONNECTIVITY

IPS often rely upon various communication and interconnectivity (C&I) techniques, in addition to hardwired connections. Local or remote I/O is acceptable as long as secure communication is provided between the I/O and the logic solver. Common C&I techniques include bus, parallel, and serial communications. C&I characteristics embedded in PE logic solvers are transparent to most users. Therefore, this discussion deals solely with C&I between systems, unless otherwise noted.

The process is controlled by sensors, logic solvers (including humans), and final control elements. The equipment comprising the control system and IPS interact with each other through hardwired systems or serial communication.

Non-redundant serial communication should not be used for safety alarms (e.g., alarms requiring operator action to reduce the risk of a hazardous event that could cause harm to people). Non-redundant serial communication is typically used to communicate information from the IPS to the BPCS, including IPF input and output signals, fault alarms, trip indications, and remote data acquisition terminals and loggers. Redundant communication links should be considered when critical information, such as alarms, trip indications, resets or manual shutdown, is being communicated between the IPS and BPCS.

The communication interface should be suitable for communication between equipment referenced to different electrical ground potentials. It should be sufficiently robust to withstand electromagnetic interference, including power surges.

The failure of a communication interface should not impact the ability of the IPS to execute its safe state action or the ability of the operator to view the state of the process. This is especially true for any communication between another system, such as the BPCS, and the SIS.

Writes to the SIS should be carefully planned and executed to ensure that bad writes do not impact the SIS operation. Writes initiated by the operator should require confirmation prior to the SIS taking action. Consideration should be given to specifying means to limit writes, such as a tolerable range or percent change for process variable setpoints or selection of specific batch recipe.

The following means are implemented for communication between the BPCS and the IPS:

- Hardwired communication (e.g., using an IPS digital or analog output to the BPCS),
- Unrestricted read from the IPS to any other system,
- Highly restricted and controlled writes from the BPCS to the IPS,

- Loss of "write" should not result in hazard, and
- Bad "write" should not cause a hazard.

If communicated information is required for the successful functioning of the IPS, the equipment used to communicate the information should be considered as part of the IPS and be designed and managed accordingly. The security of the communicated information should be assessed and managed. Refer to Appendix C.6 for more information on security.

Writes should be assessed for potentially dangerous failures, and provisions implemented to detect these failures. Writes can be controlled by various means, using limited time windows, software passwords, biometrics, or key switches. When operators are responsible for entering information to be written to the IPS, there should be a confirmation back to the operator concerning the entry value prior to the system taking action. In all cases, the design should ensure that the IPS receives any required information prior to the BPCS taking action on the change.

Bus technology is being increasingly deployed in the control of chemical processes. This technology can improve field diagnostics. Wiring costs are greatly reduced, because the need to run individual wire to each field device is eliminated. Buses provide improved communication integrity and error handling and provide time stamped alarms to enhance troubleshooting capability.

Bus technology caveats include potential for common mode faults, unknown failures (e.g., "freezes" information transfer without detection), response time limitations, and different redundancy capabilities (e.g., ring, dual ring, communication re-route after a fault). Further, bus technology is made up of many different products, each designed to handle a different plant floor need (e.g., instrumentation sensor daisy-chain connections, machine tools, discrete equipment interconnections). Extreme care should be taken to ensure the bus technology selected is user approved for the application (see Section F.1 for more information). Wireless busses are becoming available; their use requires user approval for the application and hazard and risk analysis team review and approval.

Bus technology should be approached with caution in regard to IPS. Buses are available in redundant and non-redundant versions. When redundancy is available, it is generally configured as hot standby with no voting of the redundant channels. Sometimes, it is difficult to predict inter-operability between systems.

Prior to their widespread use in IPS, bus systems must undergo more robustness testing. Stress and soak tests should be employed to determine bus reliability and fault insertion tests to verify that faults are properly identified and managed by the system hardware and software. Some testing is currently being performed, but these tests are not as extensive as those employed in the assessment of PES performance.

While buses designed to meet IEC 61508 requirements are available, it is important to note that IEC 61508 does not address system inter-operability.

Testing interoperability under all possible conditions is impossible; hence more field installation experience in control applications is essential to support user approval for SIS. Until the technology is proven in use, the BPCS functions and SIFs should be separated to ensure that bus problems do not disable both.

F.8 PRESCRIPTIVE DESIGNS

Many owner/operators use prescriptive practices to specify safety instrumented system (SIS) requirements. The use of "cookbooks" was very common in the process industry prior to the issuance of the ISA 84.01-1996. The prescriptive practices are based on historical experience and are generally intended to be sufficiently conservative that a wide range of equipment could potentially be used to implement the design. These practices generally considered only a single safe state action. The proof test interval is also specified by the practices, which should be applied unless additional analysis is performed.

This section provides examples of "cookbook" approaches and illustrates how the architectures change as successively higher safety integrity levels (SIL) are required. For all the examples, the following is assumed:

- SIS is managed throughout it lifecycle to achieve the required core attributes,
- Failure rate of the equipment is random,
- Equipment is specified to fail to the safe state on loss of power and other support systems,
- Dedicated SIS sensors are used and when redundant, the sensor are installed on separate process connections,
- Valves are specified as spring return fail to the safe state,
- Logic solver is safety-configured and fault tolerant with redundancy in the input and output modules, as well as the main processors. The logic solver is designed to achieve the required integrity level and reliability,
- If additional input subsystems or output subsystems are required, the resulting SIS may not achieve the indicated SIL,
- Proof test interval is 1 year in all cases except SIL 3 high reliability, where the 2oo2 voting solenoid operated valves are proof tested every 6 months, and
- The proof test procedure fully validates the required functionality and that inspection and preventive maintenance activities performed as part of the proof test return the equipment to the "as new" condition.

F.8.1 SIL 1

The architecture shown in Figure F.1 provides a simple, independent SIL 1 SIS. Figure F.2 provides a high reliability SIL 1 design.

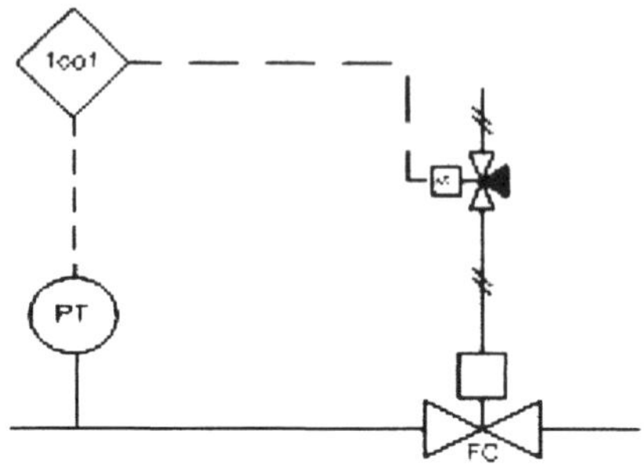

Figure F.1. Example SIL 1 SIS.

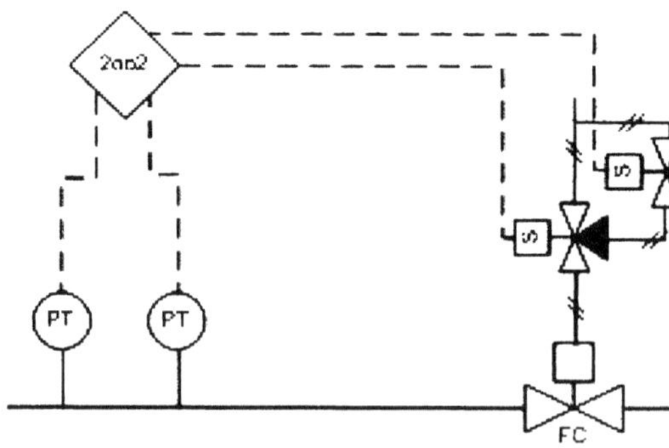

Figure F.2. Example High Reliability SIL 1 SIS.

F.8.2 SIL 2

Figure F.3 provides an SIL 2 SIS, while Figure F.4 provides a high reliability SIL 2 SIS. In Figures F.3 and F.4, the control valve or an additional block valve is used as a second means of process isolation. A solenoid operated valve is used to control the pneumatic supply to the control valve actuator. Sharing is acceptable for SIL 2 when the control valve is designed and managed to meet the SIS requirements. When the control valve cannot meet the SIS design basis (e.g., integrity, independence, leak tightness, response time, etc.), a second block valve is used instead of the control valve. This is illustrated in Figures F.3 and F.4 by the "or" and the dashed representation of the control valve and block valve.

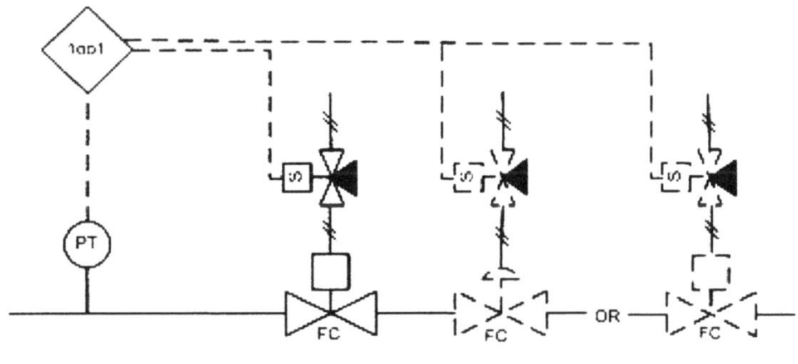

Figure F.3. Example SIL 2 SIS.

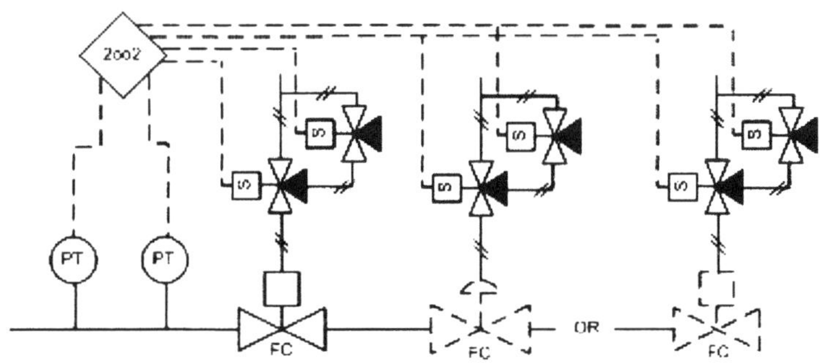

Figure F.4. Example High Reliability SIL 2 SIS.

F.8.3 fSIL 3

SIL 3 is the highest level of performance typically expected from an SIS in the process industry. For SIL 3, systematic errors must be minimized through the use of fault tolerance. Fault tolerance should be provided in the sensors, logic solver, final elements, and any required support systems. Figure F.5 provides an SIL 3 architecture that is fault tolerant against dangerous failures. Figure F.6 provides a high reliability SIL 3 architecture.

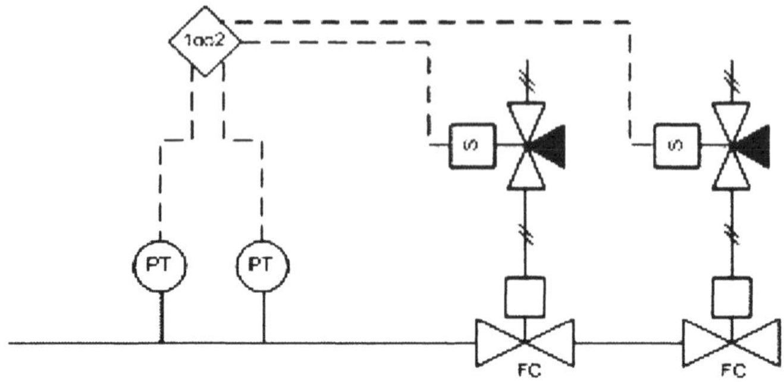

Figure F.5. Example SIL 3 SIS.

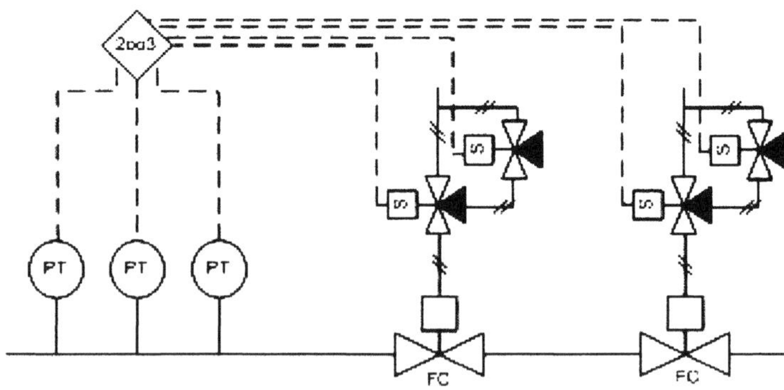

Figure F.6. Example High Reliability SIL 3 SIS.

G

REFERENCES

Alden, Edward, "Comment And Analysis: Five Fraught And Futile Years - Why America must align aims and reality," www.ft.com, Financial Times (2006).

American Petroleum Institute (API), RP 14J, "Recommended Practice for Design and Hazards Analysis for Offshore Production Facilities," Second Edition, Washington, DC (2001).

API, RP 14C, "Recommended Practice for Design, Installation, and Testing of Basic Surface Safety Systems for Offshore Production Platforms," Seventh Edition, Washington, DC (2001).

API, RP 520, "Sizing, Selection and Installation of Pressure-relieving Devices in Refineries, Part I--Sizing and Selection," 7th edition, Washington, DC (2000).

API, RP 520, "Sizing, Selection, and Installation of Pressure-relieving Devices in Refineries, Part II-Installation, 5th edition, Washington, DC (2003).

API, RP 521, Guide for Pressure Relieving and Depressuring Systems, Washington, DC (1997).

API, RP 556, Instrumentation and Control Systems for Fired Heaters and Steam Generators, First Edition, Washington, DC (1997).

American Society of Mechanical Engineers (ASME), Boiler and Pressure Vessel Code, Section VIII –Pressure Vessels, New York.

ASME, Code Case 2211-2, New York (1999).

Australian Petroleum Production and Exploration Association. Limited (APPEA), *Guidelines for Emergency Support Systems*, Canberra, Australia (2000).

Center for Chemical Process Safety (CCPS), American Institute of Chemical Engineers (AIChE), "Combined Glossary of Terms," New York (2005).

CCPS/AIChE, *Avoiding Static Ignition Hazards in Chemical Operations*, New York (1999).

CCPS/AIChE, *Deflagration and Detonation Flame Arrestors*, New York (2002).

CCPS/AIChE, Essential *Practices for Managing Chemical Reactivity Hazards*, New York (2003).

CCPS/AIChE, *Estimating the Flammable Mass of a Vapor Cloud*, New York (1999).

CCPS/AIChE, *Guidelines for Chemical Process Quantitative Risk Analysis*, Second Edition, New York (2000).

CCPS/AIChE, *Guidelines for Chemical Reactivity Evaluation +-and Application to Process Design*, New York (1995).

CCPS/AIChE, *Guidelines for Consequence Analysis of Chemical Releases*, Second Edition, New York (1999).

CCPS/AIChE, *Guidelines for Hazard Evaluation Procedures, Second Edition With Worked Examples*, New York (1992).

CCPS/AIChE, *Guidelines for Investigating Chemical Process Incidents*, Second Edition, New York (2003).

CCPS/AIChE, *Guidelines for Performing Effective Pre-Startup Safety Reviews*, New York (expected publication 2007).

CCPS/AIChE, *Guidelines for Pressure Relief and Effluent Handling Systems*, New York (1998).

CCPS/AIChE, *Guidelines for Preventing Human Error in Process Safety*, New York (1994).

CCPS/AIChE, *Guidelines for Process Equipment Reliability Data*, New York (1989).

CCPS/AIChE, *Guidelines for Process Safety in Batch Reaction Systems*, New York (1999).

CCPS/AIChE, *Guidelines for Process Safety in Outsourced Manufacturing Operations*, New York (2000).

CCPS/AIChE, *Guidelines for Safe Automation Of Chemical Processes*, New York (1993).

CCPS/AIChE, *Guidelines for Safe Warehousing Of Chemicals*, New York (1998).

CCPS/AIChE, *Guidelines for Technical Management of Chemical Process Safety*, New York (1992).

CCPS/AIChE, *Inherently Safer Chemical Processes: A Lifecycle Approach*, Concept Series, New York (1996).

CCPS/AIChE, *Layer of Protection Analysis: Simplified Process Risk Assessment*, Concept Series, New York (2001).

CCPS/AIChE, *Revalidating Process Hazards Analyses*, New York (2000).

CCPS/AIChE, *Safety Health & Loss Prevention in the Chemical Process*, Volume 2: Solutions Manual, New York (2002).

CCPS/AIChE, *Tools for Making Acute Risk Decisions with Chemical Process Safety Applications*, New York (1995).

CCPS/AIChE, *Wind Flow and Vapor Cloud Dispersion at Industrial and Urban Sites*, New York (2002).

Collins, Jim, *Good to Great: Why Some Companies Make the Leap . . . and Others Don't, Harper Business*, New York (2001).

Computer Security Institute, 600 Harrison Street, San Francisco, CA (2005).

Dowell, Arthur, "Critical Safe Operating Parameters: 'Never Exceed' Limit and 'Never Deviate' Action," Texas Chemical Council and Association of Chemical Industry of Texas, Process Safety Management Seminar, Session III: Human Factor, The Woodlands, TX (1997).

Dowell III, A. D., G. W. Carrithers, and D. C. Hendershot, "It's Never Too Late for Inherent Safety." International Conference and Workshop on Process Safety Management and Inherently Safer Processes, October 8-11, 1996, Orlando, FL, 227-241, New York: American Institute of Chemical Engineers. (1996).

Englund, Stanley, M. and Donald J. Grinwis, "Provide the Right Redundancy for Control Systems," Chemical Engineering Progress, New York (1992).

Engineering Equipment and Materials Users' Association (EEMUA), Publication 191:99, *Alarm Systems – A Guide To Design, Management And Procurement*, London, United Kingdom (1999).

EEMUA, Publication 201:02, *Process Plant Control Desks Utilising Human-Computer Interfaces-A Guide to Design, Operational, and Human Interface Issues*, London, United Kingdom (2002).

Fisher-Mayes, Laurie, works for hire, used by permission, SIS-TECH Solutions, LP, Houston, TX (2006).

Gentile, Michela and Angela Summers, "Random, Systematic, and Common Cause Failure: How do you manage them?," *Process Safety Progress* (December 2006).

Green, A. E. and A. J. Bourne, Reliability Technology, Wiley-Interscience, Bristol, England (1972).

Health and Safety Executive (HSE), *Designing and Operating Safety Chemical Reaction Processes*, HSE Books, Great Britain (2000).

HSE, *Guidance On 'As Low As Reasonably Practicable' (ALARP) Decisions In Control Of Major Accident Hazards (COMAH)*, SPC/Permissioning/12, Great Britain (2003).

HSE, *Out of Control--Why control systems go wrong and how to prevent failure*, Second Edition, HSE Books, Great Britain (2003).

HSE, *Principles And Guidelines To Assist HSE In Its Judgments That Duty Holders Have Reduced Risk As Low As Reasonably Practicable*, Great Britain (2003).

HSE, *Programmable Electronic Systems Safety Related Application PES 2 General Technical*, Great Britain (1987).

HSE, *Successful Health and Safety Management*, HSG 65, 2nd edition, HSE Books, Great Britain (1997).

Instrumentation, Systems and Automation Society (ISA), ANSI/ISA S84.01-1996, "Application of Safety Instrumented Systems (SIS) for the Process Industry," Research Triangle Park, NC (1996)

ISA, ANSI/ISA 84.00.01-2004 (IEC 61511 modified), *Functional Safety: Safety Instrumented Systems for the Process Industry Sector*, Research Triangle Park, NC (2004).

ISA, ISA TR84.00.02, *Safety Instrumented Functions (SIF) - Safety Integrity Level (SIL) Evaluation Techniques Part 1: Introduction*, Research Triangle Park, NC (2002).

ISA, ISA TR84.00.02, *Safety Instrumented Functions (SIF) - Safety Integrity Level (SIL) Evaluation Techniques Part 2: Determining the SIL of a SIF via Simplified Equations*, Research Triangle Park, NC (2002).

ISA, *ISA TR84.00.02, Safety Instrumented Functions (SIF) - Safety Integrity Level (SIL) Evaluation Techniques Part 3: Determining the SIL of an SIF via Fault Tree Analysis*, Research Triangle Park, NC (2002).

ISA, *ISA TR84.00.04, Guidelines On The Implementation Of ANSI/ISA 84.00.01-2004 (ISA 61511 Modified)*, Research Triangle Park, NC (2006).

International Electrotechnical Commission (IEC), IEC 61131 *Programmable controllers - Part 3: Programming languages*, Geneva, Switzerland (2003).

IEC, *IEC 61326, Electrical Equipment for Measurement, Control and Laboratory Use--EMC Requirements*, Geneva, Switzerland (2002).

IEC, *IEC 61508, Functional Safety of Electrical /Electronic/Programmable Electronic Safety Related Systems, Parts 1-7*, Geneva, Switzerland (1999-2001).

IEC, *IEC 61511, Functional Safety: Safety Instrumented Systems for the Process Industry Sector*, Geneva, Switzerland (2003).

International Organization for Standardization (ISO), ISO 11064, *Ergonomic Design of Control Centres*, Geneva, Switzerland (2000).

Kletz, Trevor, *Lessons from Disaster: How Organizations Have No Memory and Accidents Recur*, Institution of Chemical Engineers, United Kingdom (1993).

Kumamoto, Hiromitsu and Ernest Henley, *Probability Risk Assessment and Management for Engineering and Scientists*, Second edition, IEEE Press, New York (1996).

Lawley, H.G. and Kletz, T.A., High-pressure-trip systems for vessel protection. *Chemical Engineering*, Vol. 82 (1975)

Mannan, Sam, *Lees' Loss Prevention in the Process Industries, Volumes 1-3*, Elsevier Butterworth-Heinemann, United Kingdom (2005).

Mostia, William, "Avoid Errors," *Chemical Processing*, Putman Media, Illinois USA (October 2003).

Nimmo, Ian, User Centered Design Services, LLC, used by permission, www.mycontrolroom.com, Anthem, Arizona (2006).

National Fire Protection Association (NFPA), NFPA 85, *Boiler and Combustion Systems Hazards Codes*, Massachusetts (2001).

NFPA, NFPA 86, *Standards for Ovens and Furnaces*, Massachusetts USA (1999).

NFPA, NFPA 101, *Life Safety Code* Massachusetts USA (2006).

National Safety Council, United State of America, *Injury Facts*®, Itasca IL (2004).

Norwegian Oil Industry Association, *Application of IEC 61508 and IEC 61511 in the Norwegian Petroleum Industry*, (2004).

Idaho National Laboratory, P.O. Box 1625, Idaho Falls, Idaho (2005).

Occupational Safety and Health Administration (OSHA), "Compliance Guidelines and Recommendations for Process Safety Management (Non-mandatory), 29 CFR Part 1910 Appendix C," *Federal Register 57, 36*, Washington, DC (1992).

Offshore Reliability Data (OREDA), SINTEF Industrial Management, Det Norske Veritas, Trondheim, Norway (1992).

OSHA, correspondence, Richard Fairfax, Director, Directorate of Enforcement Program, to Lois Ferson, Manager of Standards Services, Instrumentation, Systems and Automation Society, DEP/GIE/SMK, dated November 29, 2005.

OSHA, "Process Safety Management of Highly Hazardous Chemicals; Explosives and Blasting Agents, 29 CFR Part 1910." *Federal Register 57, 36*, Washington, DC (1992).

Reason, James, *Human Error*, Cambridge University Press, Cambridge United Kingdom (1990).

Reason, James, *Managing the Risks of Organizational Accidents,* Ashgate Publishing Company, Hampshire England (1997).

Reason, James and Alan Hobbs, *Managing Maintenance Error: A Practical Guide*, Ashgate Publishing Company, Hampshire England (2003).

SIL Solver, SIS-TECH Solutions, LP, Houston, TX (2006).

Slattery, Jack, works for hire, used by permission SIS-TECH Solutions, LP, Houston, TX (2006).

Smith, David, *Reliability, Maintainability, and Risk*, Fifth Edition, Butterworth-Heinemann, United Kingdom (1997).

Summers, Angela, "Can you safely 'grandfather' your SIS?" *Chemical Processing*, Putman Media, Illinois USA (August 2005).

Summers, Angela, "Consider HIPS for Reactive Processes," *Chemical Processing*, Putman Media, www.chemicalprocessing.com, Illinois USA (March 2004).

Summers, Angela, "Fault Management Analysis," *Process Safety Progress* (September 2001).

Summers, Angela, "Software Implemented Safety Logic," *Process Safety Progress* (June 2002).

Summers, Angela, "Bridging the Safe Automation Gap Part 1," *Hydrocarbon Processing*, Gulf Publishing, Texas USA (April 2002).

Summers, Angela, "Bridging the Safe Automation Gap Part 2," *Hydrocarbon Processing*, Gulf Publishing, Texas USA (May 2002).

Summers, Angela, "The Grandfather Clause is Not a Jolly Man in a Red Suit," *Control*, Putman Media, www.controlglobal.com, Illinois USA (August 2005).

Summers, Angela, "Can you safely "grandfather" your SIS?" Putman Media, *Chemical Processing*, www.chemicalprocessing.com, Illinois USA (August 2005).

Summers, Angela and Susan Wiley, "Meet SIS 'User Approval' Mandates," *Chemical Processing*, Putman Media, www.chemicalprocessing.com, Illinois USA (May 2006).

Summers, Angela, "*Continuous Improvement in Safety Instrumented Systems*," 2007 Center for Chemical Process Safety Symposium, Houston, TX (2007).

United States, Homeland Security, Computer Emergency Readiness Team, Arlington, Virginia (2006).

United States, National Safety Council, Injury Facts® (2004).

H

ACRONYMS AND ABBREVIATIONS

AIChE	American Institute of Chemical Engineers
ALARP	As Low as Reasonably Practicable
ANSI	American National Standards Institute
API	American Petroleum Institute
APL	Asset Protection Level
ASME	American Society of Mechanical Engineers
ATEX	Atmosphères Explosibles
BMS	Burner Management System
BPCS	Basic Process Control System
C&E	Cause and Effect
CCF	Common Cause Factor
CCPS	Center for Chemical Process Safety
CENELEC	The European Committee for Electrotechnical Standardization
CFR	Code of Federal Regulations
CMOS	Complementary Metal Oxide Semiconductor
COMAH	Control of Major Accident Hazards
CPU	Central Processing Unit
DC	Diagnostic Coverage
DCS	Distributed Control System
DD	Dangerous Detected
DTT	De-energize to Trip
DU	Dangerous Undetected
EEMUA	European Equipment Manufacturers and Users Association
E/E/PES	Electrical/Electronic/Programmable Electronic System
EIL	Environmental Integrity Level
EMC	Electromagnetic Capability
EPA	Environmental Protection Agency
ESD	Emergency Shutdown System
ESS	Emergency Support System
ETA	Event Tree Analysis
ETT	Energize to Trip
FAT	Factory Acceptance Test
FM	Factory Mutual Global

FMEA	Failure Mode and Effects Analysis
FMECA	Failure Mode, Effects and Criticality Analysis
FPL	Fixed Program Language
FTA	Fault Tree Analysis
FVL	Full Variability Language
HAZOP	Hazard and Operability Study
H&RA	Hazard and Risk Analysis
HFE	Human Factors Engineering
HIPS	High Integrity Protective System
HMI	Human Machine Interface
HNIL	High Noise Immunity Logic
HRA	Human Reliability Analysis
HSE	Health and Safety Executive
IEC	International Electrotechnical Commission
IEEE	Institute of Electrical and Electronics Engineering
IL	Integrity Level
I/O	Input/Output
IPL	Independent Protection Layer
IPF	Instrumented Protective Function
IPS	Instrumented Protective System
ISA	Instrumentation, Systems, and Automation Society
ISO	International Organization for Standardization
ISPE	International Society of Pharmaceutical Engineering
ISS	Instrumented Safety System
LOPA	Layers of Protection Analysis
LVL	Limited Variability Language
MOC	Management of Change
MTBF	Mean Time Between Failure
MTTF	Mean Time To Failure
MTTR	Mean Time To Repair
NEC	National Electrical Code
NFPA	National Fire Protection Association
NRTL	Nationally Recognized Testing Laboratory
OSHA	Occupational Safety and Health Administration
PC	Personal Computer
PE	Programmable Electronic
PERD	Process Equipment Reliability Database
PES	Programmable Electronic Systems
PFD(t)	Probability of Failure as a Function of Time
PFD_{AVG}	Average Probability**Error! Bookmark not defined.** of Failure on Demand
P&IDs	Piping and Instrumentation Diagrams
PHA	Process Hazard Analysis

PLC	Programmable Logic Controller
PSAT	Pre-Startup Acceptance Test
PSM	Process Safety Management
PSSR	Pre-Startup Safety Review
QRA	Quantitative Risk Analysis
RAGAGEP	Recognized and Generally Accepted Good Engineering Practice
RMP	Risk Management Program
RRF	Risk Reduction Factor
RFI	Radio Frequency Interference
RTL	Resistor-Transistor Logic
RTD	Resistance Temperature Detector
SAT	Site Acceptance Test
SD	Safe Detected
SFF	Safe Failure Fraction
SIF	Safety Instrumented Function
SIL	Safety Integrity Level
SIS	Safety Instrumented System
SOE	Sequence of Events
SOV	Solenoid Operated Valve
SRS	Safety Requirements Specification
STR	Spurious Trip Rate
SU	Safe Undetected
TI	Test Interval
TTL	Transistor-transistor Logic
TÜV	Technisher Überwachungsverein
UPS	Uninterruptible Power Supply
1oo1	one-out-of-one
1oo2	one-out-of-two
2oo2	two-out-of-two
2oo3	two-out-of-three
2oo4	two-out-of-four
MooN	M-out-of-N

I

INDEX